MATHEMATICS IN ACTION

An Introduction to Algebraic, Graphical, and Numerical Problem Solving

FIFTH EDITION

The Consortium for Foundation Mathematics

Ralph Bertelle	Columbia-Greene Community College
Judith Bloch	University of Rochester
Roy Cameron	SUNY Cobleskill
Carolyn Curley	Erie Community College—South Campus
Ernie Danforth	Corning Community College
Brian Gray	Howard Community College
Arlene Kleinstein	SUNY Farmingdale
Kathleen Milligan	Monroe Community College
Patricia Pacitti	SUNY Oswego
Renan Sezer	Ankara University, Turkey
Patricia Shuart	Polk State College—Winter Haven, Florida
Sylvia Svitak	Queensborough Community College

D0406034

PEARSON

Boston   Columbus   Indianapolis   New York   San Francisco
Amsterdam   Cape Town   Dubai   Hoboken   London   Madrid   Milan   Munich   Paris   Montréal   Toronto
Delhi   Mexico City   São Paulo   Sydney   Hong Kong   Seoul   Singapore   Taipei   Tokyo

Editorial Director: Christine Hoag
Editor in Chief: Michael Hirsch
Assistant Editor: Matthew Summers
Project Manager: Beth Houston
Program Manager: Danielle S. Miller
Project Management Team Lead: Peter Silvia
Program Management Team Lead: Marianne Stepanian
Media Producer: Erin Carreiro
TestGen Content Manager: John Flanagan
MathXL Content Developer: Rebecca Williams
Marketing Manager: Alicia Frankel
Marketing Assistant: Emma Sarconi
Senior Author Support/Technology Specialist: Joe Vetere
Rights and Permissions Project Manager: Diahanne Lucas Dowridge
Procurement Specialist: Carol Melville
Associate Director of Design: Andrea Nix
Program Design Lead: Barbara Atkinson
Composition: Lumina Datamatics
Illustrations: Lumina Datamatics
Cover and Interior Design: Studio Montage
Cover Image: GustoImages/Science Photo Library/Getty Images

2 3 4 5 6 7 8 9 10—EBM—18 17 16 15

Library of Congress Cataloging-in-Publication Data
Mathematics in action. An introduction to algebraic, graphical, and numerical problem
 solving / The Consortium for Foundation Mathematics. — 5 [edition].
  pages cm
  ISBN 978-0-321-96993-4 (alk. paper)
   1. Mathematics. I. Consortium for Foundation Mathematics. II. Title: Introduction
to algebraic, graphical, and numerical problem solving.

QA39.3.M42 2016
512—dc23   2014036244

ISBN 13: 978-0-321-96993-4
ISBN 10:   0-321-96993-6

Contents

CHAPTER 4

An Introduction to Nonlinear Problem Solving

2. Expand and simplify the product of any two polynomials.

3. Recognize and expand the product of conjugate binomials: difference of squares.

4. Recognize and expand the product of identical binomials: perfect-square trinomials.

Objectives:

1. Evaluate quadratic functions of the form $y = ax^2$.

2. Graph quadratic functions of the form $y = ax^2$.

3. Interpret the coordinates of points on the graph of $y = ax^2$ in context.

4. Solve a quadratic equation of the form $ax^2 = c$ graphically.

5. Solve a quadratic equation of the form $ax^2 = c$ algebraically by taking square roots.

6. Solve a quadratic equation of the form $(x \pm a)^2 = c$ algebraically by taking square roots.

Note: $a \neq 0$ in Objectives 1–5.

Objectives:

1. Evaluate quadratic functions of the form $y = ax^2 + bx, a \neq 0$.

2. Graph quadratic functions of the form $y = ax^2 + bx, a \neq 0$.

3. Identify the x-intercepts of the graph of $y = ax^2 + bx$ graphically and algebraically.

4. Interpret the x-intercepts of a quadratic function in context.

5. Factor a binomial of the form $ax^2 + bx$.

6. Solve an equation of the form $ax^2 + bx = 0$ using the zero-product property.

Objectives:

1. Recognize and write a quadratic equation in standard form, $ax^2 + bx + c = 0, a \neq 0$.

2. Factor trinomials of the form $x^2 + bx + c$.

3. Solve a factorable quadratic equation of the form $x^2 + bx + c = 0$ using the zero-product property.

4. Identify a quadratic function from its algebraic form.

Objectives:

1. Use the quadratic formula to solve quadratic equations.

2. Identify the solutions of a quadratic equation with points on the corresponding graph.

APPENDIXES

Preface

Our Vision

Mathematics in Action: An Introduction to Algebraic, Graphical, and Numerical Problem Solving, Fifth Edition, is intended to help college mathematics students gain mathematical literacy in the real world and simultaneously help them build a solid foundation for future study in mathematics and other disciplines.

Our team of twelve faculty, primarily from the State University of New York and the City University of New York systems, used the AMATYC *Crossroads* Standards to develop this *Mathematics in Action* series to serve a very large population of college students who, for whatever reason, have not yet succeeded in learning mathematics. It became apparent to us that teaching the same content in the same way to students who have not previously comprehended it is not effective, and this realization motivated us to develop a new approach.

Mathematics in Action is based on the principle that students learn mathematics best by doing mathematics within a meaningful context. In keeping with this premise, students solve problems in a series of realistic situations from which the crucial need for mathematics arises. *Mathematics in Action* guides students toward developing a sense of independence and taking responsibility for their own learning. Students are encouraged to construct, reflect on, apply, and describe their own mathematical models, which they use to solve meaningful problems. We see this as the key to bridging the gap between abstraction and application and as the basis for transfer learning. Appropriate technology is integrated throughout the books, allowing students to interpret real-life data verbally, numerically, symbolically, and graphically.

We expect that by using the *Mathematics in Action* series, all students will be able to achieve the following goals:

- Develop mathematical intuition and a relevant base of mathematical knowledge.

- Gain experiences that connect classroom learning with real-world applications.

- Prepare effectively for further college work in mathematics and related disciplines.

- Learn to work in groups as well as independently.

- Increase knowledge of mathematics through explorations with appropriate technology.

- Develop a positive attitude about learning and using mathematics.

- Build techniques of reasoning for effective problem solving.

- Learn to apply and display knowledge through alternative means of assessment, such as mathematical portfolios and journal writing.

Our vision for you is to join the growing number of students using our approaches who discover that mathematics is an essential and learnable survival skill for the 21st century.

Pedagogical Features

The pedagogical core of *Mathematics in Action* is a series of guided-discovery activities in which students work in groups to discover mathematical principles embedded in realistic situations. The key principles of each activity are highlighted and summarized at the activity's conclusion. Each activity is followed by exercises that reinforce the concepts and skills revealed in the activity.

The activities are clustered within each chapter. Each cluster contains regular activities along with project and lab activities that relate to particular topics. The lab activities require more than just paper, pencil, and calculator; they also require measurements and data collection and are ideal for in-class group work. The project activities are designed to allow students to explore specific topics in greater depth, either individually or in groups. These activities are usually self-contained and have no accompanying exercises. For specific suggestions on how to use the three types of activities, we strongly encourage instructors to refer to the *Instructor's Resource Manual with Tests* that accompanies this text.

Each cluster concludes with two sections: What Have I Learned? and How Can I Practice? The What Have I Learned? exercises are designed to help students pull together the key concepts of the cluster. The How Can I Practice? exercises are designed primarily to provide additional work with the numeric, algebraic, and graphing skills of the cluster. Taken as a whole, these exercises give students the tools they need to bridge the gaps between abstraction, skills, and application.

Each chapter ends with a Summary containing a brief description of the concepts and skills discussed in the chapter, plus examples illustrating these concepts and skills. The concepts and skills are also referenced to the activity in which they appear, making the format easier to follow for those students who are unfamiliar with our approach. Each chapter also ends with a Gateway Review, providing students with an opportunity to check their understanding of the chapter's concepts and skills.

Content Changes in the Fifth Edition

The Fifth Edition retains all the features of the previous edition, with the following content changes:

- All the data-based activities and exercises have been updated to reflect the most recent information and/or replaced with more relevant topics.

- The introductory scenarios in several activities have been replaced with more robust, up-to-date situations.

- Several new real world exercises have been added in most activities.

- The exposition and treatment of topics has been carefully reviewed and revised/rewritten where necessary to provide students with a more clear and easy to understand presentation.

- Activity 1.3, "Properties of Arithmetic", has been rewritten extensively.

- Activity 1.4 has been replaced by a new Activity, "Top Chef", which provides a review of the terminology and operations associated with fractions.

- The language in Activity 1.11, "Fuel Economy", has been streamlined for a more concise presentation. A section on the metric system has been added to Activity 1.11.

- Activity 1.13, "Shedding the Extra Pounds", has been rewritten, including the addition of an expanded treatment of division by zero.

- Cluster 1 of Chapter 2, "Symbolic Rules and Expressions", has been reorganized and rewritten as follows:

- The rewritten activity, "Symbolizing Arithmetic", (previously Activity 2.4), is the new introductory Activity 2.1.

- Activity 2.2, "Blood Alcohol", (previously Activity 2.1), has been rewritten extensively.

- Previous Activity 2.2, "Earth's Temperature", has been moved to Activity 3.1.

- Activity 2.3, "College Expenses", has been modified to reflect the changes in the introductory material.

- Activities 2.6–2.9, 2.11–2.14 from the previous edition have become Activities 2.5–2.12 in the fifth edition.

- Activity 2.10, "Decoding", (previously Activity 2.12, "Math Magic"), has been rewritten.

- Cluster 1 of Chapter 3, " Function Sense", has been changed as follows:

  - Activity 3.1, "Gold, Silver, and Bronze", replaces previous Activity 3.1. The new activity gives an expanded treatment of functions and includes graphing topics from Activity 2.2 from the previous edition.

  - Activity 3.3, "Symbolically Defined Functions and their Graphs", is new.

  - Activity 3.4, "Course Grade", (previously Activity 3.2), has been rewritten to include new situations.

- The remaining Activities in Chapter 3 have been streamlined and consolidated to eliminate duplication of topics without loss of content. The number of activities have been reduced and Clusters 2, 3 and 4 in previous edition have been combined to two clusters as follows:

  - Cluster 2, "Introduction to Linear Functions", contains Activities 3.5–3.8.

  - Cluster 3, "Linear Regressions, Systems, and Inequalities", contains Activities 3.9–3.14.

- Activity 3.9, "Education Pays", replaces Activity 3.10, "Oxygen for Fish".

MyMathLab Changes in the Fifth Edition

- Fully HTML5-compatible ebook, media, and exercises are device-aware and can now be used on mobile devices.

- A new video program provides:

  - Conceptual overview of many topics at high level, to help answer the question "how will I ever use this material?"

  - A walkthrough of lessons and examples at the learning outcome level.

- Exercise coverage has been enhanced to ensure better conceptual flow, encourage conceptual thinking about math topics, and balance out the coverage of skills related questions.

- A continuously adaptive study plan monitors student work and provides customized remediation, letting them see what they have mastered and targeting where they need further practice.

- A "Ready to Go" MyMathLab course option provides students with all the same great MyMathLab features, but makes it easier for instructors to get started. Each course includes pre-assigned homework and quizzes to make creating a course even simpler.

Supplements

Instructor Supplements

Annotated Instructor's Edition

ISBN-13 978-0-321-98239-1

ISBN-10 0-321-98239-8

This special version of the student text provides answers to all exercises directly beneath each problem.

Instructor's Resource Manual with Tests

ISBN-13 978-0-321-98240-7

ISBN-10 0-321-98240-1

This valuable teaching resource includes the following materials:

- Sample syllabi suggesting ways to structure the course around core and supplemental activities and within different credit-hour options.

- Sample course outlines containing timelines for covering topics.

- Teaching notes for each chapter. These notes are ideal for those using the *Mathematics in Action* approach for the first time.

- Extra practice worksheets for topics with which students typically have difficulty.

- Sample chapter tests and final exams for in-class and take-home use by individual students and groups.

- Information about incorporating technology in the classroom, including sample graphing calculator assignments.

TestGen®

ISBN-13 978-0-321-98241-4

ISBN-10 0-321-98241-X

TestGen enables instructors to build, edit, print, and administer tests using a computerized bank of questions developed to cover all the objectives of the text. TestGen is algorithmically based, allowing instructors to create multiple but equivalent versions of the same question or test with the click of a button. Instructors can also modify test bank questions or add new questions. The software and testbank are available for download from Pearson Education's online catalog.

Student Supplements

Worksheets for Classroom or Lab Practice

ISBN-13 978-0-321-98243-8

ISBN-10 0-321-98243-6

- Extra practice exercises for every section of the text with ample space for students to show their work.

- These lab- and classroom-friendly workbooks also list the learning objectives and key vocabulary terms for every text section, along with vocabulary practice problems.

- Concept Connection exercises, similar to the What Have I Learned? exercises found in the text, assess students' conceptual understanding of the skills required to complete each worksheet.

Supplements for Instructors and Students

MathXL® Online Course (access code required)

MathXL® is the homework and assessment engine that runs MyMathLab. (MyMathLab is MathXL and a learning management system.)

With MathXL, instructors can:

- Create, edit, and assign online homework and tests using algorithmically generated exercises correlated at the objective level to the textbook.
- Create and assign their own online exercises and import TestGen tests for added flexibility.
- Maintain records of all students' work tracked in MathXL's online gradebook.

With MathXL, students can:

- Take chapter tests in MathXL and receive personalized study plans and/or personalized homework assignments based on their test results.
- Use the study plan and/or the homework to link directly to tutorial exercises for the objectives they need to study.
- Access supplemental animations and video clips directly from selected exercises.

MathXL is available to qualified adopters. For more information, visit our Web site at www.mathxl.com, or contact your Pearson representative.

MyMathLab® Online Course (access code required)

MyMathLab from Pearson is the world's leading online resource in mathematics, integrating interactive homework, assessment, and media in a flexible, easy to use format. It provides **engaging experiences** that personalize, stimulate, and measure learning for each student. Moreover, it comes from an **experienced partner** with educational expertise and an eye on the future.

To learn more about how MyMathLab combines proven learning applications with powerful assessment, visit **www.mymathlab.com** or contact your Pearson representative.

Acknowledgments

The Consortium would like to acknowledge and thank the following people for their invaluable assistance in reviewing and testing material for this text in the past and current editions:

Mark Alexander, *Kapi'olani Community College*

Kathleen Bavelas, *Manchester Community College*

Shirley J. Beil, *Normandale Community College*

Carol Bellisio, *Monmouth University*

Ann Boehmer, *East Central College*

Barbara Burke, *Hawai'i Pacific University*

San Dong Chung, *Kapi'olani Community College*

Marjorie Deutsch, *Queensboro Community College*

Jennifer Dollar, *Grand Rapids Community College*

Irene Duranczyk, *University of Minnesota*

Kristy Eisenhart, *Western Michigan University*

Mary Esteban, *Kapiolani Community College*

Brian J. Garant, *Morton College*

Thomas Grogan, *Cincinnati State Technical and Community College*

Maryann Justinger, *Erie Community College—South Campus*

Brian Karasek, *South Mountain Community College*

Jim Larson, *Lake Michigan College*

Miriam Long, *Madonna University*

Ellen Musen, *Brookdale Community College*

Roberta Pardo, *Chandler-Gilbert Community College*

Kathy Potter, *St. Ambrose University*

Cindy Pulley, *Heartland Community College*

Robbie Ray, *Sul Ross State University*

Janice Roy, *Montcalm Community College*

Andrew S. Russell, *Queensborough Community Collge*

Amy C. Salvati, *Adirondack Community College*

Philomena Sawyer, *Manchester Community College*

Brenda Shepard, *Lake Michigan College*

Karen Smith, *US Blue Ash*

Kurt Verderber, *SUNY Cobleskill*

We would also like to thank our accuracy checkers, Jon Weerts and James Lapp.

Finally, a special thank you to our families for their unwavering support and sacrifice, which enabled us to make this text a reality.

The Consortium for Foundation Mathematics

To the Student

The book in your hands is most likely very different from any mathematics textbook you have seen before. In this book, you will take an active role in developing the important ideas of arithmetic and beginning algebra. You will be expected to add your own words to the text. This will be part of your daily work, both in and out of class. It is the belief of the authors that students learn mathematics best when they are actively involved in solving problems that are meaningful to them.

The text is primarily a collection of situations drawn from real life. Each situation leads to one or more problems. By answering a series of questions and solving each part of the problem, you will be led to use one or more ideas of introductory college mathematics. Sometimes, these will be basic skills that build on your knowledge of arithmetic. Other times, they will be new concepts that are more general and far reaching. The important point is that you won't be asked to master a skill until you see a real need for that skill as part of solving a realistic application.

Another important aspect of this text and the course you are taking is the benefit gained by collaborating with your classmates. Much of your work in class will result from being a member of a team. Working in small groups, you will help each other work through a problem situation. While you may feel uncomfortable working this way at first, there are several reasons we believe it is appropriate in this course. First, it is part of the learning-by-doing philosophy. You will be talking about mathematics, needing to express your thoughts in words. This is a key to learning. Secondly, you will be developing skills that will be very valuable when you leave the classroom. Currently, many jobs and careers require the ability to collaborate within a team environment. Your instructor will provide you with more specific information about this collaboration.

One more fundamental part of this course is that you will have access to appropriate technology at all times. You will have access to calculators and some form of graphics tool— either a calculator or computer. Technology is a part of our modern world, and learning to use technology goes hand in hand with learning mathematics. Your work in this course will help prepare you for whatever you pursue in your working life.

This course will help you develop both the mathematical and general skills necessary in today's workplace, such as organization, problem solving, communication, and collaborative skills. By keeping up with your work and following the suggested organization of the text, you will gain a valuable resource that will serve you well in the future. With hard work and dedication you will be ready for the next step.

The Consortium for Foundation Mathematics

Chapter

1

Number Sense

Your goal in this chapter is to use the numerical mathematical skills you already have—and those you will learn or relearn—to solve problems. Chapter activities are based on practical, real-world situations that you may encounter in your daily life and work. Before you begin the activities in Chapter 1, think about your previous encounters with mathematics and choose one word to describe those experiences.

Cluster 1 — Introduction to Problem Solving

Activity 1.1

The Bookstore

Objectives

1. Practice communication skills.

2. Organize information.

3. Write a solution in sentences.

4. Develop problem-solving skills.

By 11 A.M., a line has formed outside the crowded bookstore. You ask the guard at the gate how long you can expect to wait. She provides you with the following information: She is permitted to let 6 people into the bookstore only after 6 people have left; students are leaving at the rate of 2 students per minute; and she has just let 6 new students in. Also, each student spends an average of 15 minutes gathering books and supplies and 10 minutes waiting in line to check out.

Currently, 38 people are ahead of you in line. You know that it is a 10-minute walk to your noon class. Can you buy your books and still expect to make it to your noon class on time? Use the following questions to guide you in solving this problem.

1. What was your initial reaction after reading the problem?

2. Have you ever worked a problem such as this before?

3. Organizing the information will help you solve the problem.

   a. How many students must leave the bookstore before the guard allows more to enter?

   b. How many students per minute leave the bookstore?

   c. How many minutes are there between groups of students entering the bookstore?

   d. How long will you stand in line outside the bookstore?

  e. Now finish solving the problem and answer the question: How early or late for class will you be?

  4. In complete sentences, write what you did to solve this problem. Then explain your solution to a classmate.

SUMMARY: ACTIVITY 1.1

Steps in Problem Solving

1. Sort out the relevant information and organize it.

2. Discuss the problem with others to increase your understanding of the problem.

3. Write your solution in complete sentences to review your steps, and check your answer.

EXERCISES: ACTIVITY 1.1

1. Think about the various approaches you and your classmates used to solve Activity 1.1, The Bookstore. Choose the approach that is best for you, and describe it in complete sentences.

2. What mathematical operations and skills did you use?

Activity 1.2

The Classroom

Objectives

1. Organize information.

2. Develop problem-solving strategies.
  - Draw a picture.
  - Recognize a pattern.
  - Do a simpler problem.

3. Communicate problem-solving ideas.

The Handshake

This algebra course involves working with other students in the class, so form a group of 3, 4, or 5 students. Introduce yourself to every other student in your group with a firm handshake. Share some information about yourself with the other members of your group.

1. How many people are in your group?

2. How many handshakes in all were there in your group?

3. Discuss how your group determined the number of handshakes. Be sure everyone understands and agrees with the method and the answer. Write the explanation of the method here.

4. Share your findings with the other groups, and fill in the table.

NUMBER OF STUDENTS IN GROUP	NUMBER OF HANDSHAKES
2	
3	
4	
5	
6	
7	

5. a. Describe a rule for determining the number of handshakes in a group of 7 students.

b. Describe a rule for determining the number of handshakes in a class of n students.

6. If each student shakes hands with each other student, how many handshakes will be needed in your algebra class?

7. Is shaking hands during class time a practical way for students to introduce themselves? Explain.

George Polya's book *How to Solve It* outlines a four-step process for solving problems.

  i. Understand the problem (determine what is involved).

  ii. Devise a plan (look for connections to obtain the idea of a solution).

  iii. Carry out the plan.

  iv. Look back at the completed solution (review and discuss it).

8. Describe how your experiences with the handshake problem correspond with Polya's suggestions.

The Classroom

Suppose the tables in your classroom have square tops. Four students can comfortably sit at each table with ample working space. Putting tables together in clusters as shown will allow students to work in larger groups.

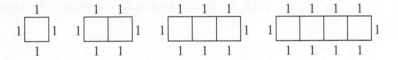

9. Construct a table of values for the number of tables and the corresponding total number of students.

NUMBER OF SQUARE TABLES IN EACH CLUSTER	TOTAL NUMBER OF STUDENTS
1	4
2	6

10. How many students can sit around a cluster of 7 square tables?

11. Describe the pattern that connects the number of square tables in a cluster and the total number of students that can be seated. Write a rule (as a complete sentence) that will determine the total number of students who can sit in a cluster of a given number of square tables.

12. There are 24 students in a math course at your college.

a. How many tables must be placed together to seat a group of 6 students?

b. How many clusters of tables are needed?

13. Discuss the best way to arrange the square tables into clusters given the number of students in your class.

SUMMARY: ACTIVITY 1.2

1. Problem-solving strategies include:

- discussing the problem
- organizing information
- drawing a picture
- recognizing patterns
- doing a simpler problem

2. George Polya's book *How to Solve It* outlines **a four-step process for solving problems**.

i. Understand the problem (determine what is involved).

ii. Devise a plan (look for connections to obtain the idea of a solution).

iii. Carry out the plan.

iv. Look back at the completed solution (review and discuss it).

EXERCISES: ACTIVITY 1.2

1. At the opening session of the United States Supreme Court, each justice shakes hands with all the others.

a. How many justices are there?

b. How many handshakes do they make?

Exercise numbers appearing in color are answered in the Selected Answers appendix.

2. Identify how the numbers are generated in this triangular arrangement, known as Pascal's triangle. Fill in the missing numbers.

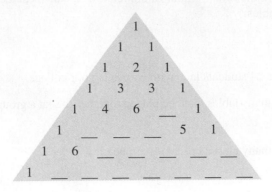

3. An **arithmetic sequence** is a list of numbers in which consecutive numbers share a common difference. Each number after the first is calculated by adding the common difference to the preceding number. For example, the arithmetic sequence 1, 4, 7, 10, . . . has 3 as its common difference. Identify the common difference in each arithmetic sequence that follows.

   a. 2, 4, 6, 8, 10, . . .

   b. 1, 3, 5, 7, 9, 11, . . .

   c. 26, 31, 36, 41, 46, . . .

4. A **geometric sequence** is a list of numbers in which consecutive numbers share a common ratio. Each number after the first is calculated by multiplying the preceding number by the common ratio. For example, 1, 3, 9, 27, . . . has 3 as its common ratio. Identify the common ratio in each geometric sequence that follows.

   a. 2, 4, 8, 16, 32, . . .

   b. 1, 5, 25, 125, 625, . . .

5. The operations needed to get from one number to the next in a sequence can be more complex. Describe a relationship shared by consecutive numbers in the following sequences.

   a. 2, 4, 16, 256, . . .

   b. 2, 5, 11, 23, 47, . . .

   c. 1, 2, 5, 14, 41, 122, . . .

6. In biology lab, you conduct the following experiment. You put two rabbits in a large caged area. In the first month, the pair produces no offspring (rabbits need a month to reach adulthood). At the end of the second month the pair produces exactly one new pair of rabbits (one male and one female). The result makes you wonder how many male/female pairs you might have if you continue the experiment and each existing pair of rabbits produces a new pair each month, starting after their first month. The numbers for the first 4 months are calculated and recorded for you in the following table. The arrows in the table illustrate that the number of pairs produced in a given month equals the number of pairs that existed at the beginning of the preceding month. Continue the pattern, and fill in the rest of the table.

MONTH	NUMBER OF PAIRS AT THE BEGINNING OF THE MONTH	NUMBER OF NEW PAIRS PRODUCED	TOTAL NUMBER OF PAIRS AT THE END OF THE MONTH
1	1	0	1
2	1	1	2
3	2	1	3
4	3	2	5
5			
6			
7			
8			

The list of numbers in the second column is called the **Fibonacci sequence**. This problem on the reproduction of rabbits first appeared in 1202 in the mathematics text *Liber Abaci*, written by Leonardo of Pisa (nicknamed Fibonacci). Using the first two numbers, 1 and 1, as a starting point, describe how the next number is generated. Your rule should generate the rest of the numbers shown in the sequence in column 2.

7. If you shift all the numbers in Pascal's triangle so that all the 1s are in the same column, you get the following triangle.

   a. Add the numbers crossed by each arrow. Put the sums at the tip of each arrow.

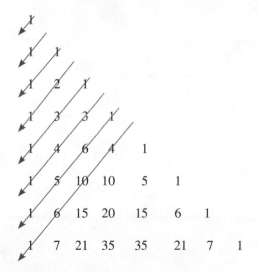

   b. What is the name of the sequence formed by these sums?

8. There are some interesting patterns within the Fibonacci sequence itself. Take any number in the sequence, and multiply it by itself; then subtract the product of the number immediately before it and the number immediately after it. What is the result? Pick another number, and follow the same procedure. What result do you obtain? Try two more numbers in the sequence.

For example, choose 5.

$5 \cdot 5 = 25$

$3 \cdot 8 = 24$

$25 - 24 = 1$

For example, choose 3.

$3 \cdot 3 = 9$

$2 \cdot 5 = 10$

$9 - 10 = -1$

Activity 1.3

Properties of Arithmetic

Objectives

1. Identify and use the commutative property in calculations.

2. Use the associative property to evaluate arithmetic expressions.

3. Use the order of operations convention to evaluate arithmetic expressions.

4. Identify and use the properties of exponents in calculations.

5. Convert numbers to and from scientific notation.

Counting, addition, and then subtraction are the first tools you learned to describe quantitative situations—from childhood games to tomorrow's weather forecast. By including multiplication and division as well, a person possesses the basic quantitative skills so fundamental to everyday life. In this activity, you will review the properties and vocabulary for arithmetic calculations and discover a compact way of writing very large numbers.

Terminology

The four operations of arithmetic—addition, subtraction, multiplication, and division—are called **binary** operations because they are all ways for combining two numbers at a time.

- When two numbers are *added*, the numbers are called **terms** and the result is called the **sum**.

- When two numbers are *subtracted*, the numbers are also called **terms** and the result is called the **difference**.

- When two numbers are *multiplied*, the numbers are called **factors** and the result is called the **product**.

- When two numbers are *divided*, the number being divided is called the **dividend**, the number it is divided by is called the **divisor**, and the result is called the **quotient**.

Commutative Property

1. Suppose 12 male and 15 female students are enrolled in your class.

   a. What arithmetic operation must you use to determine the total number of students in the class?

   b. How many students are in the class?

   c. Does it matter in which order you perform this operation?

   The fact that the sum of 12 and 15 is the same regardless of the order in which they are added illustrates the commutative property of addition.

> When the order in which two numbers are added is reversed, the sum remains the same. This property is called the **commutative property of addition** and can be written symbolically as
>
> $$a + b = b + a.$$

2. Is the commutative property true for the operation of subtraction? Multiplication? Division? Explain by giving examples for each operation.

SUBTRACTION	MULTIPLICATION	DIVISION

When the order in which two numbers (factors) are multiplied is reversed, the product remains the same. This property is called the **commutative property of multiplication** and can be written symbolically as

$$a \cdot b = b \cdot a.$$

Associative Property

3. What if more than two numbers are added? For example, consider the sum $3 + 5 + 7$.

    a. Perhaps the most obvious way to proceed is to add $3 + 5$ and then add the 7. What is the final result?

    b. Suppose you first add $5 + 7$ and then add this value to 3. What is the final result?

Problem 3 demonstrates that however you choose to group the numbers, $(3 + 5) + 7$ or $3 + (5 + 7)$, the sum remains the same. This property is called the **associative property of addition** and can be written symbolically as

$$(a + b) + c = a + (b + c).$$

Note that the parentheses are symbols used to group numbers. Operations contained within parentheses are performed first.

The associative property of addition combined with the commutative property of addition allows you to regroup and reorder the terms in any sum.

4. Explain how you might reorder and regroup the terms in the following sum to perform the calculation in your head easily and quickly: $15 + 6 + 5 + 4 + 7$.

5. What if more than two numbers are multiplied? Consider the product of three factors such as $7 \cdot 5 \cdot 2$?

    a. Calculate the value by first determining the product $7 \cdot 5$ and then multiplying by 2.

    b. Calculate the value by multiplying 7 by the product $5 \cdot 2$.

    c. How do your results from parts a and b compare?

Problem 5 demonstrates that however you choose to group the factors, $(7 \cdot 5) \cdot 2$ or $7 \cdot (5 \cdot 2)$, the product remains the same. This property is called the **associative property of multiplication** and can be written symbolically as.

$$(a \cdot b) \cdot c = a \cdot (b \cdot c).$$

The associative property of multiplication combined with the commutative property of multiplication allows you to regroup and reorder the factors in any product.

6. Explain how you might reorder and regroup the following factors to perform the calculation in your head easily and quickly: $2 \cdot 4 \cdot 5 \cdot 6$.

7. Is there an associative property of subtraction? For example, is $(12 - 5) - 3$ the same as $12 - (5 - 3)$?

8. Is there an associative property of division? For example, is $(24 \div 6) \div 2$ the same as $24 \div (6 \div 2)$?

Order of Operations

Arithmetic computations involving more than one operation may cause confusion. Therefore, a straightforward computational rule has been universally adopted that eliminates any uncertainty. It is called the **order of operations** convention.

> **Order of Operations**
>
> **1.** Multiplications and divisions are performed from left to right exactly in the order in which they appear. For example,
>
>   **a.** $24 \div 6 \div 2$ is evaluated as $(24 \div 6) \div 2 = 4 \div 2 = 2$.
>
>   **b.** $24 \div 6 \cdot 2$ is evaluated as $(24 \div 6) \cdot 2 = 4 \cdot 2 = 8$.
>
> **2.** Additions and subtractions are performed from left to right exactly in the order in which they appear. For example,
>
>   **a.** $12 - 5 - 3$ is evaluated as $(12 - 5) - 3 = 7 - 3 = 4$.
>
>   **b.** $12 - 5 + 3$ is evaluated as $(12 - 5) + 3 = 7 + 3 = 10$.
>
> **3.** The operations of multiplication and division have priority over addition and subtraction. Therefore, multiplication and division operations are performed first, then the additions and subtractions are done. For example,
>
> $20 - 12 \div 4 + 1$ is evaluated as $20 - (12 \div 4) + 1 = 20 - 3 + 1 = (20 - 3) + 1 = 18$.

9. Perform the following calculations without a calculator.

   a. $42 \div 3 + 4$ **b.** $42 + 8 \div 4$

   c. $24 \div 6 - 2 \cdot 2$ **d.** $4 + 8 \div 2 \cdot 3 - 9$

10. a. Describe in words the order of operations required to correctly calculate $\dfrac{24}{2+6}$, and then perform the calculation.

b. In horizontal format, can the expression $\dfrac{24}{2+6}$ be written as $24 \div 2 + 6$? Explain.

Note that $\dfrac{24}{2+6}$ is the same as the quotient $\dfrac{24}{8}$. Therefore, the addition in the denominator is done first, followed by the division. To write $\dfrac{24}{2+6}$ in a horizontal format, you must use parentheses to group the expression in the denominator to indicate that the addition is performed first. That is, write $\dfrac{24}{2+6}$ as $24 \div (2+6)$.

When you write a meaningful combination of numbers and arithmetic operations, the result is called an **arithmetic expression**. If the expression contains more than one operation, then the last operation to be performed classifies the expression as a sum, difference, product, or quotient. For example, the expression in Problem 10 above is called a quotient because division is performed last.

Example 1   *Evaluate $2(3 + 4 \cdot 5)$, and classify this expression as a sum, difference, product, or quotient.*

SOLUTION

$2(3 + 4 \cdot 5)$ **First evaluate the arithmetic expression in parentheses using the order of operations indicated. That is, multiply $4 \cdot 5$ and then add 3.**

$= 2(3 + 20)$

$= 2 \cdot 23$ **Finally, multiply 23 by 2.**

$= 46$

The expression is a product.

---

11. Evaluate the following, and classify the expression as a sum, difference, product, or quotient.

a. $\dfrac{6}{3+3}$ **b.** $\dfrac{2+8}{4-2}$

c. $5 + 2(4 \div 2 + 3)$ **d.** $10 - (12 - 3 \cdot 2) \div 3$

Exponential Expressions

Exponents and exponential notation are extremely useful in simplifying an expression containing a repeated multiplication such as $3 \cdot 3$ or $5 \cdot 5 \cdot 5 \cdot 5$.

In the first case, $3 \cdot 3$, the factor 3 occurs 2 times: The repeated factor, 3, is called the **base;** the number of times it occurs is called the **exponent;** and the product $3 \cdot 3$ can be written as 3^2 using **exponential notation.** This notation is read as "3 raised to the second power," or simply "3 to the second power." Note that the exponent always appears as a superscript to the right of the base.

In the second case, $5 \cdot 5 \cdot 5 \cdot 5$, the factor 5 occurs 4 times and the product $5 \cdot 5 \cdot 5 \cdot 5$ can be written as 5^4. This notation is read as "5 to the fourth power." When the exponents are either 2 or 3, the expressions are more commonly called "squared" or "cubed," respectively. That is, 4^2 and 4^3 are usually read as "4 squared" and "4 cubed," respectively.

The exponents considered here are **natural numbers** (the familiar positive counting integers $1, 2, 3, \ldots$). Later in the textbook, you will encounter and learn the meanings of other numerical exponents.

Square Roots

Closely related to raising a number to the second power is its "inverse" operation of taking a square root. For example, because $5^2 = 25$ (read "5 **squared**" $= 25$), you say that the "**square root**" of 25 is 5. The radical sign $\sqrt{}$ is used to signify a square root, and you write $\sqrt{25} - 5$.

Like parentheses, the radical sign is a grouping symbol indicating that all operations inside the radical are performed *before* determining the square root. For example,

$$\sqrt{36 + 64} = \sqrt{100} = 10 \quad \text{and} \quad \sqrt{\frac{50}{8 - 3 \cdot 2}} = \sqrt{\frac{50}{2}} = \sqrt{25} = 5.$$

Numbers such as 100 and 25 are called **perfect squares** because they can each be expressed as the square of an integer, 10^2 and 5^2 respectively.

Definition

A number is called a **perfect square** if it can be expressed as the square of an integer.

Procedure

Order of operations convention for evaluating arithmetic expressions

Perform operations inside all grouping symbols first. This includes expressions in the numerator and denominator of a fraction, inside a radical, and within a pair of parentheses. All operations are performed in the following order.

1. Evaluate all exponential expressions as you read the expression from left to right.

2. Perform all multiplication and division as you read the expression from left to right.

3. Perform all addition and subtraction as you read the expression from left to right.

Example 2   *Evaluate the expression* $20 - 2 \cdot 3^2$.

SOLUTION

$20 - 2 \cdot 3^2$	Evaluate all exponential expressions as you read the arithmetic expression from left to right.
$= 20 - 2 \cdot 9$	Perform all multiplications and divisions as you read the expression from left to right.
$= 20 - 18$	Perform all additions and subtractions as you read the expression from left to right.
$= 2$	

The expression is a difference.

12. Evaluate the following. Classify each expression as a sum, difference, product, or quotient.

    a. $6 + 3 \cdot 4^3$ **b.** $2 \cdot 3^4 - 5^3$

    c. $(12 - 2^3) \cdot (9 - 6 \div 3)$ **d.** $\dfrac{128}{16 - 2^3}$

Using a Calculator to Evaluate Arithmetic Expressions

The calculator is a powerful tool in the problem-solving process. A quick look at the desks around you probably shows that calculators come in many sizes and shapes and with various capabilities.

- Some calculators perform only the four basic arithmetic operations of addition, subtraction, multiplication, and division (and perhaps square roots). These calculators are *not recommended* for use in this course; they are meant for calculations containing a single operation. They perform arithmetic operations one at a time exactly in the order in which you enter them and *do not* follow the order of operations convention.

- Other calculators contain parentheses keys and handle operations with exponents, trigonometry, and fractions. Some even have statistical capabilities. These are classified as **scientific calculators**. They are designed to follow the order of operations convention and are recommended for use in this course.

- Enhanced scientific calculators graph equations and generate tables of values. Some even manipulate algebraic symbols. These are called **graphing calculators** and are also acceptable for use in this course. **Note:** This course requires that you have access to at least a scientific calculator. If you have a TI-84 Plus or Plus C graphing calculator, you can find detailed information on many of its features in Appendix E.

- Many reliable online graphing calculator simulators are available on the Internet. In addition, powerful graphing calculator apps that you may find helpful are available on smartphones and tablets.

When evaluating arithmetic expressions containing more than one operation, you should enter the complete expression into the calculator before you press the ☐ key. This guarantees that any necessary rounding is done at the very end, giving you the most accurate result. When you evaluate the expression in pieces you are also more likely to introduce key stroke errors. Entering an expression into your calculator requires the same notation used to write the expression in a horizontal format. In particular, you often will need to include parentheses—especially when fractions or square roots are involved. For example,

$\dfrac{12 + 4}{2 + 6}$ must be entered as $(12 + 4) \div (2 + 6)$ and $\sqrt{25 + 11}$ must be entered as $\sqrt{(25 + 11)}$.

13. Evaluate the following expressions by hand, and then confirm the results on your calculator.

   a. $\dfrac{30}{2 + 3}$ **b.** $\dfrac{32}{4 \cdot 2}$ **c.** $3\sqrt{9} + 16$ **d.** $\sqrt{25 + 24} + 8$

14. Use your calculator to evaluate the following expressions, and round your results to 2 decimal places.

   a. $\dfrac{2.5}{4.2 + 3.6}$ **b.** $\dfrac{18}{2\pi}$ **c.** $\dfrac{2}{3}\sqrt{75}$ **d.** $\sqrt{28 - 5 \cdot 2}$

An exponent key is found on all scientific and graphing calculators, but the notation on the face of the key and the required keystrokes can differ among calculators.

- Some calculators have a key labeled x^2. As indicated, this key only performs the squaring operation. For example, to calculate 5^2, input 5 and then press the x^2 key.

- Many calculators have a caret key labeled ☐^. This key performs any exponential operation using the keystroke sequence [base] ☐^ [exponent] . For example, to calculate 3^4, use the keystroke sequence (3) (^) (4).

- Other calculators have a key labeled y^x or x^y. This key also performs any exponential operation. For example, to calculate 3^4, use the keystroke sequence (3) (y^x) (4).

15. Use your calculator to evaluate the following expressions. Round your results to 2 decimal places.

   a. $\dfrac{4 \cdot 3^2}{21 - 2^3}$ **b.** $\sqrt{6^2 + 7^2}$ **c.** $6.5\,(1.23)^4$ **d.** $\dfrac{128}{2.5^3}$

16. **a.** Use the exponent key to evaluate the following powers of 10: 10^2, 10^3, 10^4, and 10^5. What relationship do you notice between the exponent and the number of zeros in the result?

   b. Evaluate 10^6. Is the result what you expected? How many zeros are in the result?

c. Complete the following table.

EXPONENTIAL FORM	EXPANDED FORM	VALUE
10^5	$10 \cdot 10 \cdot 10 \cdot 10 \cdot 10$	
10^4	$10 \cdot 10 \cdot 10 \cdot 10$	
10^3		
10^2		
10^1		

17. a. Follow the patterns in the preceding table and determine a reasonable value for 10^0.

b. Use your calculator to evaluate 10^0. Is the result what you expected? How many zeros are in the result?

c. Use the exponent key to evaluate the following powers: 3^0, 8^0, 23^0, and 526^0.

d. Evaluate other nonzero numbers with a zero exponent.

e. Write a sentence describing the result of raising a nonzero number to a zero exponent.

Scientific Notation

Very large numbers appear in a variety of situations. For example, in the year 2012, the total personal income of all Americans was estimated to be $13,401,868,693,000. Because this number contains so many digits, it is difficult to read and to use in computations.

18. a. Use place values to write the 2012 total personal income of all Americans in words.

b. Round this number to the nearest million and to the nearest billion.

Scientific notation presents a convenient way to write this and other very large numbers.

19. a. The number 5000 is said to be written in **standard notation**. Because 5000 is 5×1000, the number 5000 can also be written as 5×10^3. Use your calculator to verify this by evaluating 5×10^3.

b. Your calculator provides another way to evaluate a number multiplied by a power of 10, such as 5×10^3. Find the key labeled (EE) or (E) or (EXP); sometimes it is a second function. This key takes the place of the (×) (1) (0) and (^) keystroke sequence. To evaluate 5×10^3, enter 5, press the (EE) key, and then enter 3. Try this now and verify the result.

A number written as the product of a decimal number between 1 and 10 and a power of 10 is said to be written in **scientific notation**.

Example 3 *Write 9,420,000 in scientific notation.*

SOLUTION

9,420,000 can be written in scientific notation as follows:

Place a decimal point immediately to the right of the first digit of the number: 9.420000.

Count the number of digits to the right of the decimal point. This number will be the exponent. In this example, these digits are 420000, that is, 6 digits.

Drop all the trailing zeros in the decimal number, and then multiply by 10^6.

Therefore, $9,420,000 = 9.42 \times 10^6$.

The number 9,420,000 can be entered into your calculator in scientific notation using the $\boxed{\text{EE}}$ or $\boxed{\text{E}}$ key as follows: 9.42 $\boxed{\text{EE}}$ 6. Try it.

20. Write each of the following numbers in scientific notation.

    a. 5120   **b.** 2,600,000

21. **a.** Write each of the following numbers in standard notation.

    i. 4.72×10^5 **ii.** 3.5×10^{11}

    b. Describe the process you used to convert a number from scientific notation to standard notation.

22. Input 4.23 $\boxed{\text{E}}$ 2 into your calculator, press the $\boxed{\text{ENTER}}$ or $\boxed{=}$ key, and see if you obtain the result that you expect.

Note: Scientific notation can also be used in writing very small numbers such as 0.00000027. This is discussed in Activity 1.14.

Computations Using Scientific Notation

One advantage of using scientific notation becomes evident when you must perform computations involving very large numbers.

23. **a.** The average distance from Earth to the Sun is 93,000,000 miles. Write this number in words and in scientific notation. Then use the $\boxed{\text{EE}}$ key to enter it into your calculator.

b. Estimate 5 times the average distance from Earth to the Sun.

c. Use your calculator to verify your estimate from part b. Write the answer in standard notation, in scientific notation, and in words.

24. There are heavenly bodies that are thousands of times farther away from Earth than is the Sun. Multiply 93,000,000 miles by 1000. Write the result in scientific notation, in standard notation, and in words.

25. **a.** Write the total U.S. personal income in 2012, estimated to be $13,402,000,000,000 rounded to the nearest billion, in scientific notation.

    b. The population of the United States in 2012 is estimated to have been 314,000,000. Write this number in scientific notation.

    c. Determine the per capita personal income (total income divided by population) of the United States in 2012 in both standard and scientific notation.

26. Use your calculator to perform each of the following calculations. Then express each result in standard notation.

    a. $(3.26 \times 10^4)(5.87 \times 10^3)$

    b. $\dfrac{750 \times 10^{17}}{25 \times 10^{10}}$

    c. $(25 \times 10^3) + (750 \times 10^2)$

SUMMARY: ACTIVITY 1.3

1. The four operations of arithmetic—addition, subtraction, multiplication, and division—are called **binary** operations because they are rules for combining two numbers at a time.

   • When two numbers are *added*, the numbers are called **terms** and the result is called the **sum**.

   • When two numbers are *subtracted*, the numbers are also called **terms** and the result is called the **difference**.

- When two numbers are *multiplied*, the numbers are called **factors** and the result is called the **product**.

- When two numbers are *divided*, the number being divided is called the **dividend**, the number it is divided by is called the **divisor**, and the result is called the **quotient**.

2. The **commutative property** states that changing the order in which you add or multiply two numbers does not change the result. The **commutative property of addition** is written symbolically as

$$a + b = b + a,$$

and the **commutative property of multiplication** is written symbolically as

$$a \cdot b = b \cdot a.$$

The commutative property does not hold for subtraction or division.

3. The **associative property of addition** states that however you choose to group the terms of a sum, the result remains the same. This property is written symbolically as

$$(a + b) + c = a + (b + c).$$

The **associative property of multiplication** states that however you choose to group the factors of a product, the result remains the same. This property is written symbolically as

$$(a \cdot b) \cdot c = a \cdot (b \cdot c).$$

The associative property does not hold for subtraction or division.

4. A repeated multiplication, such as $5 \cdot 5 \cdot 5$, can be written as 5^3 using **exponential notation**. This notation is read as "5 to the third power." The repeated factor, 5, is called the **base**, and 3, the number of times this factor appears, is called the **exponent**. When the exponent is either 2 or 3, the expression is commonly called **squared** or **cubed**, respectively.

5. An exponent of 0 has its own special definition. Any number, except zero, raised to the zero power is defined to have the value 1. That is, for any number $a \neq 0$, $a^0 = 1$. However, 0^0 has no numerical value—it is undefined.

6. For any positive number, a, the symbol $\sqrt{a}$ is called the **square root** of a.

7. An **arithmetic expression** consists of a meaningful combination of numbers and arithmetic operations, including exponents and radical signs.

8. **Order of operations convention** for evaluating arithmetic expressions:

   Perform operations inside all grouping symbols first. This includes expressions in the numerator and denominator of a fraction, inside a radical, and within a pair of parentheses. All operations are performed in the following order.

   a. Evaluate all exponential expressions as you read the expression from left to right.

   b. Perform all multiplications and divisions as you read the expression from left to right.

   c. Perform all additions and subtractions as you read the expression from left to right.

9. A number is expressed in **scientific notation** when it is written as the product of a decimal number between 1 and 10 and a power of 10.

EXERCISES: ACTIVITY 1.3

1. Evaluate each expression, and use your calculator to check your answers.

a. $7(20 + 5)$ **b.** $7 \cdot 20 + 5$ **c.** $20 + 7 \cdot 5$

2. Evaluate each expression, and use your calculator to check your answers.

$20(100 - 2)$ **b.** $20 \cdot 100 - 2$ **c.** $100 - 20 \cdot 2$

3. Use the order of operations convention to evaluate each expression.

a. $17(50 - 2)$ **b.** $(90 - 7)5$

4. Perform the following calculations without a calculator. Then use your calculator to verify your result.

a. $45 \div 3 + 12$ **b.** $54 \div 9 - 2 \cdot 3$ **c.** $26 + 2 \cdot 7 - 12 \div 4$

5. a. Explain why the result of $72 \div 8 + 4$ is 13.

b. Explain why the result of $72 \div (8 + 4)$ is 6.

6. Evaluate the following, and verify on your calculator.

a. $48 \div (4 + 4)$ $\dfrac{8 + 12}{6 - 2}$

c. $120 \div (6 + 4)$ **d.** $64 \div (6 - 2) \cdot 2$

e. $(16 + 84) \div (4 \cdot 3 - 2)$ **f.** $(6 + 2) \cdot 20 - 12 \div 3$

g. $39 + 3 \cdot (8 \div 2 + 3)$ **h.** $100 - (81 - 27 \cdot 3) \div 3$

7. Evaluate the following. Use a calculator to verify your answers.

a. $5 \cdot 2^4 - 3^3$ **b.** $8 + 12 \div 2 \cdot 2^2$ **c.** $8 + 2 \cdot 6 - 3$

Exercise numbers appearing in color are answered in the Selected Answers appendix.

8. Evaluate each of the following arithmetic expressions by performing the operations in the appropriate order. Use a calculator to check your results.

a. $4^2 + 3^2$

b. $(4 + 3)^2$

c. $(3^2 - 4)^2$

d. $\dfrac{4^2 - 6}{2}$

e. $4^2 - 6 \div 2$

f. $36 \div 3 \cdot 2 + 1$

g. $36 \div 3 \cdot (2 + 1)$

h. $12 \div 2^2 \cdot 3$

i. $5 + 2 \cdot 3^2$

9. Evaluate the following expressions by hand, and then confirm the results on your calculator.

a. $\dfrac{54}{6 \cdot 3}$

b. $\dfrac{48}{4 + 2}$

c. $4\sqrt{64 + 36}$

10. Use your calculator to evaluate the following expressions. Round your results to 2 decimal places.

a. $\dfrac{28}{4 + 7}$

b. $\dfrac{3}{8}\sqrt{32}$

c. $\sqrt{45 - 5 + 2 \cdot 7}$

d. $\dfrac{3 \cdot 5^3}{(12 - 7.5)^2}$

e. $\sqrt{8^2 + 9^2}$

f. $4.2(1.15)^6$

11. The following numbers are written in standard notation. Convert each number to scientific notation.

a. 213,040,000,000

b. 555,140,500,000,000

12. The following numbers are written in scientific notation. Convert each number to standard notation.

a. 4.532×10^{11}

b. 4.532×10^7

13. The distance that light travels in 1 second is 186,000 miles. How far will light travel in 1 year? (There are approximately 31,500,000 seconds in 1 year.) Change the numbers to scientific notation and then perform the appropriate operations.

14. The total amount of taxes collected by the U.S. government in 2012 was $2.35 trillion. At that time, the U.S. population was approximately 314 million. If the total tax collections were evenly divided among all Americans, how much would each citizen have paid (rounded to the nearest dollar)? This amount is called the tax per capita.

15. The human heart beats approximately 70 times per minute. If you live for 80 years, determine how many times your heart beats during your lifetime. Express your answer in scientific notation, where the decimal factor is rounded to 2 decimal places.

16. The land area of Alaska is approximately 3.66×10^8 acres. The state was purchased from Russia in 1867 for $7.2 million. Determine the price per acre to the nearest cent that the United States paid Russia.

Internet Exploration

17. The Web search engine Google is named after a *googol*. Sergey Brin, cofounder of Google Inc. and a mathematics major, chose the name Google to describe the extent of the power of the search engine. How many zeroes are in a *googol* ? Write the number in scientific notation.

18. a. What is engineering notation? Is there an advantage of using engineering notation over scientific notation? Explain.

b. Complete the following table.

NUMBER	SCIENTIFIC NOTATION	ENGINEERING NOTATION
37,000		
47,900,000,000		

Cluster 1 | What Have I Learned?

Activities 1.1 and 1.2 gave you an opportunity to develop some problem-solving strategies. Apply the skills you used in this cluster to solve the following problems.

1. As you settle down to read a chapter for tomorrow's class, you notice a group of students forming a circle outside and beginning to randomly kick an odd-looking ball from person to person. You notice that 12 students are in the circle and that they are able to keep the object in the air as they kick it. Sometimes they kick it to the person next to them; other times they kick it to someone across the circle.

You later learn that this is a refined version of the original Hacky Sack game invented in 1972. Hacky Sack has regained popularity internationally and online with blogs and Facebook groups dedicated to promoting a resurgence of the game.

a. Suppose each student kicks the Hacky Sack exactly once to each of the others in the circle. How many total kicks would that take?

b. One student in the circle invites you and another student to join them for a total of 14. How many kicks will it take now if each student kicks the Hacky Sack exactly once to each of the other players? How do you arrive at your answer?

c. George Polya's book *How to Solve It* outlines a four-step process for solving problems.

  i. Understand the problem (determine what is involved).

  ii. Devise a plan (look for connections to obtain the idea of a solution).

  iii. Carry out the plan.

  iv. Look back at the completed solution (review and discuss it).

Describe how your procedures in parts a and b correspond with Polya's suggestions.

2. You are assigned to read *War and Peace* for your literature class. The edition you have contains 1232 pages. You time yourself and estimate that you can read 12 pages in 1 hour. You have 5 days before your exam on this book. Will you be able to finish reading it before the exam?

3. You purchase 3 lbs of American cheese at $4 per pound, 4 lbs of Swiss cheese at $5 per pound and 2 lbs of imported Gruyere cheese at $7 per pound.

   a. Write and evaluate an arithmetic expression that represents the amount (lbs) of cheese that you bought.

   b. Write and evaluate an arithmetic expression that represents the total cost of the cheese.

   c. Write and evaluate an arithmetic expression that represents how much more the Swiss cheese order cost you than did the Gruyere.

   d. You hand the cashier 3 ten-dollar bills and 4 five-dollar bills. Write and evaluate an arithmetic expression that represents your change.

4. Your part-time job pays $9 per hour, $3 per hour less than your older brother earns. How much does your brother earn working a 20-hour week?

5. At the flea market you sold 18 tee shirts and 25 tank tops, earning $294 for the day. If the tank tops sold for $6 apiece, what did you charge for the tee shirts?

6. Given the algebraic expression $24 \div 3 \cdot 2^2 + 4$:

   a. Indicate the order in which you must perform the operations.

   b. Evaluate the expression.

   c. What is the value of the expression if parentheses are inserted as follows:

   i. $24 \div 3 \cdot (2^2 + 4)$

   ii. $24 \div (3 \cdot 2^2) + 4$

Cluster 1   How Can I Practice?

1. Describe the relationship shared by consecutive numbers in the following sequences.

   a. $1, 4, 7, 10, \ldots$

   b. $1, 2, 4, 8, \ldots$

   c. $1, 3, 7, 15, 31, 63, \ldots$

2. Evaluate each of the following arithmetic expressions by performing the operations in the appropriate order. Use your calculator to check your results.

   a. $4(6 + 3) - 9 \cdot 2$

   b. $5 \cdot 9 \div 3 - 3 \cdot 4 \div 6$

   c. $2 + 3 \cdot 4^3$

   d. $\dfrac{256}{28 + 6^2}$

   e. $7 - 3(8 - 2 \cdot 3) + 2^2$

   f. $3 \cdot 2^5 + 2 \cdot 5^2$

   g. $144 \div (24 - 2^3)$

   h. $\dfrac{36 - 2 \cdot 9}{6}$

   i. $9 \cdot 5 - 5 \cdot 2^3 + 5$

   j. $1^5 \cdot 5^1$

   k. $2^3 \cdot 2^0$

   l. $7^2 + 7^2$

   m. $(3^2 - 4 \cdot 0)^2$

   n. $(\sqrt{25 - 9})^2$

   o. $\sqrt{5^2 + 12^2}$

   p. $\sqrt{\dfrac{16}{9}}$

   q. $\dfrac{\sqrt{16}}{9}$

   r. $\dfrac{48}{\sqrt{12 + 24}}$

Exercise numbers appearing in color are answered in the Selected Answers appendix.

3. Write each number in standard notation.

    a. 8^0 **b.** 12^2 **c.** 3^4 **d.** 2^6

4. a. Write 214,000,000,000 in scientific notation.

    b. Write 7.83×10^4 in standard notation.

5. A newly discovered binary star, Shuart 1, is located 185 light-years from Earth. One light-year is 9,460,000,000,000 kilometers. Express the distance to Shuart 1 in kilometers. Write the answer in scientific notation.

6. The human brain contains about 100,000,000,000 nerve cells. If the average cell contains about 200,000,000,000,000 molecules, determine the number of molecules in the nerve cells of the human brain.

7. Identify the arithmetic property expressed by each numerical statement.

    a. $25 \cdot 30 = 30 \cdot 25$

    b. $(15 \cdot 9) \cdot 3 = 15 \cdot (9 \cdot 3)$

8. Determine whether each of the following numerical statements is true or false. In each case, justify your answer.

    a. $7(20 + 2) = 7 \cdot 22$

    b. $25 - 10 - 4 = 25 - (10 - 4)$

9. In a recent promotion for a hot chocolate drink, the following question was featured on the box of the mix: On average, how many mini marshmallows are in one serving? The answer is

$$3 + 2 \cdot 4 \div 2 - 3 \cdot 7 - 4 + 48.$$

    a. The box gave 93 as the answer. What is the correct answer?

    b. Explain the error that somebody at the company made in calculating the answer.

| **Cluster 2** | **Problem Solving with Fractions and Decimals (Rational Numbers)** |

Activity 1.4

Top Chef

Objectives

1. Add and subtract fractions.

2. Multiply and divide fractions.

Fractions are everywhere in everyday life. A store in the mall advertises a half-off sale. A friend divides a homemade pizza into six equal portions to share with the group. Your band requests one-quarter of the space on the dance floor. Perhaps nowhere else is the use of fractions more prevalent than in the kitchen. Cooking has always been a favorite family activity. Today, television cooking shows with chefs such as Bobby Flay and Giada De Laurentiis are very popular.

The purpose of this activity is to review the terminology and operations associated with fractions. This discussion on fractions can be supplemented by the material in Appendix A, which contains detailed explanations, definitions, examples, and practice exercises with worked-out solutions.

Rational Numbers

Appendix

You decide to have some friends over to watch a video. You are eager to try a new pizza recipe that you saw on the Food Network. You plan to make three large pizzas of equal size. Because some friends are hungrier than others, you cut the first pizza into 4 equal-sized large pieces, the second pizza into 8 equal-sized medium pieces, and the third pizza into 16 equal-sized small pieces.

Pizza 1: Large pieces   Pizza 2: Medium pieces   Pizza 3: Small pieces

4
Slices

8
Slices

16
Slices

A fraction is classified by its **denominator**. The denominator indicates the total number of equal parts into which a unit or whole has been divided. The **numerator** indicates the number of parts being counted. Fractions with the same denominator are called **like fractions**.

1. a. If you eat five small pieces, what fractional part of the whole pizza did you eat? Explain.

b. What is the denominator of this fraction? What does the denominator represent in this context?

c. What is the numerator of this fraction? What does this number represent?

The fraction $\frac{5}{16}$ indicates that the unit has been divided into 16 equal parts, each representing $\frac{1}{16}$ of the unit, and that you are counting 5 of these parts. So $\frac{5}{16}$ can also be thought of as $5 \cdot \frac{1}{16}$.

A fraction that has an integer numerator and a nonzero integer denominator is called a **rational number**.

Definition

A **rational number** is a number that can be written in the form $\frac{a}{b}$, where a and b are integers and b is not zero. Every integer is also a rational number because any integer a can be written as $\frac{a}{1}$.

Note that the denominator of a fraction cannot be zero. Division by 0 is discussed in more detail in Activity 1.13, page 121.

2. As you are cutting the pizzas, you notice that two medium pieces together are the same size as one large piece. Therefore, $\frac{2}{8}$ of the pizza represents the same portion as $\frac{1}{4}$.

   a. How many small pieces represent the same portion as one large piece?

   b. What fraction of the pizza does the number of small pieces in part a represent?

In Problem 2, three different fractions were used to represent the same portion of a whole pizza. These fractions, $\frac{1}{4}$, $\frac{2}{8}$, and $\frac{4}{16}$, are called **equivalent fractions**. They represent the same quantity.

The fraction $\frac{1}{4}$ is said to be in **lowest terms** because the largest factor common to both the numerator and denominator of the fraction is 1. For the fraction $\frac{4}{16}$, the largest factor common to both 4 and 16 is 4. To reduce a fraction to lowest terms, divide both the numerator and denominator by the greatest common factor.

3. Reduce the fraction $\frac{4}{16}$ to an equivalent fraction in lowest terms.

Adding and Subtracting Like Fractions

You want to bake zucchini bread and cookies for snacking. The recipes call for $\frac{1}{4}$ teaspoon of baking powder for the bread and $\frac{3}{4}$ teaspoon of baking powder for the cookies. How much baking powder do you need for both recipes? The amount of baking powder in this situation involves the addition of like fractions.

Adding like fractions means counting the number of parts contained in all the fractions. For example, consider the sum $\frac{3}{5} + \frac{4}{5}$. Because the first fraction contains 3 parts $\left(\frac{1}{5}s\right)$ and the second contains 4 parts $\left(\frac{1}{5}s\right)$, together there are 7 parts $\left(\frac{1}{5}s\right)$ so that $\frac{3}{5} + \frac{4}{5} = \frac{7}{5}$, or $1\frac{2}{5}$.

4. Now, determine the amount of baking powder you need for the bread and cookie recipes $\left(\frac{1}{4}\text{ teaspoon for bread and }\frac{3}{4}\text{ teaspoon for cookies}\right)$. Explain the procedure you used.

Adding Fractions having Different Denominators

Now suppose you use a different recipe for the bread that calls for $1\frac{1}{2}$ teaspoons of baking powder. The number $1\frac{1}{2}$ is called a **mixed number** and represents the sum of $1 + \frac{1}{2}$. Because $1 = \frac{2}{2}$, you have

$$1 + \frac{1}{2} = \frac{2}{2} + \frac{1}{2} = \frac{3}{2}.$$

Note: The number $\frac{3}{2}$ is called an **improper fraction**.

Therefore, the total amount of baking soda needed for the new bread recipe and the cookies is the sum

$$1\frac{1}{2} + \frac{3}{4} = \frac{3}{2} + \frac{3}{4}.$$

5. How is the addition problem $\frac{3}{2} + \frac{3}{4}$ different from the addition exercise in Problem 4?

If the fractions to be added have different denominators, both fractions must be expressed as equivalent fractions with identical denominators prior to addition. The common denominator for a set of fractions is determined by finding the smallest number, called the **least common denominator** (LCD), that is divisible by all the denominators. For example, to determine the sum

$$\frac{1}{4} + \frac{5}{6},$$

Step 1. Calculate the LCD. Because the smallest number that is divisible by 4 and 6 is 12, the LCD is 12.

Step 2. Write each fraction as an equivalent fraction having 12 as its denominator.

$$\frac{1}{4} = \frac{1 \cdot 3}{4 \cdot 3} = \frac{3}{12} \text{ and } \frac{5}{6} = \frac{5 \cdot 2}{6 \cdot 2} = \frac{10}{12}$$

Step 3. Add the equivalent fractions.

$$\frac{1}{4} + \frac{5}{6} = \frac{3}{12} + \frac{10}{12} = \frac{13}{12}, \text{ or } 1\frac{1}{12}.$$

6. Now, determine the amount of baking powder you need for the new bread recipe $\left(\dfrac{3}{2}\text{ teaspoons}\right)$ and the cookie recipe $\left(\dfrac{3}{4}\text{ teaspoons}\right)$.

7. Determine the following sums, expressing each sum as a fraction in lowest terms.

a. $\dfrac{1}{3} + \dfrac{1}{2}$ **b.** $\dfrac{2}{7} + \dfrac{1}{3}$

c. $\dfrac{3}{10} + \dfrac{1}{4}$ **d.** $\dfrac{3}{8} + \dfrac{1}{20}$

e. $\dfrac{1}{4} + \dfrac{1}{6} + \dfrac{3}{8}$ **f.** $\dfrac{5}{12} + \dfrac{1}{8} + \dfrac{3}{16}$

g. $\dfrac{2}{11} + \dfrac{1}{2} + \dfrac{1}{4}$ **h.** $\dfrac{1}{5} + \dfrac{1}{4} + \dfrac{1}{3}$

i. $\dfrac{1}{7} + \dfrac{1}{6} + \dfrac{1}{3}$

Multiplying Fractions

The recipe for chocolate chip cookies includes the following ingredients:

$$2\frac{1}{4}\text{ cups flour, } \frac{1}{2}\text{ tsp. salt, } 1\frac{1}{3}\text{ tsp. baking soda}$$

$$\frac{3}{4}\text{ cup sugar, and } 1\frac{2}{3}\text{ cups chocolate chips}$$

How much of each ingredient would you need to double or triple the recipe? This calculation involves multiplication of fractions and can be done by rearranging the factors.

For example, consider the product $\dfrac{2}{3} \cdot \dfrac{5}{4}$. Because the factor $\dfrac{2}{3}$ can be thought of as $2 \cdot \dfrac{1}{3}$ and factor $\dfrac{5}{4}$ can be thought of as $5 \cdot \dfrac{1}{4}$, the product $\dfrac{2}{3} \cdot \dfrac{5}{4} = \left(2 \cdot \dfrac{1}{3}\right) \cdot \left(5 \cdot \dfrac{1}{4}\right)$. By rearranging the factors, this product can be rewritten as $(2 \cdot 5) \cdot \left(\dfrac{1}{3} \cdot \dfrac{1}{4}\right) = 10 \cdot \dfrac{1}{12} = \dfrac{10}{12} = \dfrac{5}{6}$.

Therefore, $\dfrac{2}{3} \cdot \dfrac{5}{4} = \dfrac{10}{12} = \dfrac{5}{6}$.

8. Describe a method to calculate $\frac{2}{3} \cdot \frac{5}{4}$ that does not require rearranging the factors.

$$\frac{2}{3} \cdot \frac{5}{4} = \frac{10}{12} = \frac{5}{6}$$

Because multiplication of fractions involves multiplying the numerators and multiplying the denominators, common factors in the numerator and denominator can be divided out before or after performing these multiplications.

9. Multiply each of the following, and write your answer in lowest terms.

a. $\frac{4}{7} \cdot \frac{3}{5}$

b. $\frac{5}{6} \cdot \frac{2}{3}$

c. $\frac{9}{4} \cdot \frac{7}{8}$

d. $1\frac{3}{10} \cdot \frac{5}{9}$

e. $2\frac{3}{4} \cdot \frac{7}{8}$

10. Determine the amount of each ingredient for double the chocolate chip cookie recipe. Write 2 as $\frac{2}{1}$. Fill in the blanks below.

_____ cups flour

_____ tsp. salt

_____ tsp. baking soda

_____ cups sugar

_____ cups chocolate chips

11. Suppose you want to use one-half the amount of sugar in the new cookie recipe in Problem 10. How much sugar do you need? Note that one-half the amount of an ingredient is $\frac{1}{2}$ times the amount.

Dividing Fractions

It takes $1\frac{1}{3}$ teaspoons of baking soda to make one batch of chocolate chip cookies. If you have 8 teaspoons available, how many batches of cookies can you make? This situation involves division of fractions.

Division can be thought of as dividing into equal parts. For example, 20 cookies split equally among 4 friends can be expressed as 20 cookies ÷ 4 friends = 5 cookies/friend, or simply $20 \div 4 = 5$. In vertical form, this can be written as

$$\frac{20}{4} = 5, \text{ or equivalently as } 20 \cdot \frac{1}{4} = 5.$$

Therefore, the division $20 \div 4 = 5$ is equivalent to the multiplication $20 \cdot \frac{1}{4} = 5$. Division can also be thought of as asking how many groupings of the divisor fit into the dividend. For example, $20 \div 4 = \mathbf{5}$ indicates that $\mathbf{5}$ groups of 4 make 20. The division $20 \div \frac{1}{2}$ is asking how many groups of size $\frac{1}{2}$ make 20. Because 2 groups of $\frac{1}{2}$ will make 1 and 4 groups of $\frac{1}{2}$ will make 2, it follows that 40 groups of $\frac{1}{2}$ will make 20. Therefore, the division $20 \div \frac{1}{2} = 40$ is equivalent to the multiplication $20 \cdot 2 = 40$.

The two examples above help illustrate why **division by any nonzero number is equivalent to multiplication by its reciprocal.**

Recall that the reciprocal of any nonzero number n can be written as $\frac{1}{n}$; the product of any number and its reciprocal is always 1: $n \cdot \frac{1}{n} = 1$. If the original number is a fraction, $\frac{a}{b}$, its reciprocal can be written as the fraction $\frac{b}{a}$ because the product $\frac{a}{b} \cdot \frac{b}{a} = 1$.

12. Determine the reciprocals of the following numbers.

    a. 7   **b.** $\frac{1}{3}$ **c.** $\frac{2}{9}$ **d.** $\frac{16}{3}$

    e. $2\frac{3}{4}$ **f.** $1\frac{1}{2}$ **g.** $3\frac{5}{8}$

13. Because division by a fraction can be understood as multiplication by its reciprocal, rewrite each division as an equivalent multiplication and then perform the multiplication.

    a. $4 \div \frac{2}{5}$ **b.** $\dfrac{\frac{2}{3}}{8}$ **c.** $9 \div 2\frac{3}{4}$

    d. $5\frac{1}{3} \div \frac{8}{9}$ **e.** $7\frac{1}{2} \div 1\frac{1}{4}$

14. If you have 8 teaspoons of baking soda, how many batches of chocolate chip cookies can you make?

Additional Applications

Crab Supreme

4 small (6 oz.) cans crabmeat	2 dashes of Tabasco sauce
1 egg, hard-boiled and mashed	$3\frac{1}{2}$ tbsp. chopped fresh chives
$\frac{1}{2}$ cup mayonnaise	$\frac{1}{4}$ tsp. salt
$2\frac{1}{2}$ tbsp. chopped onion	$\frac{1}{2}$ tsp. garlic powder
$3\frac{2}{3}$ tbsp. plain yogurt	1 tsp. lemon juice

Drain and rinse crab in cold water. Mash crab and egg together. Add all remaining ingredients except chives. Stir well. Chill and serve with chips or crackers. SERVES 6.

15. Determine the ingredients for one-half of the crab recipe. Fill in the blanks below.

_____ small (6 oz.) cans crabmeat   _____ dash Tabasco

_____ egg, hard-boiled and mashed   _____ tbsp. chives

_____ cup(s) mayonnaise   _____ tsp. salt

_____ tbsp. chopped onion   _____ tsp. garlic powder

_____ tbsp. plain yogurt   _____ tsp. lemon juice

16. List the ingredients needed for the crab recipe if 18 people attend the party.

_____ small (6 oz.) cans crabmeat   _____ dashes Tabasco

_____ eggs, hard-boiled and mashed   _____ tbsp. chives

_____ cup(s) mayonnaise   _____ tsp. salt

_____ tbsp. chopped onion   _____ tsp. garlic powder

_____ tbsp. plain yogurt   _____ tsp. lemon juice

17. If a container of yogurt holds 1 cup, how many full batches of crab appetizer can you make with one container (1 cup = 16 tbsp.)?

18. If each person drinks $2\frac{2}{3}$ cups of soda, how many cups of soda will be needed for 18 people?

Apple Crisp

4 cups tart apples	$\frac{1}{3}$ *cup softened butter*
peeled, cored, and sliced	$\frac{1}{2}$ *tsp. salt*
$\frac{2}{3}$ *cup packed brown sugar*	$\frac{3}{4}$ *tsp. cinnamon*
$\frac{1}{4}$ *cup rolled oats*	$\frac{1}{8}$ *tsp. allspice or nutmeg*
$\frac{1}{2}$ *cup flour*	

Preheat oven to 375°F. Place apples in a greased 8-inch square pan. Blend remaining ingredients until crumbly, and spread over the apples. Bake approximately 30 minutes uncovered until the topping is golden and the apples are tender. SERVES 4.

---

19. List the ingredients needed for the apple crisp recipe if 18 people attend the party.

20. How many times would you need to fill a $\frac{2}{3}$-cup container to measure 4 cups of apples?

21. If it takes $\frac{3}{4}$ teaspoon of cinnamon to make one batch of apple crisp and you have only 6 teaspoons of cinnamon left in the cupboard, how many batches can you make?

Potato Pancakes	
6 cups potato	$\frac{1}{3}$ cup flour
(pared and grated)	$3\frac{3}{8}$ tsp. salt
9 eggs	$2\frac{1}{4}$ tbsp. grated onion
Drain the potatoes well. Beat eggs and stir into the potatoes. Combine and sift the flour and salt; then stir in the onions. Add to the potato mixture. Shape into patties and sauté in hot fat. Best served hot with applesauce. MAKES 36 three-inch pancakes.	

22. If you were to make one batch each of the crab supreme, apple crisp, and potato pancake recipes, how much salt would you need? How much flour? How much onion?

23. If you have 2 cups of flour in the cupboard before you start cooking for the party and you make one batch of each recipe, how much flour will you have left?

EXERCISES: ACTIVITY 1.4

1. The year that you enter college, your freshman class consists of 760 students. According to statistical studies, about $\frac{4}{7}$ of these students will actually graduate. Approximately how many of your classmates will receive their degrees?

Exercise numbers appearing in color are answered in the Selected Answers appendix.

2. You rent an apartment for the academic year (two semesters) with three of your college friends. The rent for the entire academic year is $10,000. Each semester you receive a bill for your share of the rent. If you and your friends divide the rent equally, how much must you pay each semester?

3. Your residence hall has been designated a quiet building. This means that there is a no-noise rule from 10 P.M. every night to noon the next day. During what fraction of a 24-hour period is one allowed to make noise?

4. You are a member of the college chorus and have three practices scheduled before an upcoming concert. If the group practices $1\frac{1}{4}$ hours, $2\frac{1}{2}$ hours, and $3\frac{2}{3}$ hours, what is your total practice time before the concert?

5. You are planning a summer cookout and decide to serve quarter-pound hamburgers. If you buy $5\frac{1}{2}$ pounds of hamburger meat, how many burgers can you make?

6. Your favorite muffin recipe calls for $2\frac{2}{3}$ cups of flour, 1 cup of sugar, $\frac{1}{2}$ cup of crushed cashews, and $\frac{5}{8}$ cup of milk, plus assorted spices. How many cups of mixture do you have?

7. You must take medicine in 4 equal doses each day. Each day's medicine comes in a single container and measures $3\frac{1}{5}$ tablespoons. How much medicine is in each dose?

8. Perform the indicated operations.

a. $4\dfrac{2}{3} - 1\dfrac{6}{7}$

b. $5\dfrac{1}{2} + 2\dfrac{1}{3}$

c. $2\dfrac{1}{6} \cdot 4\dfrac{1}{2}$

d. $2\dfrac{3}{7} + \dfrac{14}{5}$

e. $\dfrac{4}{5} \div \dfrac{8}{3}$

f. $4\dfrac{1}{5} \div \dfrac{10}{3}$

9. You need $5\dfrac{1}{3}$ tablespoons of butter to bake a small apple tart. You have one stick of butter, which is 8 tablespoons. How much butter will you have left after you make the tart?

10. At one point in 2011, there were approximately 2.1 billion Internet users worldwide.

a. If approximately $\dfrac{9}{40}$ of the users lived in Europe, estimate the number of Internet users in Europe at this time.

b. How many Internet users were in North America if approximately $\dfrac{13}{100}$ of the worldwide users lived in North America?

11. A distributor shipped only $\dfrac{3}{4}$ of a cheese order that was to be distributed equally between three pizza restaurants. What fractional part of the original order will each restaurant receive?

Activity 1.5

Course Grades and Your GPA

Objectives

1. Recognize and calculate a weighted average.

2. Express fractions in decimal format.

You are a college freshman, and the end of the semester is approaching. You are concerned about keeping a B− (80 through 83) average in your English literature class. Your grade will be determined by computing the **simple average**, or **mean**, of your exam scores. To calculate a simple average, you add all your scores and divide the sum by the number of exams. So far, you have scores of 82, 75, 85, and 93 on four exams. Each exam has a maximum score of 100.

1. What is your current average for the four exams?

2. There is another way you can view simple averages that will lead to the important concept of weighted average. Note that the 4 in the denominator of the fraction $\dfrac{82 + 75 + 85 + 93}{4}$ divides each term of the numerator. Therefore, you can write

$$\frac{82 + 75 + 85 + 93}{4} = \frac{82}{4} + \frac{75}{4} + \frac{85}{4} + \frac{93}{4}.$$

   a. The sum $\dfrac{82}{4} + \dfrac{75}{4} + \dfrac{85}{4} + \dfrac{93}{4}$ can be calculated by performing the four divisions and then adding the results to obtain the average score 83.75. Do this calculation.

   b. Part a shows that the average 83.75 is the sum of 4 values, each of which is $\dfrac{1}{4}$ of a test. This means that each test contributed $\dfrac{1}{4}$ of its value to the average, and you will note that each test contributed equally to the average. For example, the first test score 82 contributed 20.5 to the average. How much did each of the other tests contribute?

3. This calculation of averages can also be viewed in a way that allows you to understand a weighted average. Recall that dividing by 4 is the same as multiplying by its reciprocal $\dfrac{1}{4}$. So you can rewrite the sum from Problem 2a as follows:

$$\frac{82}{4} + \frac{75}{4} + \frac{85}{4} + \frac{93}{4} = \frac{1}{4} \cdot 82 + \frac{1}{4} \cdot 75 + \frac{1}{4} \cdot 85 + \frac{1}{4} \cdot 93$$

   Calculate the expression $\dfrac{1}{4} \cdot 82 + \dfrac{1}{4} \cdot 75 + \dfrac{1}{4} \cdot 85 + \dfrac{1}{4} \cdot 93$ by multiplying each test score by $\dfrac{1}{4}$ and then adding the terms to get the average 83.75.

From the point of view in Problems 2 and 3, the ratio $\dfrac{1}{4}$ is called the **weight** of a test score. It means that a test with a weight of $\dfrac{1}{4}$ contributes $\dfrac{1}{4}$ (or 0.25, or 25%) of its score to the overall average. In the preceding problems, each test had the same weight. However, there are many other situations where not every score contributes the same weight. One such situation occurs when your college determines an average for your semester's work. The average is called a **grade point average**, or **GPA**.

Weighted Averages

The semester finally ended, and your transcript just arrived in the mail. As a part-time student this semester, you took 8 credits—a 3-credit English literature course and a 5-credit biology course. You open the transcript and discover that you earned an A (numerically equivalent to 4.0) in biology and a B (numerically 3.0) in English literature.

   4. Do you think that your biology grade should contribute more to your semester average (GPA) than your English literature grade? Discuss this with your classmates, and give a reason for your answer.

Because it is a 5-credit course, your grade in biology will count more (be more heavily weighted) than your grade in the 3-credit English course. Problems 5 and 6 will show you how this is done.

   5. a. The way to give your 5-credit biology course more weight in your GPA than your literature course is to determine what part of your 8 total credits your biology course represents. Write that ratio in fractional as well as decimal form.

   b. Write the ratio that represents the weight your 3-credit literature grade contributes to your GPA.

   c. Calculate the sum of the two weights from part a and part b.

As you continue working with weighted averages, you will notice that the sum of the weights for a set of scores will always total 1, as it did in Problem 5c.

   6. In Problem 3, you multiplied each test score by its weight and then summed the products to obtain your test average. Do the same kind of calculation here to obtain your GPA; that is, multiply each grade (4.0 for biology and 3.0 for literature) by its respective weight (from Problem 5) and then sum the products. Write your GPA to the nearest hundredth.

Problem 6 illustrates the procedure for computing a **weighted average**. The general procedure is given below.

Procedure

Computing the Weighted Average of Several Data Values

1. Multiply each data value by its respective weight.

2. Sum these weighted data values.

> **Example 1** *You will have a 12-credit load next semester as a full-time student. You will have a 2-credit course, two 3-credit courses, and a 4-credit course. Determine the weight of each course.*

SOLUTION

The 2-credit course will have a weight of $\dfrac{2}{12} = \dfrac{1}{6}$; each of the 3-credit courses will have a weight of $\dfrac{3}{12} = \dfrac{1}{4}$; the 4-credit course will have a weight of $\dfrac{4}{12} = \dfrac{1}{3}$.

7. Do the weights in Example 1 sum to 1? Explain.

Your letter grade for a course translates into a numerical equivalent according to the following table.

Letter Grade	A	A−	B+	B	B−	C+	C	C−	D+	D	D−	F
Numerical Equivalent	4.00	3.67	3.33	3.00	2.67	2.33	2.00	1.67	1.33	1.00	0.67	0.00

Suppose you took 17 credit-hours this past semester, your third semester in college. You earned an A− in psychology (3 hours), a C+ in economics (3 hours), a B+ in chemistry (4 hours), a B in English (3 hours), and a B− in mathematics (4 hours).

8. a. Use the first four columns of the following table to record the information regarding the courses you took. As a guide, the information for your psychology course has been recorded for you.

1 COURSE	2 LETTER GRADE	3 NUMERICAL EQUIVALENT	4 CREDIT HOURS	5 WEIGHT	6 CONTRIBUTION TO GPA
Psychology	A−	3.67	3	$\dfrac{3}{17}$	$\dfrac{3}{17} \cdot 3.67 \approx 0.648$

b. Calculate the weight for each course, and enter it in column 5.

c. For each course, multiply your numerical grade (column 3) by the course's weight (column 5). Round to 3 decimal places, and enter this product in column 6, the course's contribution to your GPA.

d. You can now calculate your semester's GPA by summing the contributions of all your courses. What is your semester GPA?

9. ESR Manufacturing Corporation of Tampa, Florida, makes brass desk lamps each of which requires 10 man-hours split among three levels of labor to make and finish. The table shows the number of man-hours, skill levels and salaries of workers required to produce each lamp.

LEVEL OF LABOR	LABOR HOURS REQUIRED	HOURLY WAGE ($)
Skilled	6	13
Semiskilled	3	11
Unskilled	1	9
Total:	10	

a. Determine the weights for the time commitment (hours) required by each level of labor to produce a single lamp.

b. Determine the weighted hourly salary for producing a single lamp.

SUMMARY: ACTIVITY 1.5

1. To calculate a **simple average** (also called a **mean**), add all the values and divide the sum by the number of values.

2. To compute a **weighted average** of several data values:

   i. Multiply each data value by its respective weight.

   ii. Sum the weighted data values.

3. The sum of the weights used to compute a weighted average will always equal 1.

EXERCISES: ACTIVITY 1.5

1. A grade of W is given if you withdraw from a course before a certain date. The W appears on your transcript but is not included in your grade point average. Suppose that instead of a C+ in economics in Problem 8, you receive a W. Use this new grade to recalculate your GPA.

2. Now suppose that you earn an F in economics. The F is included in your grade point average. Recalculate your GPA from Problem 8.

3. In your first semester in college, you took 13 credit hours and earned a GPA of 2.13. In your second semester, your GPA of 2.34 was based on 12 credit hours. You calculated the third semester's GPA in Problem 8.

   a. Explain why the calculation of your overall GPA for the three semesters requires a weighted average.

   b. Calculate your overall GPA for the three semesters.

SEMESTER	GPA	CREDITS	WEIGHT	NUMERICAL EQUIVALENT
1				
2				
3				
				Total:

4. You are concerned about passing your economics class with a C–(70) average. Your grade is determined by averaging your exam scores. So far, you have scores of 78, 66, 87, and 59 on four exams. Each exam is based on 100 points. Your economics teacher uses the simple average method to determine your average.

   a. What is your current average for the four exams?

   b. What is the lowest score you can achieve on the fifth exam to have at least a 70 average?

5. Suppose you took 15 credit hours last semester. You earned an A– in English (3 hours), a B in mathematics (4 hours), a C+ in chemistry (3 hours), a B+ in health (2 hours), and a B– in history (3 hours). Calculate your GPA for the semester.

COURSE	LETTER GRADE	NUMERICAL EQUIVALENT	CREDIT HOURS	WEIGHT	WEIGHT × NUM. EQUIV.

Cluster 2   What Have I Learned?

I. To add or subtract fractions, you must write them in equivalent form with common denominators. However, to multiply or divide fractions, you do not need a common denominator. Why is this reasonable?

2. The operation of division can be viewed from several different points of view. For example, $24 \div 3$ has at least two meanings:

- Write 24 as the sum of some number of 3s.

- Divide 24 into three equal-sized parts, whose size you must determine.

These interpretations can be applied to fractions as well as to whole numbers.

a. Calculate $2 \div \frac{1}{2}$ by answering this question: 2 can be written as the sum of how many $\frac{1}{2}$s?

b. Calculate $\frac{1}{5} \div 2$ by answering this question: If you divide $\frac{1}{5}$ into two equal parts, how large is each part?

c. Do your answers to parts a and b agree with the results you would obtain by using the procedures for dividing fractions reviewed in this cluster? Explain.

d. Use these ideas to interpret and calculate the following quotients.

   i. $4 \div \frac{1}{3}$

   ii. $\frac{2}{3} \div 2$

Cluster 2   How Can I Practice?

1. You are in a golf tournament. There is a prize for the person who drives the ball closest to the green on the sixth hole. You drive the ball to within 4 feet $2\frac{3}{8}$ inches of the hole, and your nearest competitor is 4 feet $5\frac{1}{4}$ inches from the hole. By how many inches do you win?

2. One of your jobs as the assistant to a weather forecaster is to determine the average thickness of the ice in a bay on the St. Lawrence River. Ice fishermen use this report to determine whether the ice is safe for fishing. You must chop holes in five different areas, measure the thickness of the ice, and take the average. During the first week in January, you record the following measurements: $2\frac{3}{8}, 5\frac{1}{2}, 6\frac{3}{4}, 4$, and $5\frac{7}{8}$ inches. What do you report as the average thickness of the ice in this area?

3. You and two others in your family will divide 119 shares of computer stock left by a relative who died. The stock is worth $12 per share. When the stock is sold, what will be your share of the proceeds?

4. You are about to purchase a rug for your college dorm room. The rug's length is perfect for your room. The width of the rug you want to purchase is $6\frac{1}{2}$ feet. If you center the rug in the middle of your room, which is 10 feet wide, how much of the floor will show on each side of the rug?

5. You are making some repairs in your apartment. You have $12\frac{1}{2}$ feet of plastic pipe. You use $3\frac{2}{3}$ feet for the sink line and $5\frac{3}{4}$ feet for the washing machine. You need approximately $3\frac{1}{2}$ feet for a disposal. Do you have enough pipe left for a disposal?

6. Perform the following operations. Write the result in simplest terms or as a mixed number.

a. $\dfrac{5}{7} + \dfrac{2}{7}$

b. $\dfrac{3}{4} + \dfrac{3}{8}$

c. $\dfrac{3}{8} + \dfrac{1}{12}$

d. $\dfrac{4}{5} + \dfrac{5}{6}$

e. $\dfrac{1}{2} + \dfrac{3}{5} + \dfrac{4}{15}$

f. $\dfrac{11}{12} - \dfrac{5}{12}$

g. $\dfrac{7}{9} - \dfrac{5}{12}$

h. $\dfrac{2}{3} - \dfrac{1}{4}$

i. $\dfrac{7}{30} - \dfrac{3}{20}$

j. $\dfrac{4}{5} - \dfrac{3}{4} + \dfrac{1}{2}$

k. $\dfrac{3}{5} \cdot \dfrac{1}{2}$

l. $\dfrac{2}{3} \cdot \dfrac{7}{8}$

m. $\dfrac{15}{8} \cdot \dfrac{24}{5}$

n. $5 \cdot \dfrac{3}{10}$

o. $\dfrac{3}{8} \div \dfrac{3}{4}$

p. $8 \div \dfrac{1}{2}$

q. $\dfrac{5}{7} \div \dfrac{20}{21}$

r. $4\dfrac{5}{6} + 3\dfrac{2}{9}$

s. $12\dfrac{5}{12} - 4\dfrac{1}{6}$

t. $6\dfrac{2}{13} - 4\dfrac{7}{26}$

u. $2\dfrac{1}{4} \cdot 5\dfrac{2}{3}$

v. $6\dfrac{3}{4} \div 1\dfrac{2}{7}$

7. At the end of the semester, the bookstore buys back books that will be used again in courses the next semester. Usually, the store gives you one-sixth of the original cost of the book. If you spend $643 on books this semester and the bookstore buys back all your books, how much money can you expect to receive?

8. a. You are driving from Buffalo, New York, to Orlando, Florida, during spring break. The driving distance is approximately 1178 miles. If your average speed is 58.5 miles per hour, calculate the total driving time.

   b. The driving distance between Erie, Pennsylvania, and Daytona Beach, Florida, is approximately 1048 miles. If your average speed is 68.2 miles per hour, calculate the total driving time. (Round to the nearest tenths place.)

   c. If you need to make the trip in part b in 14 hours, calculate the average speed needed.

Cluster 3 | Comparisons and Proportional Reasoning

Activity 1.6

Everything Is Relative

Objectives

1. Distinguish between absolute and relative measure.

2. Write ratios in fraction, decimal, and percent formats.

3. Determine equivalence of ratios.

During the 2012–2013 National Basketball Association (NBA) season, Carmelo Anthony of the New York Knicks, Kevin Durant of the Oklahoma City Thunder, Kobe Bryant of the Los Angeles Lakers, and LeBron James of the Miami Heat were among the NBA's highest scorers. The statistics in the table represent each player's 3-point field goal totals for the entire season.

PLAYER	NUMBER OF 3-POINT FIELD GOALS MADE	NUMBER OF 3-POINT FIELD GOALS ATTEMPTED
Carmelo Anthony, NYK	157	414
Kobe Bryant, LAL	132	407
Kevin Durant, OKC	139	334
LeBron James, MIA	103	254

1. Using only the data in column 2, Number of 3-Point Field Goals Made, rank the players from best to worst according to their field goal performance.

You can also rank the players using the data from both column 2 and column 3. For example, Carmelo Anthony made 157 three-point field goals out of the 414 he attempted. The 157 successful baskets can be compared with the 414 attempts by dividing 157 by 414.

You can represent that comparison numerically as the fraction $\frac{157}{414}$, or equivalently as the decimal 0.379 (rounded to thousandths).

2. Complete the following table, and use your results to determine another ranking of the four players from best to worst performance.

PLAYER	NUMBER OF 3-POINT FIELD GOALS MADE	NUMBER OF 3-POINT FIELD GOALS ATTEMPTED	RELATIVE PERFORMANCE		
			VERBAL	FRACTIONAL	DECIMAL
Anthony	157	414	157 out of 414	$\frac{157}{414}$	0.379
Bryant	132	407			
Durant	139	334			
James	103	254			

The two sets of rankings are based on two different points of view. The first ranking (in Problem 1) takes an **absolute** viewpoint in which you count the actual number of 3-point field goals made. The second ranking (in Problem 2) takes a **relative** perspective in which you take into account the number of successes *relative* to the number of attempts.

3. Does one measure, either absolute or relative, better describe field goal performance than the other? Explain.

4. Identify which statements refer to an absolute measure and which refer to a relative measure. Explain your answers.

   a. I got seven answers wrong.

   b. I guessed on four answers.

   c. Two-thirds of the class failed.

   d. I saved $10.

   e. I saved 40%.

   f. Four out of five students work to help pay tuition.

   g. Johnny Damon's batting average in 2005, his final year with the Red Sox, was .316.

   h. Ichiro Suzuki got 262 hits in 2004, a new major league baseball record.

   i. In 2012, the U.S. Census Bureau estimated that 53,000,000 members of the U.S. population were Hispanic (of any race).

   j. In 2012, the U.S. Census Bureau indicated that 17% of the nation's population was Hispanic.

5. What mathematical notation or verbal phrases in Problem 4 indicate a relative measure?

Definition

Relative measure is a quotient that compares two similar quantities, often a "part" and a "total." The part is *divided* by the total.

Ratio is the term used to describe the relative measure quotient. Ratios can be expressed in several forms—words (verbal form), fractions, decimals, and percents.

6. Use the free-throw statistics from the NBA 2012–2013 regular season to express each player's relative performance as a ratio in verbal, fractional, and decimal form. The data for Carmelo Anthony is completed for you. Which player had the best relative performance?

PLAYER	NUMBER OF FREE THROWS MADE	NUMBER OF FREE THROWS ATTEMPTED	RELATIVE PERFORMANCE		
			VERBAL	FRACTION	DECIMAL
Anthony	425	512	425 out of 512	$\frac{425}{512}$	0.830
Bryant	525	626			
Durant	679	750			
James	403	535			

7. a. Three friends, shooting baskets in the schoolyard, kept track of their performance. Andy made 9 out of 15 shots, Pat made 28 out of 40, and Val made 15 out of 24. Rank their relative performance.

   b. Which ratio form (fractional, decimal, or other) did you use to determine the ranking?

8. How would you determine whether two ratios such as "12 out of 20" and "21 out of 35" were equivalent?

9. a. Match each ratio from column A with the equivalent ratio in column B.

COLUMN A	COLUMN B
15 out of 25	84 out of 100
42 out of 60	65 out of 100
63 out of 75	60 out of 100
52 out of 80	70 out of 100

   b. Which ratio in each matched pair is more useful in comparing and ranking the four pairs? Why?

Percents

Relative measure based on 100 is familiar and seems natural. There are 100 cents in a dollar and 100 points on a test. You have probably been using a ranking scale from 0 to 100 since childhood. You most likely possess an instinctive understanding of ratios relative to 100. A ratio such as 40 out of 100 can be expressed as 40 **per** 100, or more commonly as 40 **percent**, written as 40%. Percent always indicates a ratio, number of parts out of 100.

10. Express each ratio in column B of Problem 9 as a percent, using the symbol %.

Each ratio in Problem 10 is already a ratio "out of 100," so you just need to replace the phrase "out of 100" with the % symbol. However, suppose you need to write a ratio such as 21 out of 25 in percent format. You may recognize that the denominator, 25, is a factor of 100 ($25 \cdot 4 = 100$). Then the fraction $\frac{21}{25}$ can be written equivalently as $\frac{21 \cdot 4}{25 \cdot 4} = \frac{84}{100}$, which is 84%.

A more general method used to convert a ratio such as 21 out of 25 into percent format is described next.

Procedure

Converting a Fraction to a Percent

1. Obtain the equivalent decimal form by dividing the numerator of the fraction by the denominator.

2. Move the decimal point in the quotient from step 1 two places to the right, inserting placeholder zeros, if necessary.

3. Attach the % symbol to the right of the number.

---

11. Write the following ratios in percent format.

    a. 35 out of 100   **b.** 16 out of 50

    c. 8 out of 20   **d.** 7 out of 8

In many applications, you will also need to convert a percent into decimal form. This can be done by replacing the % symbol with its equivalent meaning "out of 100," $\frac{1}{100}$.

Procedure

Converting a Percent to a Decimal

1. Locate the decimal point in the number preceding the % symbol.

2. Move the decimal point 2 places to the left, inserting placeholding zeros if needed. (Note that moving the decimal point 2 places to the left is the same as dividing by 100.)

3. Delete the % symbol.

---

12. Write the following percents in decimal format.

    a. 75%   **b.** 3.5%   **c.** 200%   **d.** 0.75%

13. Use the field goal statistics from the regular 2012–2013 NBA season to express each player's relative performance as a ratio in verbal, fractional, decimal, and percent form. Round the decimal to the nearest hundredth.

PLAYER	NUMBER OF FIELD GOALS MADE	NUMBER OF FIELD GOALS ATTEMPTED	RELATIVE PERFORMANCE			
			VERBAL	FRACTION	DECIMAL	PERCENT
Anthony	669	1489	669 out of 1489	$\dfrac{669}{1489}$	0.45	45%
Bryant	738	1595				
Durant	731	1433				
James	765	1354				

SUMMARY: ACTIVITY 1.6

1. Relative measure is a quotient that compares two similar quantities, often a "part" and a "total." The part is divided by the total.

2. Ratio is the term used to describe a relative measure.

3. Ratios can be expressed in several forms: **verbal** (4 out of 5), **fractional** $\left(\dfrac{4}{5}\right)$, **decimal** (0.8), and as a **percent** (80%).

4. Percent always indicates a ratio out of 100.

5. To convert a fraction or decimal to a percent:

    i. Convert the fraction to decimal form by dividing the numerator by the denominator.

    ii. Move the decimal point 2 places to the right and then attach the % symbol.

6. To convert a percent to a decimal:

    i. Locate the decimal point in the number preceding the % symbol.

    ii. Move the decimal point 2 places to the left and then drop the % symbol.

7. Two ratios are **equivalent** if their decimal or reduced-fraction forms are equal.

EXERCISES: ACTIVITY 1.6

1. Complete the table below by representing each ratio in all four formats. Round decimals to thousandths and percents to tenths.

VERBAL FORM	REDUCED-FRACTION FORM	DECIMAL FORM	PERCENT FORM
	$\frac{1}{3}$		
	$\frac{2}{5}$		
	$\frac{18}{25}$		
	$\frac{8}{9}$		
3 out of 8			
25 out of 45			
120 out of 40			
		0.75	
		0.675	
		0.6	
		$0.\overline{6} \approx 0.667$	
			80%
			0.50%
			200%

2. a. Match each ratio from column I with the equivalent ratio in column II.

	COLUMN I		COLUMN II
1.	12 out of 27	A.	60 out of 75
2.	28 out of 36	B.	25 out of 40
3.	45 out of 75	C.	21 out of 27
4.	64 out of 80	D.	42 out of 70
5.	35 out of 56	E.	20 out of 45

Exercise numbers appearing in color are answered in the Selected Answers appendix.

b. Write each matched pair of equivalent ratios as a percent.

3. Your biology instructor returned three quizzes today. On which quiz did you perform best? Explain how you determined the best score.

Quiz 1: 18 out of 25   Quiz 2: 32 out of 40   Quiz 3: 14 out of 20

4. Baseball batting averages are the ratios of hits to "at bats." They are reported as three-digit decimals. Determine the batting averages and ranking of three players with the following records.

a. 16 hits out of 54 at bats

b. 25 hits out of 80 at bats

c. 32 hits out of 98 at bats

5. There are 1720 women among the 3200 students at the local community college. Express this ratio in each of the following forms.

a. fraction

b. reduced fraction

c. decimal

d. percent

6. At the state university campus, 2304 women and 2196 men are enrolled. In which school—the community college in Exercise 5 or the state university—is the relative number of women greater? Explain your reasoning.

7. In the 2013 Major League Baseball season, the world champion Boston Red Sox ended the regular season with a 97–65 win-loss record. During the playoff season, the team's win-loss record was 11–5. Did the Red Sox play better in the regular season or in the postseason? Justify your answer mathematically.

8. A random check of 150 Southwest Airlines flights last month identified that 113 of them arrived on time. What "on-time" percent does this represent?

9. In the following table, the admissions office has organized data describing the number of men and women who are currently enrolled in your college in order to formulate recruitment and retention efforts for the coming academic year. A student is matriculated if he or she is enrolled in a program that leads to a degree (such as associate or baccalaureate).

	MATRICULATION BY GENDER		
	FULL-TIME MATRICULATED **(≥12 CREDITS)**	**PART-TIME MATRICULATED** **(< 12 CREDITS)**	**PART-TIME NONMATRICULATED** **(< 12 CREDITS)**
Men	214	174	65
Women	262	87	29

  a. How many men attend your college?

  b. What percent of the men are full-time students?

  c. How many women attend your college?

  d. What percent of the women are full-time students?

  e. How many students are enrolled full-time?

  f. How many students are enrolled part-time?

  g. Women constitute what percent of the full-time students?

  h. Women constitute what percent of the part-time students?

  i. How many students are nonmatriculated?

  j. What percent of the student body is nonmatriculated?

Activity 1.7

The Devastation of AIDS in Africa

Objective

1. Use proportional reasoning to apply a known ratio to a given piece of information.

International public health experts are desperately seeking to stem the spread of the AIDS epidemic in sub-Saharan Africa, especially in the small nation of Botswana. According to a recent United Nations report, approximately 24.8% of Botswana's estimated 1,200,000 adults (ages 15–49) were infected with the AIDS virus.

Botswana

The data in the paragraph above is typical of the kind of numerical information you will find in reading virtually any print or electronic document, report, or article.

1. What *relative* data (ratio) appears in the opening paragraph? What phrase or symbol identifies it as a relative measure?

2. Express the ratio in Problem 1 in fractional, decimal, and verbal form.

Recall from Activity 1.6 that a ratio represents a relative measure, the quotient of a part divided by a total.

3. The opening paragraph contains a second piece of information, namely, 1,200,000. Does this number represent a part or a total in this situation?

Proportional Reasoning

Once you know two of the three values in the basic relationship

$$ratio = \frac{part}{total},$$

the third unknown value can be determined using **proportional reasoning**. This basic relationship can be rewritten in one of two equivalent forms:

$$part = total \cdot ratio \quad or \quad total = \frac{part}{ratio}$$

Procedure

Applying Proportional Reasoning

Step 1. Identify the known ratio.

Step 2. Identify whether the second known piece of information represents a total or a part.

Step 3. • If the given information represents a *total*, then the part is unknown. You can determine the unknown part by *multiplying* the known total by the known ratio as follows:

$$unknown\ part = total \cdot ratio$$

• If the given information represents a *part*, then the total is the unknown.

$$unknown\ total = \frac{part}{ratio}$$

In this case, you *divide* the known part by the known ratio to determine the total.

Examples 1 and 2 demonstrate this procedure.

> **Example 1**   *How many adults in Botswana are infected with AIDS (refer to the introductory paragraph)?*

SOLUTION

Proportional reasoning is needed to determine the actual number of adults in Botswana who are infected with AIDS.

Step 1. The known ratio is 24.8%, or 0.248 in decimal form—that is, 248 out of every 1000 adults.

Step 2. The given piece of information is Botswana's total adult population, 1,200,000.

Step 3. The unknown piece of information is the actual number, or part, of Botswana's adults infected with AIDS. Therefore,

$$\text{unknown part} = \text{total} \cdot \text{ratio}$$

$$\text{unknown part} = 1{,}200{,}000 \cdot 0.248 = 297{,}600.$$

The number of adults in Botswana infected with AIDS is 297,600.

> **Example 2**   *You do further research about this disturbing health problem and discover that in 2011, approximately 22,900,000 adults and children were living with HIV/AIDS in sub-Saharan Africa. This represented 5% of the total population there. Use proportional reasoning to determine the total population of sub-Saharan Africa in 2011.*

SOLUTION

The known ratio is 5%, or 0.05.

The given information, 22,900,000 adults and children with HIV/AIDS, is a part of the total population.

$$\text{unknown total} = \frac{\text{part}}{\text{ratio}},$$

$$\text{unknown total} = \frac{22{,}900{,}000}{0.05} = 458{,}000{,}000.$$

The total population of sub-Saharan Africa in 2011 was approximately 458 million people.

4. In 2011, sub-Saharan Africa was home to approximately 22,900,000 people living with HIV/AIDS. This represented an estimated 69% of all people living with HIV/AIDS. From this, estimate the number of people in the world living with HIV/AIDS in 2011.

5. a. In 2011, the U.S. Census Bureau estimated the U.S. population at approximately 312 million. Approximately 23.7% of this population was under the age of 18 years. How many people were under the age of 18 in the United States in 2011?

b. The U.S. Census Bureau projects that in 2020, the number of adults aged 65 years and over will be 54.8 million. If this represents approximately 16% of the total population, determine the projected total U.S. population in 2020.

6. According to the U.S. Census Bureau, approximately 110 million American adults used at least one social networking Web site in 2010. This was about 47% of all American adults. Determine to the nearest million the number of adults in the United States in 2010.

7. In June 2013, the American Medical Association voted to classify obesity as a disease. Although this announcement did not receive the full support of the medical community, the general consensus was that obesity is "a major health concern" and "a complex disorder."
   a. It was estimated that 34.9% of adults in the United States age 20 years and older were obese in 2012. If the total U.S. adult population in 2012 was approximately 225 million, how many adults (to the nearest million) were obese in 2012?

   b. Approximately 12.5 million of children and adolescents ages 2–19 years were obese in 2013. If this represents 17% of the total number of children and adolescents, determine the total population of ages 2–19 in the United States in 2013. Round to the nearest million.

SUMMARY: ACTIVITY 1.7

1. **Proportional reasoning** is the process by which you apply a known ratio to one piece of information (part or total) to determine a related, yet unknown, piece of information (total or part).

2. Given the total, *multiply* the total by the ratio to obtain the unknown part:

$$\text{total} \cdot \text{ratio} = \text{unknown part}$$

3. Given the part, *divide* the part by the ratio to obtain the unknown total:

$$\frac{\text{part}}{\text{ratio}} = \text{unknown total}$$

EXERCISES: ACTIVITY 1.7

1. Many people around the world are living in poverty, as defined by The World Bank and determined by both income and available assets. In 2010, Nigeria had 134.9 million people living on less than $2 per day. This was 84.5% of the total Nigerian population.

   a. Identify the given quantities as a ratio, total, or part.

   b. Identify the unknown quantity.

   c. Determine the total population of Nigeria in 2010.

2. One of The World Bank's stated goals is to eradicate extreme poverty worldwide. Extreme poverty is defined as living on less than $1.25 per day. In 1994, the population of India was 938.8 million, when 81.7% of the population lived in extreme poverty. By 2010, India's population had grown to 1.205 billion people, with 68.8% living in extreme poverty. The ratio of those in extreme poverty clearly decreased. Determine the actual number of people living in extreme poverty in both 1994 and 2010. Compare these numbers.

3. In 2011, approximately 44% of the worldwide Internet users were living in the Asian region of the world. If there were approximately 2.1 billion Internet users worldwide, determine the number of Internet users living in the Asian region in 2011.

   a. Identify the known ratio.

   b. Identify the given piece of information. Does this represent a part or a total?

   c. What is the number of Internet users living in the Asian region?

4. The number of smartphone users in the United States in 2010 was estimated to be 49.1 million. If this represents 15.7% of the U.S. population, approximate the population of the United States in 2010.

5. The customary tip on waiter service in New York City restaurants is approximately 20% of the food and beverage cost. This means that your server is counting on a tip of $0.20 for each dollar you spent on your meal. What is the expected tip on a dinner for two costing $65?

6. You purchase a new car for $22,500. The state sales tax rate is 8%. How much sales tax will you pay on the car?

7. The sales tax on a new laptop computer is $51. If the sales tax rate is 7.5%, how much did the computer cost?

8. In your recent school board elections, only 45% of the registered voters went to the polls. Approximately how many voters are registered in your school district if 22,000 votes were cast?

9. In 2013, approximately 15,600,000 new cars and trucks were sold in the United States. The market share breakdown among the top five manufacturers was as follows:

General Motors	18%
Ford	16%
Toyota	14.4%
Chrysler	11.5%
Honda	9.6%

a. What percent of total U.S. car and truck sales in 2013 was attributed to the top five manufacturers?

b. Approximately how many new vehicles did each of these five companies sell in 2013?

c. Nissan sold 1,248,420 new vehicles in 2013. What was its market share?

10. For a sociology class, you conduct a survey in the residence halls regarding students' daily use of various forms of social media. The results of the survey are given in the following table.

SOCIAL MEDIA	PERCENT OF STUDENTS (Nearest Tenth)	NUMBER OF STUDENTS
Facebook	68.1	
Twitter	52.6	
LinkedIn	7.4	
Google+	15.6	
YouTube	25.2	

   a. If 71 students reported that they use Twitter daily, how many students were in the survey?

   b. Record 71 in the appropriate location in the table. Use the result from part a to complete the table.

11. In a very disappointing season, your softball team won only 40% of the games it played this year. If your team won 8 games, how many games did it play? How many games did the team lose?

12. In a typical telemarketing campaign, 5% of the people contacted agree to purchase the product. If your quota is to sign up 50 people, approximately how many phone calls do you anticipate making?

13. Of the world's 6.7 billion people, 3.1 billion live on less than $2 per day. Calculate the percent of the world's population that live on less than $2 per day.

14. Approximately three-fourths of the teacher education students in your college pass the licensing exam on their first attempt. This year 92 students will sit for the exam. How many are expected to pass?

15. You paid $140 for your economics textbook. At the end of the semester, the bookstore will buy back your book for 40% of the purchase price. The book sells used for $90. How much money does the bookstore net on the resale?

Activity 1.8

Who Really Did Better?

Objectives

1. Define actual and relative change.

2. Distinguish between actual and relative change.

3. Calculate relative change as a percent increase or percent decrease.

You have a discussion in your business class regarding investing. Your discussion includes topics such as common and preferred stocks; treasury, municipal, and corporate bonds; and certificates of deposit (CDs). As a class project, you are asked to select a stock to purchase and then track the performance of the stock. After researching several companies, you decide to choose a technology stock and purchase virtual shares worth $200. Your classmate selects an energy stock and purchases virtual shares worth $400. After several weeks of significant gains in the stock market, your stock is worth $280, whereas your classmate's stock is valued at $500.

1. How much did the value of your investment change?

2. How much did the value of your classmate's investment change?

3. Who earned more money?

Your answers to Problems 1 and 2 represent **actual change**, the actual numerical value by which a quantity has changed. When a quantity increases in value, such as your stock, the actual change is positive. When a quantity decreases in value, the actual change is negative.

4. Keeping in mind the amount each of you originally invested, whose investment do you believe did better? Explain your answer.

Relative Change

In answering Problem 4, you considered the actual change relative to the original amount invested.

Definition

The ratio formed by comparing the **actual change** to the **original value** is called the **relative change**. Relative change is written in fraction form, with the actual change placed in the numerator and the original (or earlier) amount placed in the denominator.

$$\text{relative change} = \frac{\text{actual change}}{\text{original value}}$$

---

5. Determine the relative change in your $200 investment. Write your answer as a fraction and as a percent.

6. Determine the relative change in your classmate's $400 investment. Write your answer as a fraction and as a percent.

7. In relative terms, whose stock performed better?

Percent Change

Because *relative change* (increase or decrease) is frequently reported as a percent, it is often called *percent change*. The two terms are interchangeable.

8. Determine the actual change and the percent change of the following quantities.

a. You paid $20 for a rare baseball card that is now worth $30.

b. Last year's graduating class had 180 students; this year's class size is 225.

c. Ann invested $300 in an energy stock that is now worth $540.

d. Bob invested $300 in a technology stock that is now worth only $225.

e. Pat invested $300 in a risky dot-com venture, and his stock is now worth $60.

When someone is speaking about percent change, a percent greater than 100 makes sense. For example, suppose your $200 stock investment tripled to $600. The actual increase in your investment is $600 − $200 = $400. Therefore, the relative change is

$$\frac{\text{actual increase}}{\text{original value}} = \frac{\$400}{\$200} = 2.00.$$

Converting 2.00 from a decimal to percent, your investment has increased by 200%.

9. Determine the actual change and the percent change of the following quantities.

a. In 1970, your parents purchased their home for $40,000. Recently, they sold it for $200,000.

b. After a new convention center was built, the number of hotel rooms in the city increased from 1500 to 6000.

Actual Change Versus Relative Change

Actual change and relative (percent) change provide two perspectives for understanding change. Actual change looks only at the size of the change (ignoring the original value), whereas relative change indicates the significance or importance of the change by comparing it to the original value.

The following problems will help to clarify these two perspectives.

10. Two of your friends in the business class were bragging about their respective stock performances. Each of their stocks earned $250 last year. Then it slipped out that the original investment of one friend was $500, whereas the other friend invested $2000 in her stock.

    a. What was the actual increase in each investment?

    b. Calculate the relative increase (in percent format) for each investment.

    c. Whose investment performance was more impressive? Why?

11. During the first week of classes, campus clubs actively recruit members. The Ultimate Frisbee Club signed up 8 new members, reaching an all-time high membership of 48 students. The Yoga Club also attracted 8 new members and now boasts 18 members.

    a. What was the membership of the Ultimate Frisbee Club before recruitment week?

    b. By what percent did the Frisbee Club increase its membership?

    c. What was the membership of the Yoga Club before recruitment week?

    d. By what percent did the Yoga Club increase its membership?

SUMMARY: ACTIVITY 1.8

1. The **relative change** in a quantity such as a stock price or a utility bill can be determined by dividing the actual change in the quantity by the original value of the quantity:

$$\text{relative change} = \frac{\text{actual change}}{\text{original value}}.$$

EXERCISES: ACTIVITY 1.8

1. Last year the full-time enrollment in your college was 3200 students. This year 3560 full-time students are on campus.

   a. Determine the actual increase in full-time enrollment.

   b. Determine the relative increase (as a percent) in full-time enrollment.

2. The table below indicates the average annual pump price of gasoline in the United States from 2008 through 2013. Determine the actual increase (decrease) and the percent increase (decrease) in the price of gasoline from year to year. Record your results in the appropriate place in the table.

YEAR	2008	2009	2010	2011	2012	2013
Price/Gallon	$3.23	$2.35	$2.78	$3.51	$3.61	$3.49
Actual Change in Price						
Percent Change in Price						

3. The table below indicates the legal minimum hourly wage in San Francisco from 2006 through 2014.

YEAR	2006	2007	2008	2009	2010	2011	2012	2013	2014
Minimum Wage	$8.82	$9.14	$9.36	$9.79	$9.79	$9.92	$10.24	$10.55	$10.74

   a. Between which two consecutive years did the actual minimum wage increase the most?

Exercise numbers appearing in color are answered in the Selected Answers appendix.

b. What percent increase does this represent?

c. What has been the percent increase from 2006 through 2014?

4. Last month two large companies announced that they were forced to lay off 500 workers each. The first company's workforce is now 1000. The second company currently has 4500 employees. What percent of its workforce did each company lay off?

5. You are now totally committed to dieting. You have decided to cut your daily caloric intake from 2400 to 1800 calories. By what percent are you cutting back on calories?

6. a. You paid $10 per share for a promising technology stock. Within a year, the stock price skyrocketed to $50 per share. By what percent did your stock value increase?

b. You held on to the stock a little too long. The price has just fallen from its $50 high back to $10. By what percent did your stock decrease?

7. Perform the following calculations.

a. Start with a value of 10, double it, and then calculate the percent increase.

b. Start with a value of 25, double it, and then calculate the percent increase.

   c. Start with a value of 40, double it, and then calculate the percent increase.

   d. When any quantity doubles in size, by what percent does it increase?

8. Perform a set of calculations similar to the ones you did in Exercise 7 to help determine the percent increase of any quantity that triples in size.

Activity 1.9

Going Shopping

Objectives

1. Define growth factor.

2. Determine growth factors from percent increases.

3. Apply growth factors to problems involving percent increases.

4. Define decay factor.

5. Determine decay factors from percent decreases.

6. Apply decay factors to problems involving percent decreases.

To collect revenue (income), many state and local governments require merchants to collect sales tax on the items they sell. In several localities, the sales tax is assessed at as much as 8% of the selling price and is passed on directly to the purchaser.

1. Determine the total cost (including the 8% sales tax) to the customer of the following items. Include each step of your calculation in your answer.

   a. A greeting card selling for $3.50

   b. A Blu-ray player priced at $120.

Many people correctly determine the total costs for Problem 1 in two steps:

- They first compute the sales tax on the item.

- Then they add the sales tax to the selling price to obtain the total cost.

It is also possible to compute the total cost in one step by using the idea of a **growth factor**.

Growth Factors

2. If a quantity increases by 50%, how does its new value compare with its original value? That is, what is the ratio of the new value to the original value? The first row of the following table is completed for you. Complete the remaining rows to confirm or discover your answer.

ORIGINAL VALUE	NEW VALUE (INCREASED BY 50%)	RATIO OF NEW VALUE TO ORIGINAL VALUE		
		FRACTIONAL FORM	DECIMAL FORM	PERCENT FORM
20	10 + 20 = 30	$\frac{30}{20} = \frac{3}{2} = 1\frac{1}{2}$	1.50	150%
50				
100				
Your choice of an original value				

3. Use the results from the preceding table to answer the question, What is the ratio of the new value to the original value of any quantity that increases by 50%? Express this ratio as a reduced fraction, a mixed number, a decimal, and a percent.

The ratio $\dfrac{3}{2} = \dfrac{\text{new value}}{\text{original value}}$ that you calculated in the table and observed in Problem 3 is called the **growth factor** for a 50% increase in any quantity. The definition of a growth factor for any percent increase is given below.

> **Definition**
>
> For a specified percent increase, no matter what the original value may be, the ratio of the new value to the original value is always the same. This ratio is called the **growth factor** associated with the specified percent increase. The formula is
>
> $$\text{growth factor} = \dfrac{\text{new value}}{\text{original value}}.$$

The growth factor is most often written in decimal form. As you determined in Problems 2 and 3, the growth factor, in decimal form, associated with a 50% increase is 1.50. Because 150% is the same as 100% + 50%, the growth factor is the sum of 100% and 50%. Because 150% = 1.50 in decimal form, the growth factor is also the sum of 1.00 and 0.50. This leads to a second way to determine a growth factor.

> **Procedure**
>
> **Determining a Growth Factor from a Percent Increase**
>
> The growth factor is determined by adding the specified percent increase to 100% and then changing this percent to its decimal form.

4. In each of the following, determine the growth factor of any quantity that increases by the given percent. Write your result as a percent and as a decimal.

   a. 30%   **b.** 75%

   c. 15%   **d.** 7.5%

5. A growth factor can always be written in decimal form. Explain why this number will always be greater than 1.

Using Growth Factors in Percent Increase Problems

In Problems 2 and 3, you saw that a growth factor is given by the formula

$$\text{growth factor} = \dfrac{\text{new value}}{\text{original value}}.$$

This formula is equivalent to the formula

original value · growth factor = new value.

For example, $1.5 = \dfrac{75}{50}$ is the same as $50 \cdot 1.5 = 75$. This observation leads to a procedure for determining a new value using growth factors.

Procedure

Determining the New Value Using Growth Factors

When a quantity increases by a specified percent, its new value can be obtained by multiplying the original value by the corresponding growth factor. That is,

original value · growth factor = new value.

---

6. a. Use the growth factor 1.50 to determine the new value of a stock portfolio that increased 50% over its original value of $400.

   b. Use the growth factor 1.50 to determine the population of a town that has grown 50% over its previous size of 120,000 residents.

7. a. Determine the growth factor for any quantity that increases 20%.

   b. Use the growth factor to determine this year's budget, which has increased 20% over last year's budget of $75,000.

8. a. Determine the growth factor of any quantity that increases 8%.

   b. Use the growth factor to determine the total cost of each item in Problem 1.

In the previous problems, you were given an original (earlier) value and a percent increase and were asked to determine the new value. Suppose you are asked a "reverse" question: Sales in your small retail business have increased 20% over last year. This year you had gross sales receipts of $75,000. How much did you gross last year?

Using the growth factor 1.20, you can write

original value · 1.20 = 75,000.

To determine last year's sales receipts, divide both sides of the equation by 1.20:

original value = 75,000 ÷ 1.20

Therefore, last year's sales receipts totaled $62,500.

> **Procedure**
>
> **Determining the Original Value Using Growth Factors**
>
> When a quantity has *already increased* by a specified percent, its original value is obtained by dividing the new value by the corresponding growth factor. That is,
>
> new value ÷ growth factor = original value.

9. You determined a growth factor for an 8% increase in Problem 8. Use it to answer the following questions.

   a. The cash register receipt for your new coat, which includes the 8% sales tax, totals $243. What was the ticketed price of the coat?

   b. The credit-card receipt for your new sofa, which includes the 8% sales tax, totals $1620. What was the ticketed price of the sofa?

Decay Factors

It's the sale you have been waiting for all season. Everything in the store is 40% off the original ticket price.

10. Determine the discounted price of a pair of sunglasses originally selling for $25.

Many people correctly determine the discounted prices for Problem 10 in two steps:

- They first compute the actual dollar discount on the item.

- Then they subtract the dollar discount from the original price to obtain the sale price.

It is also possible to compute the sale price in one step by using the idea of a **decay factor**.

11. a. If a quantity decreases 20%, how does its new value compare with its original value? That is, what is the ratio of the new value to the original value? Complete the rows in the following table. The first row has been done for you. Notice that the new values are less than the originals, so the ratios of new values to original values will be less than 1.

ORIGINAL VALUE	NEW VALUE (DECREASED BY 20%)	RATIO OF NEW VALUE TO ORIGINAL VALUE		
		FRACTIONAL FORM	DECIMAL FORM	PERCENT FORM
20	16	$\frac{16}{20} = \frac{4}{5}$	0.80	80%
50				
100				
Choose any value				

b. Use the results from the preceding table to answer the question, What is the ratio of the new value to the original value of any quantity that decreases 20%? Express this ratio in reduced-fraction, decimal, and percent forms.

The ratio $\dfrac{4}{5} = \dfrac{\text{new value}}{\text{original value}}$ that you calculated in the table and observed in Problem 11 is called the **decay factor** for a 20% decrease in any quantity. The definition of a decay factor for any percent decrease is given below.

Definition

For a specified percent *decrease*, no matter what the original value may be, the ratio of the new value to the original value is always the same. This ratio is called the **decay factor** associated with the specified percent decrease. The formula is

$$\text{decay factor} = \frac{\text{new value}}{\text{original value}}.$$

As you have seen, the decay factor associated with a 20% decrease is 80%. For calculation purposes, a decay factor is usually written in decimal form; so the decay factor 80% is expressed as 0.80.

Procedure

Determining a Decay Factor from a Percent Decrease

Note that a percent decrease describes the percent that has been *removed* (such as the discount taken off an original price), but the corresponding decay factor always represents the percent remaining (the portion that you still must pay). Therefore, a decay factor is formed by *subtracting* the specified percent decrease from 100% and then changing this percent to its decimal form.

Example 1   *The decay factor corresponding to a 10% decrease is*

$$100\% - 10\% = 90\%.$$

If you change 90% to a decimal, the decay factor is 0.90.

12. Determine the decay factor of any quantity that decreases by the following given percents.

  a. 30%   **b.** 75%

  c. 15%   **d.** 7.5%

Remember that multiplying the original value by the decay factor will always result in the amount remaining, not the amount that was removed.

13. Explain why every decay factor will have a decimal format that is less than 1.

> ## Procedure
>
> **Determining the New Value Using Decay Factors**
>
> When a quantity *decreases* by a specified percent, its new value is obtained by multiplying the original value by the corresponding decay factor. That is,
>
> $$\text{original value} \cdot \text{decay factor} = \text{new value}.$$

14. Use the decay factor 0.80 to determine the new value of a stock portfolio that has lost 20% of its original value of $400.

In the previous questions, you were given an original value and a percent decrease and were asked to determine the new (*smaller*) value. Suppose you are asked a "reverse" question: Voter turnout in the local school board elections declined 40% from last year. This year only 3600 eligible voters went to the polls. How many people voted in last year's school board elections?

Using the decay factor 0.60 you can write

$$\text{original value} \cdot 0.60 = 3600.$$

To determine last year's voter turnout, divide both sides of the equation by 0.60:

$$\text{original value} = 3600 \div 0.60.$$

Therefore, last year 6000 people voted in the school board elections.

> ## Procedure
>
> **Determining the Original Value Using Decay Factors**
>
> When a quantity has *already decreased* by a specified percent, its original value is obtained by dividing the new value by the corresponding decay factor. That is,
>
> $$\text{new value} \div \text{decay factor} = \text{original value}.$$

15. a. You purchased a cell phone on sale for $150. The discount was 40%. What was the original price?

b. Determine the original price of a treadmill that is on sale for $1140. Assume that the store has reduced all merchandise by 25%.

SUMMARY: ACTIVITY 1.9

1. When a quantity increases by a specified percent, the ratio of its new value to its original value is called the **growth factor**. The growth factor is determined by adding the specified percent increase to 100% and then changing this percent to its decimal form.

2. a. When a quantity increases by a specified percent, its new value can be obtained by multiplying the original value by the corresponding growth factor. So

$$\text{original value} \cdot \text{growth factor} = \text{new value.}$$

 b. When a quantity has already increased by a specified percent, its original value can be obtained by dividing the new value by the corresponding growth factor. So

$$\text{new value} \div \text{growth factor} = \text{original value.}$$

3. When a quantity decreases by a specified percent, the ratio of its new value to its original value is called the **decay factor** associated with the specified percent decrease. The decay factor is determined by subtracting the specified percent decrease from 100% and then changing this percent to its decimal form.

4. a. When a quantity decreases by a specified percent, its new value can be obtained by multiplying the original value by the corresponding decay factor. So

$$\text{original value} \cdot \text{decay factor} = \text{new value.}$$

 b. When a quantity has already decreased by a specified percent, its original value can be obtained by dividing the new value by the corresponding decay factor. So

$$\text{new value} \div \text{decay factor} = \text{original value.}$$

EXERCISES: ACTIVITY 1.9

1. Complete the following table.

Percent Increase	5%		15%		100%	300%	
Growth factor		1.35		1.045			11.00

2. You are purchasing a new car. The price you negotiated with a car dealer in Staten Island, New York, is $19,744, excluding sales tax. The sales tax rate is 8.875%.

 a. What growth factor is associated with the sales tax rate?

 b. Use the growth factor to determine the total cost of the car.

3. The cost of a first-class stamp in the United States increased approximately 6.5% in January 2014 to 49¢. What was the pre-January cost of the first-class stamp? Round your answer to the nearest penny.

4. A state university in the western United States reported that its enrollment increased from 11,952 students in 2012–2013 to 12,144 students in 2013–2014.

   a. What is the ratio of the number of students in 2013–2014 to the number of students in 2012–2013?

   b. From your result in part a, determine the growth factor. Write it in decimal form to the nearest thousandth.

   c. By what percent did the enrollment increase?

5. **a.** The 2010 Census showed that from 2000 to 2010, the U.S. population increased by approximately 9.7% to 308.7 million. Use this information to estimate the U.S. population in 2000.

   b. The 2010 Census report also showed that in the same period, Texas's population increased from 20,851,820 to 25,145,561. How does the increase in Texas's population compare with the increase in the total U.S. population over this decade?

6. In 1967, your grandparents bought a house for $26,000. The U.S. Bureau of Labor Statistics reports that due to inflation, the cost of the same house in 2005 was $152,030.

   a. What was the inflation growth factor of housing from 1967 to 2005?

   b. What was the percent increase to the nearest whole number percent for housing from 1967 to 2005?

   c. In 2005, your grandparents sold their house for $245,000. What profit did they make in terms of 2005 dollars?

7. Your friend plans to move from Newark, New Jersey, to Anchorage, Alaska. She earns $50,000 a year in Newark. How much must she earn in Anchorage to maintain the same standard of living if the cost of living in Anchorage is 35% higher?

8. You decide to invest $3000 in a 1-year certificate of deposit that earns an annual percentage yield of 1.85%. How much will your investment be worth in a year?

9. You live in Conway, Arkansas, and have a good job with an annual salary of $40,000. For each city listed in the table, the second column indicates the salary needed to maintain the same standard of living that you enjoy in Conway.

Fill in the third column to indicate by what percent your Conway salary would need to increase so that you could maintain the same standard of living in the other four cities.

CITY	SALARY	% INCREASE IN SALARY NEEDED
Conway, Arkansas	$40,000	0%
Baltimore, Maryland	$50,600	
Las Vegas, Nevada	$44,200	
Boston, Massachusetts	$62,200	
Manchester, New Hampshire	$53,800	

10. Since 2007, the International Data Corporation (IDC) has been keeping track of the amount of digital information created and replicated on computers and the Internet. IDC calls this set of information the digital universe. In 2012, IDC reported that the size of the digital universe was 2.8 trillion gigabytes. In 2011, that number was 1.8 trillion gigabytes.

   a. What was the growth factor in the amount of digital data from 2011 to 2012? Round your answer to the nearest hundredth.

   b. Use the growth factor from part a to predict the size of the digital universe in 2013, assuming that the growth factor remains the same. Round your answer to the nearest trillion.

11. Complete the following table.

Percent Decrease	5%		15%		12.5%
Decay Factor		0.45		0.94	

12. You wrote an 8-page article that will be published in a journal. The editor has asked you to revise the article and reduce the number of pages to 6. By what percent must you reduce the length of your article?

13. In 1995, California recorded 551,226 live births. In 2000, the number of live births dropped to 531,285. For the 5-year period from 1995 to 2000, what is the decay factor for live births? What is the percent decrease?

14. A car dealer will sell you the car you want for $18,194, which is just $200 over the dealer invoice price (the price the dealer pays the manufacturer for the car). You tell him that you will think about it. The dealer is anxious to meet his monthly quota of sales, so he calls the next day to offer you the car for $17,994 if you agree to buy it tomorrow. You decide to accept the deal.

   a. What is the decay factor associated with the decrease in the price? Write your answer in decimal form to the nearest thousandth.

b. What is the percent decrease of the price to the nearest tenth of a percent?

c. The sales tax is 6.5%. How much did you save on sales tax by taking the dealer's second offer?

15. Your company is moving you from Boston, Massachusetts, to one of the other cities listed in the following table. For each city, the table below lists the cost of goods and services that would cost $100 in Boston, Massachusetts.

CITY	COST OF GOODS AND SERVICES ($)	DECAY FACTOR IN COST OF LIVING	PERCENT DECREASE IN COST OF LIVING
Boston, MA	100		
Bogota, Colombia	81		
Budapest, Hungary	94		
Kolkata, India	88		
Medan, Indonesia	68		
Krakow, Poland	63		
Cape Town, South Africa	79		

a. Determine the decay factor for each cost in the table, and list it in the third column.

b. Determine the percent decrease in the cost of living for each city compared with living in Boston, Massachusetts. List the percent decrease in the fourth column of the table.

c. You earn a salary of $45,000 in Boston, Massachusetts. If you move to Cape Town, South Africa, what salary will you need to maintain your standard of living there?

d. How much of your salary can you save if you move to Budapest, Hungary, and your salary remains $45,000?

16. You need a fax machine but plan to wait for a sale. Yesterday the model you want was reduced by 30%. It originally cost $129.95. Use the decay factor to determine the sale price of the fax machine.

17. You are on a weight-reducing diet. Your doctor advises you that losing 10% of your body weight will significantly improve your health.

  a. You weigh 175 pounds. Determine and use the decay factor to calculate your goal weight to the nearest pound.

  b. After 6 months, you reach your goal weight. However, you still have more weight to lose to reach your ideal weight range, which is 125–150 pounds. If you lose 10% of your new weight, will you be in your ideal weight range? Explain.

18. Aspirin is typically absorbed into the bloodstream from the duodenum (the beginning portion of the small intestine). In general, 75% of a dose of aspirin is eliminated from the bloodstream in an hour. A patient is given a dose of 650 milligrams. How much aspirin remains in the patient after 1 hour?

19. Airlines often encourage their customers to book online by offering a 5% discount on tickets. You are traveling from Kansas City to Denver. A fully refundable fare is $558.60. A restricted, nonrefundable fare is $273.60. Determine and use the decay factor to calculate the cost of each fare if you book online.

20. The annual inflation rate is calculated as the percent increase of the Consumer Price Index (CPI). The table below shows the average annual CPI for 2010, 2011, 2012, and 2013 as reported by the U.S. Bureau of Labor Statistics. Use the CPI measures to determine the annual inflation rate for 2011, 2012, and 2013. Record your answer in the appropriate place in the table.

Year	2010	2011	2012	2013
CPI	218.056	224.939	229.594	232.957
Annual Inflation Rate	—			

21. You are taking a photography class at the local community college and need to purchase a digital camera. The college bookstore is having a big sale on digital cameras. After a 40% price reduction, you purchase a digital camera for $342. What was the price of the camera before the reduction?

22. The college bookstore marks up the price of a textbook by 35%. If the selling price of a book is $135, how much did the bookstore pay the publisher for the textbook?

Activity 1.10

Take an Additional 20% Off

Objectives

1. Define consecutive growth and decay factors.

2. Determine a consecutive growth or decay factor from two or more consecutive percent changes.

3. Apply consecutive growth and/or decay factors to solve problems involving percent changes.

Cumulative Discounts

Your friend arrives at your house. Today's newspaper contains a 20%-off coupon at Old Navy. The $100 jacket she had been eyeing all season was already reduced by 40%. She clipped the coupon, drove to the store, selected her jacket, and walked up to the register. The cashier brought up a price of $48; your friend insisted that the price should have been only $40. The store manager arrived and reentered the transaction, and again, the register displayed $48. Your friend left the store without purchasing the jacket and drove to your house.

1. How do you think your friend calculated a price of $40?

2. After your friend calms down, you grab a pencil and start your own calculation. First, you determine the ticketed price that reflects the 40% reduction. At what price is Old Navy selling the jacket? Explain how you calculated this price.

3. To what price does the 20%-off coupon apply?

4. Apply the 20% discount to determine the final price of the jacket.

You are now curious: Could you justify a better price by applying the discounts in the reverse order—that is, by applying the 20%-off coupon, followed by a 40% reduction? You start a new set of calculations.

5. Starting with the list price, determine the sale price after taking the 20% reduction.

6. Apply the 40% discount to the intermediate sale price.

7. Which sequence of discounts gives a better sale price?

The important point to remember here is that when multiple discounts are given, they are applied sequentially, one after the other—never all at once.

In the following example, you will see how to use decay factors to simplify calculations such as the ones above.

> **Example 1** *A stunning $2000 gold and diamond necklace you saw was far too expensive to even consider purchasing. However, over several weeks, you tracked the following successive discounts: 20% off list, 30% off marked price, and an additional 40% off every item. Determine the selling price after each of the discounts is taken.*

SOLUTION

Step 1. To calculate the first reduced price, apply the decay factor corresponding to a 20% discount. Recall that you form a decay factor by subtracting the 20% discount from 100% to obtain 80%, or 0.80 as a decimal. Apply the 20% reduction by multiplying the original price by the decay factor.

The first sale price = $2000 · 0.80; the selling price is now $1600.

Step 2. Determine the decay factor corresponding to a 30% discount by subtracting 30% from 100% to obtain 70%, or 0.70 as a decimal. Apply the 30% reduction by multiplying the already discounted price by the decay factor.

The second sale price = $1600 · 0.70; the selling price is now $1120.

Step 3. Determine the decay factor corresponding to a 40% discount by subtracting 40% from 100% to obtain 60%, or 0.60 as a decimal. Apply the 40% reduction by multiplying the most current discounted price by the decay factor.

$1120 · 0.60; the final selling price is $672.

You may have noticed that it is possible to *calculate the final sale price from the original price using a single chain of multiplications:*

$$2000 · 0.80 · 0.70 · 0.60 = \$672$$

The final sale price is $672.

8. Use a chain of multiplications as in Example 1 to determine the final sale price of the necklace if the discounts are taken in the reverse order (40%, 30%, and 20%).

You can form a single decay factor that represents the cumulative effect of applying the three consecutive percent decreases; the single decay factor is the *product* of the three decay factors.

In Example 1, the effective decay factor is given by the product 0.80 · 0.70 · 0.60, which equals 0.336 (or 33.6%). The effective discount is calculated by subtracting the decay factor (in percent form) from 100% to obtain 66.4%. Therefore, the effect of applying 20%, 30%, and 40% consecutive discounts is identical to a single discount of 66.4%.

9. **a.** Determine a single decay factor that represents the cumulative effect of consecutively applying Old Navy's 40% and 20% discounts.

b. Use this decay factor to determine the effective discount on your friend's jacket.

10. **a.** Determine a single decay factor that represents the cumulative effect of consecutively applying discounts of 40% and 50%.

b. Use this decay factor to determine the effective discount.

Cumulative Increases

Your family's fast-food franchise is growing faster than you had ever imagined. Last year you had 100 employees statewide. This year you opened several additional locations and increased the number of workers by 30%. With demand so high, next year you will open new stores nationwide and plan to increase your number of employees by an additional 50%.

To determine the projected number of employees next year, you can use growth factors to simplify the calculations.

11. a. Determine the growth factor corresponding to a 30% increase.

    b. Apply this growth factor to calculate your current workforce.

12. a. Determine the growth factor corresponding to a 50% increase.

    b. Apply this growth factor to your current workforce to determine the projected number of employees next year.

13. Starting from last year's workforce of 100, write a single chain of multiplications to calculate the projected number of employees.

> You can form a single growth factor that represents the cumulative effect of applying the two consecutive percent increases; the single growth factor is the *product* of the two growth factors.

In Problem 13, the effective growth factor is given by the product $1.30 \cdot 1.50$, which equals 1.95 (an increase of 95%), nearly double the number of employees last year.

What Goes Up Often Comes Down

You purchased $1000 of a recommended stock last year and watched gleefully as it rose quickly by 30%. Unfortunately, the economy turned downward, and your stock recently fell 30% from last year's high. Have you made or lost money on your investment? The answer might surprise you. Find out by solving the following sequence of problems.

14. a. Determine the growth factor corresponding to a 30% increase.

    b. Determine the decay factor corresponding to a 30% decrease.

You can form a single factor that represents the cumulative effect of applying the consecutive percent increase and decrease—the single factor is the *product* of the growth and decay factors. In this example, the effective factor is given by the product $1.30 \cdot 0.70$, which equals 0.91.

15. Does 0.91 represent a growth factor or a decay factor? How can you tell?

16. a. What is the current value of your stock?

    b. What is the cumulative effect (as a percent change) of applying a 30% increase followed by a 30% decrease?

17. What is the cumulative effect if the 30% decrease had been applied first, followed by the 30% increase?

SUMMARY: ACTIVITY 1.10

1. The cumulative effect of a sequence of percent changes is the *product* of the associated growth or decay factors. For example,

   a. To calculate the effect of consecutively applying 20% and 50% increases, form the respective growth factors and multiply. The effective growth factor is

   $$1.20 \cdot 1.50 = 1.80,$$

   which represents an effective increase of 80%.

   b. To calculate the effect of applying a 25% *increase* followed by a 20% *decrease*, form the respective growth and decay factors and then multiply them. The effective factor is $1.25 \cdot 0.80 = 1.00$, which indicates neither growth nor decay. That is, the quantity has returned to its original value.

2. The cumulative effect of a sequence of percent changes is the same, regardless of the order in which the changes are applied. For example, the cumulative effect of applying a 20% *increase* followed by a 50% *decrease* is equivalent to having first applied the 50% *decrease*, followed by the 20% *increase*.

EXERCISES: ACTIVITY 1.10

1. A college raises its tuition in each of three consecutive years. If the percent increase was 5%, 9%, and 7%, determine the overall percent increase at the end of the three-year period.

2. A pair of running shoes is on sale for 25% off at a local sporting goods store. You receive an online offer from the store for an additional 40% off. If the original price of the shoes is $260, what is the final selling price of the shoes after the sale and online discounts are applied?

3. A $300 leather jacket is on sale for 30% off. You present a coupon at the cash register for an additional 20% off.

   a. Determine the decay factor corresponding to each percent decrease.

   b. Use these decay factors to determine the price you paid for the jacket.

Exercise numbers appearing in color are answered in the Selected Answers appendix.

4. A local labor union has just negotiated a 3-year contract containing annual raises of 3%, 4%, and 5% during the term of contract. If the current salary of your brother is $42,000, what will he be earning in 3 years?

5. A toy manufacturer anticipated a large demand for a popular toy and increased its inventory of 1600 by 25%. If 75% of the inventory was sold, how many toys remain?

6. You anticipate that a $2000 investment in a local business will pay an average of 4% interest in each of the next 5 years. Determine to the nearest dollar the amount of the investment at the end of 5 years.

7. Budget cuts have severely crippled your department over the last few years. Your previous operating budget of $600,000 has decreased 5% in each of the last 3 years. What is your current operating budget?

8. A digital camera with an original price tag of $500 was marked down by 40%. You have a coupon good for an additional 30% off.

   a. What is the decay factor for the first discount?

   b. What is the decay factor for the additional discount?

   c. What is the effective decay factor?

   d. What is the final cost of the camera?

   e. What is the effective percent discount?

9. Your friend took a job 3 years ago that started at $30,000 a year. Last year she got a 5% raise. She stayed at the job for another year but decided to make a career change and took another job that paid 5% less than her current salary. Was the starting salary at the new job more, less, or the same as her starting salary at her previous job?

10. You wait for the price to drop on a diamond-studded watch at Macy's. Originally, it cost $2500. The first discount was 20%, and the second discount is 50%.

   a. What are the decay factors for each of the discounts?

   b. What is the effective decay factor for the two discounts?

c. Use the effective decay factor in part b to determine how much you will pay for the watch.

11. Determine whether the following statement is true or false. If true, confirm it with a valid computation. If false, determine the correct percent change.

A 50% increase followed by a 50% decrease results in no net change.

Activity 1.11

Fuel Economy

Objectives

1. Apply rates directly to solve problems.

2. Use unit or dimensional analysis to solve problems that involve consecutive rates.

You are excited about purchasing a new car for your commute to college and a part-time job. Concerned about the cost of driving, you do some research on the Internet and come across a website that lists fuel efficiency, in miles per gallon (mpg), for five cars that you are considering. You record the mpg for city and highway driving in the table below.

1. **a.** For each of the cars listed in the following table, how many city miles can you travel per week on 5 gallons of gasoline? Explain the calculation you will do to obtain the answers. Record your answers in the third column of the table.

   b. The drive to your college, part-time job, and back home is 38 city miles, which you do 4 days a week. Which of the cars would allow you to complete this trip each week on 5 gallons of gas?

FUEL ECONOMY GUIDE MODEL YEAR 2013					
MAKE/MODEL	CITY MPG	CITY MILES ON 5 GAL OF GAS	HIGHWAY MPG	GALLONS NEEDED TO DRIVE 304 HIGHWAY MILES	FUEL TANK CAPACITY IN GAL
Chevrolet Cruze	32		42		14.0
Ford Focus	27		36		13.2
Honda Civic	29		41		13.2
Hyundai Accent	28		38		11.4
Toyota Corolla	34		41		11.9

In Problem 1a, you can write miles per gallon in fraction form, $\dfrac{\text{number of miles}}{1 \text{ gallon}}$. For example, the city fuel economy rating for the Chevrolet Cruze is $\dfrac{32 \text{ mi.}}{1 \text{ gal.}}$. To determine how many city miles you can travel on 5 gallons of gasoline, you multiply the known miles per gallon by 5 gallons as follows:

$$\frac{32 \text{ mi.}}{1 \text{ gal.}} \cdot 5 \text{ gal.} = 160 \text{ mi.}$$

Notice that the unit of measurement, gallon, occurs in both the numerator and the denominator. You divide out common units of measurement the same way you divide out common numerical factors.

Therefore, for the Chevrolet Cruze, you can drive 160 miles on 5 gallons of gas.

2. Suppose your round-trip commute to a summer job would be 304 highway miles each week. How many gallons of gas would each of the cars require? Explain the calculation needed to obtain the answers. Record your answers to the nearest tenth in the fifth column of the above table.

In Problem 2, you were asked to determine how many gallons of gas were required to drive 304 highway miles. You may have known to divide the total miles by the miles per gallon. For example, in the case of the Honda Civic, $\frac{304}{41} = 7.4$ gallons.

One way to be sure the problem is set up correctly is to use the units of measurement. In Problem 2, you are determining the number of gallons. Set up the calculation so that miles will divide out and the remaining measurement unit will be gallons.

$$304 \text{ mi.} \cdot \frac{1 \text{ gal.}}{41 \text{ mi.}} = \frac{304}{41} \text{ gal.} = 7.4 \text{ gal.}$$

Rates and Unit Analysis

Miles per gallon (mpg) is an example of a rate. Mathematically, a **rate** is a comparison of two quantities that have different units of measurement. Numerically, you calculate with rates as you would with ratios. Paying attention to what happens to the units of measurement during the calculation is critical to determining the unit of the result. This process is called **unit analysis**.

> **Definition**
>
> A **rate** is a comparison, expressed as a quotient, of two quantities that have different units of measurement.
>
> **Unit analysis** (sometimes called *dimensional analysis*) uses units of measurement as a guide in setting up a calculation involving one or more rates. When the calculation is set up properly, the result is an answer with the appropriate units of measurement.

A rate may be expressed in two ways. For miles and gallons, the rate can be expressed as $\frac{miles}{gallon}$ or $\frac{gallon}{miles}$. Therefore, the highway mpg of the Chevrolet Cruze, 42 miles per gallon, can be written as $\frac{42 \ miles}{1 \ gallon}$ or $\frac{1 \ gallon}{42 \ miles}$. To determine how many gallons of gas were required to drive the Chevrolet Cruze (see Problem 2) 304 highway miles, its highway mpg would be expressed as $\frac{1 \ gal}{42 \ miles}$. Thus, the multiplication by 304 miles would result in the unit miles divided out, and the remaining measurement in the numerator will be gallons.

$$304 \ miles \cdot \frac{1 \ gal.}{42 \ mi.} = \frac{304 \ gal.}{42} \approx 7.2 \ gal.$$

3. a. The gas tank of a Ford Focus holds 13.2 gallons. How many highway miles can you travel on a full tank of gas?

   b. The Toyota Corolla's gas tank holds 11.9 gallons. Is it possible to travel as far on the highway in this car as in the Ford Focus?

4. After you purchase your new car, you would like to take a trip to see a good friend in another state. The highway distance is approximately 560 miles.

   a. If you bought the Hyundai Accent, how many gallons of gas would you need to make the round-trip?

   b. How many tanks of gas would you need for the trip?

5. You are on a part of a 1500-mile trip where gas stations are far apart. Your car is a hybrid model that is averaging 50 miles per gallon and you are traveling at 60 miles per hour (mph). The fuel tank holds 12 gallons of gas, and you just filled the tank. How many hours can you drive before you must fill the tank again?

You can solve this problem in two parts, as follows.

   a. Determine how many miles you can travel on one tank of gas.

   b. Use the result of part a to determine how many hours you can drive before you must fill the tank again.

Problem 5 is an example of applying consecutive rates to solve a problem. Part a was the first step; part b, the second step. Alternatively, you can determine the answer in a *single calculation* by considering what the measurement unit of the answer should be. In this case, it is hours per tank. Therefore, you will need to set up the calculation so that you can divide out miles and gallons.

$$\frac{50 \text{ mi.}}{1 \text{ gal.}} \cdot \frac{12 \text{ gal.}}{1 \text{ tank}} \cdot \frac{1 \text{ hr.}}{60 \text{ mi.}} = \frac{10 \text{ hr.}}{1 \text{ tank}}$$

Notice that miles and gallons divide out to leave hours per tank as the measurement unit of the answer.

Procedure

Using Unit Analysis to Solve Problems

To apply consecutive rates:

1. Identify the measurement unit of the result.

2. Set up the sequence of multiplications so that the appropriate units divide out, leaving the appropriate measurement unit of the result.

3. Multiply and divide the numbers as indicated to obtain the numerical part of the result.

4. Check that the appropriate measurement units divide out, leaving the expected unit for the result.

6. You have been driving for several hours and notice that your car's 13.2-gallon fuel tank registers half empty. How many more miles can you travel if your car is averaging 45 miles per gallon?

Metric System

The Old English system of measurement (e.g., feet for length and pounds for weight) used throughout most of the United States was not designed for its computational efficiency. It evolved over time, before the decimal system was adopted by Western European cultures. The metric system of measurement, which you may already know something about, was devised to take advantage of the decimal numeration system.

In general, the metric system of measurement relies on each unit being divided into 10 equal subunits, which, in turn, are divided into 10 equal subunits, and so on. The prefix added to the base unit signifies how far the unit has been divided. Starting with the meter as the basic unit for measuring distance, the following smaller units are defined by Latin prefixes.

UNIT OF LENGTH	ABBREVIATION	FRACTIONAL PART OF A METER
meter	m	1 meter
decimeter	dm	0.1 meter
centimeter	cm	0.01 meter
millimeter	mm	0.0001 meter

Larger metric units are defined using Greek prefixes. The most commonly used is kilo, meaning 1000 times. Therefore, a kilometer is 1000 meters.

The metric system is also used to measure the mass of an object or volume of a liquid. Grams are the basic unit for measuring the mass of an object (basically, how heavy it is). Liters are the basic unit for measuring volume (usually of a liquid). The same prefixes are used to indicate larger and smaller units of measurement. To learn more about changing measurements to the metric system, visit *www.metrication.com*.

Unit Conversion

In most countries, distance is measured in kilometers (km) and gasoline in liters (L). A mile is equivalent to 1.609 kilometers, and a liter is equivalent to 0.264 gallon. Each of these equivalences can be treated as a rate and written in fraction form. Thus, the fact that a mile is equivalent to 1.609 kilometers is written as $\frac{1.609 \text{ km}}{1 \text{ mi.}}$ or $\frac{1 \text{ mi.}}{1.609 \text{ km}}$. Using the fraction form, you can convert one measurement unit to another by applying unit analysis.

7. Your friend joins you on a trip through Canada, where gasoline is measured in liters and distance in kilometers.

a. Write the equivalence of liters and gallons in fraction form.

b. If you bought 20 liters of gas, how many gallons did you buy?

8. To keep track of mileage and fuel needs in Canada, your friend suggests that you convert your car's miles per gallon into kilometers per liter (km/L). Your car's highway fuel efficiency is 45 miles per gallon. What is its fuel efficiency in kilometers per liter?

SUMMARY: ACTIVITY 1.11

1. A **rate** is a comparison, expressed as a quotient, of two quantities that have different units of measurement.

2. **Unit analysis** (sometimes called *dimensional analysis*) uses units of measurement as a guide in setting up a calculation or writing an equation involving one or more rates. When the calculation is set up properly, the result is an answer with the appropriate units of measurement.

3. Using unit analysis to solve problems involving rates:

   i. Identify the measurement unit of the result.

   ii. Set up the sequence of multiplications so the appropriate measurement units divide out, leaving the unit of the result.

   iii. Multiply and divide the numbers as usual to obtain the numerical part of the result.

   iv. Check that the appropriate measurement units divide out, leaving the expected unit of the result.

4. The basic metric units for measuring distance, mass, and volume are meter, gram, and liter, respectively.

5. Larger and smaller units in the metric system are based on the decimal system. The prefix before the basic unit determines the size of the unit, as shown in the table.

UNIT PREFIX (BEFORE METER, GRAM, OR LITER)	ABBREVIATION	SIZE IN BASIC UNITS (METERS, GRAMS, OR LITERS)
milli-	mm, mg, mL	0.001
centi-	cm, cg, cL	0.01
deci-	dm, dg, dL	0.1
deka-	dam, dag, daL	10
hecto-	hm, hg, hL	100
kilo-	km, kg, kL	1000

EXERCISES: ACTIVITY 1.11

Use the conversion tables on the inside front cover of the textbook for conversion equivalencies.

1. You have entered a 10 km race to raise money for a local charity. How many miles will you be running?

2. Your family is vacationing in Hawaii. While on a tour of Hawaii Volcanoes National Park on the Big Island, you learn that the leading edge of a lava flow on a steep slope can travel as fast as 2 kilometers per hour. How fast is the rate in miles per day?

3. If you send an average of 70 text messages a day, determine the average number of text messages sent in a year.

4. You have been selected to study for a semester in London next year. You also have received a $1000 grant for books and expenses.

   a. What is the value of the 1000 U.S. dollars in British pounds? Many websites (e.g., www.gocurrency.com) provide up-to-the minute currency exchange rates. Use the following exchange rates obtained on January 8, 2014.

	USD	GBP	EUR	JPY
1 USD =	1	0.61	0.734	104.413
1 GBP =	1.64	1	1.204	171.281
1 EUR =	1.363	0.831	1	142.271
1 JPY =	0.01	0.006	0.007	1

   b. You plan to visit Paris, France, which uses euros (EUR) as its unit of currency. Determine the value of 100 USD in euros.

   c. Convert 75 euros to British pounds.

   d. A group of students from Japan is attending your college for the academic year. The unit of currency in Japan is the yen. Convert 1000 yen to U.S. dollars.

   e. Which is the largest unit of currency—a U.S. dollar, a British pound, a European euro, or a Japanese yen? Explain.

Exercise numbers appearing in color are answered in the Selected Answers appendix.

f. Convert $100 from U.S. dollars to British pounds, to euros, and to yen?

5. Suppose a faucet leaks 1 drop per second. You want to determine how many gallons per month are wasted. First, you experiment with a teaspoon and estimate that there are approximately 75 drops per teaspoon. If there are 3 teaspoons per tablespoon, 16 tablespoons per cup, 4 cups per quart, and 4 quarts per gallon, determine the number of gallons of water wasted per month.

6. The distance between New York City, New York, and Los Angeles, California, is approximately 4485 kilometers. What is the distance between these two major cities in miles?

7. The aorta is the largest artery in the human body. In the average adult, the aorta attains a maximum diameter of about 1.18 inches where it adjoins the heart. What is the maximum diameter of the aorta in centimeters? In millimeters?

8. In the Himalayan Mountains along the border of Tibet and Nepal, Mt. Everest reaches a record height of 29,035 feet. How high is Mt. Everest in miles? In kilometers? In meters?

9. The average weight of a mature human brain is approximately 1400 grams. What is the equivalent weight in kilograms? In pounds?

10. Approximately 4.5 liters of blood circulates in the body of the average human adult. How many quarts of blood does the average person have? How many pints?

11. How many seconds are in a day? In a week? In a year?

12. The following places are three of the wettest locations on Earth. For each site, determine the rainfall in centimeters per year. Record your results, rounded to the nearest centimeter in the third column of the table.

LOCATIONS	ANNUAL RAINFALL (IN INCHES)	ANNUAL RAINFALL (IN CM)
Mawsynram, Meghalaya, India	467	
Tutenendo, Colombia	463.5	
Mount Waialeale, Kauai, Hawaii	410	

13. Jewelry is commonly weighed in carats. Five carats are equivalent to 1 gram. What is the weight of a 24-carat diamond in grams? In ounces?

14. Tissues of living organisms consist primarily of organic compounds; that is, they contain carbon molecules known as proteins, carbohydrates, and fats. A healthy human body is approximately 18% carbon by weight. Determine how many pounds of carbon your own body contains. How many kilograms?

15. Nurses who administer medications must ensure that each dosage is customized to the patient. Units of measurement in the following dosage problems are g, grams and mg, milligrams. In each problem, determine how many tablets need to be administered.

a. The dose to be given is 750 milligrams. The stock strength is 0.25 gram per tablet.

b. The dose to be given is 0.1 gram. The stock strength is 50 milligrams per tablet.

In each of the following, determine how many milliliters need to be administered.

c. The dose to be given is 100 milligrams. The stock solution contains 0.2 gram per 5 milliliters.

d. The dose to be given is 3 grams. The stock solution contains 120 milligrams per 5 milliliters.

16. You recently rented a Toyota Prius with a full tank of gas and 6570 miles on the odometer. After 4 days, you returned the car with 7205 miles on the odometer. If the rental agency charged you $98.76 for the rental and it took 12 gallons of gas to fill up the tank, determine the following rates.

a. The car's rate of gas mileage, in miles per gallon

b. The average cost of the rental, in dollars per day

c. The car's rate of travel, in miles per day

Cluster 3   What Have I Learned?

1. On a 40-question practice test for this course, you answered 32 questions correctly. On the test itself, you correctly answered 16 out of 20 questions. Does this mean that you did better on the practice test than you did on the test itself? Explain your answer using the concepts of actual and relative comparison.

2. **a.** Ratios are often written as a fraction in the form $\frac{a}{b}$. List the other ways you can express a ratio.

   b. Thirty-three of the 108 colleges and universities in Michigan are two-year institutions. Write this ratio in each of the ways you listed in part a.

3. According to the U.S. Census Bureau, in 2000, Florida's total population was approximately 15,982,400, of which 2,924,800 individuals were aged 65 and older. In 2010, the total population had increased to 18,249,100 and the 65 and older population was 3,194,000.

   a. What percent of Florida's population was represented by seniors (65 and older) in 2000?

   b. What percent of Florida's population was represented by seniors (65 and older) in 2010?

   c. By what percent did Florida's population increase in the decade from 2000 to 2010?

d. By what percent did Florida's senior population increase in the decade from 2000 to 2010?

4. A 12-ounce cup of coffee in the campus center vending machine costs $1.25. In a local franchise coffee shop, you will pay $1.85. Explain how to determine the percent increase you will pay if you purchase your coffee at the franchise shop.

5. a. Think of an example, or find one in a newspaper or magazine or on the Internet, and use it to show how to determine the growth factor associated with a percent increase. For instance, you might think of or find growth factors associated with yearly interest rates for a savings account.

b. Show how to use the growth factor to apply a percent increase twice, then three times.

6. Current health research shows that losing just 10% of body weight produces significant health benefits, including a reduced risk for a heart attack. Suppose a relative who weighs 199 pounds begins a diet to reach a goal weight of 145 pounds.

a. Use the idea of a decay factor to show your relative how much he will weigh after losing the first 10% of his body weight.

b. Explain to your relative how he can use the decay factor to determine how many times he must lose 10% of his body weight to reach his goal weight of 145 pounds.

7. a. Write the conversion, 1 quart = 0.946 liters in two different fraction formats.

b. Write the conversion 1 liter = 1.057 quarts in two different fraction formats.

c. Are the two conversions given in parts a and b equivalent? Explain.

Cluster 3   How Can I Practice?

1. Write the following percents in decimal format.

   a. 25%   b. 87.5%   c. 6%

   d. 3.5%   e. 250%   f. 0.3%

2. An error in a measurement is the difference between the measured value and the true value. The relative error in measurement is the ratio of the error to the true value. That is,

$$\text{relative error} = \frac{\text{error}}{\text{true value}}.$$

   Determine the actual and relative errors of the measurements in the following table.

MEASUREMENT	TRUE VALUE	ACTUAL ERROR	RELATIVE ERROR (AS A PERCENT)
107 in.	100 in.		
5.7 oz.	5.0 oz.		
4.3 g	5.0 g		
11.5 cm	12.5 cm		

3. A study compared three different treatment programs for cocaine addiction. Subjects in the study were divided into three groups. One group was given the standard therapy, the second group was treated with an antidepressant, and the third group was given a placebo. The groups were tracked for 3 years. Data from the study is given in the following table.

TREATMENT THERAPY	NO. OF SUBJECTS RELAPSED	NO. OF SUBJECTS STILL SUBSTANCE-FREE
Standard therapy	36	20
Antidepressant therapy	27	18
Placebo	30	9

   a. Interpret this data to determine which, if any, of the three therapies was most effective. Explain the calculations you would do and how you use them to interpret the study.

b. Complete the following table. Which therapy appears most effective?

TREATMENT THERAPY	TOTAL NUMBER OF SUBJECTS	PERCENT OF SUBJECTS SUBSTANCE-FREE
Standard therapy		
Antidepressant therapy		
Placebo		

4. The 2619 women students on a campus constitute 54% of the entire student body. What is the total enrollment of the college?

5. JCPenney employs approximately 116,000 workers. In January 2014, it announced that it would be laying off 2000 workers and closing 33 stores, thereby reducing the total number of stores by 3%, in an attempt to save over $65 million each year.

   a. What percent of the JCPenney work force will lose jobs in this layoff?

   b. How many stores did JCPenney operate prior to the announced closings?

6. Over a 10-year period the Cajun community in Louisiana dwindled from 407,000 to 44,000. What percent decrease does that represent?

7. Hourly workers at a fiber-optics cable plant saw their workweek cut from 60 hours to 36. What percent decrease in income does this represent?

8. A suit with an original price of $300 was marked down 30%. You have a coupon good for an additional 20% off.

   a. What is the suit's final cost?

   b. What is the equivalent percent discount?

9. In recent contract negotiations between a transit workers union and a Metropolitan Transit Authority, the workers rejected a 3-year contract, with 3%, 4%, and 3.5% wage increases, respectively, over each year of the contract. At the end of this 3-year period, what would have been the salary of a motorman who had been earning $48,000 at the start of the contract?

10. In 2013, the NYC Rent Guidelines Board approved rent increases of 4% on one-year-leases, and 7.75% on leases of two years.

   a. You have been living in a studio apartment in Brooklyn for a few years. At the end of 2013, you signed a 1-year lease effective 2014. In 2014 your rent was $860 each month. What was your rent in 2013?

   b. Your brother has been living in a two bedroom apartment in Queens. In 2013, after the approval of rent guidelines, he signed a 2-year lease for the apartment he was living in. In this new lease, his 2014 and 2015 rent was $1200 per month. How much was his 2013 rent?

11. In a recent survey, 70% of the 1400 female students and 30% of the 1000 make students on campus indicated that using Facebook was their favorite leisure activity. How many students placed Facebook use at the top of their list? What percent of the entire student body does this represent?

12. The 2010 U.S. Census report showed that from 2000 to 2010, Nevada's population increased from 1,998,257 to 2,700,551. Determine the actual and relative increases in Nevada's population over this decade.

13. Genworth's 2013 Cost of Care Survey, based on data from nearly 15,000 long-term care providers, indicated that from 2008 to 2013, the median annual cost of private nursing home care jumped 24% to $83,950.

   a. Determine the median annual cost of nursing home care in 2008.

   b. If the cost jumps 24% over the next 5 years as well, determine the predicted median annual cost of nursing home care in 2018.

14. In 2011, the world's population reached approximately 7 billion people. In that year, The World Bank estimated that 88.78% of the population had access to an improved water source. Approximately how many people did not have access to an adequate water supply?

15. The Mariana Trench in the South Pacific Ocean near the Mariana Islands extends down to a maximum depth of 35,827 feet. How deep is this deepest point in miles? In kilometers? In meters?

Skills Check 1 in Appendix C

Cluster 4 · Problem Solving with Signed Numbers

Activity 1.12

Celsius Thermometers

Objectives

1. Identify signed numbers.

2. Use signed numbers to represent quantities in real-world situations.

3. Compare signed numbers.

4. Calculate the absolute value of numbers.

5. Identify and use properties of addition and subtraction of signed numbers.

6. Add and subtract signed numbers using absolute value.

1. What are some situations in which you have encountered negative numbers?

Definition

The collection of positive counting numbers, negative counting numbers, and zero is basic to our number system. This collection is called the set of **integers.**

A good technique for visualizing the relationship between positive and negative numbers is to use a number line, scaled with integers, much like an outdoor thermometer's scale.

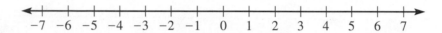

On a number line, 0 separates the positive and negative numbers: 0 is neither positive nor negative. Recall that numbers increase in value as you move to the right. Thus, -1 has a value greater than -4, written as the inequality $-1 > -4$. This relationship between -1 and -4 makes sense in the thermometer model because $-1°$ represents a (higher) warmer temperature than $-4°$.

2. Insert the appropriate inequality symbol between the two numbers in each of the following pairs.

   a. 2 ▒ 6   **b.** -2 ▒ -6 **c.** 2 ▒ -3

   d. -5 ▒ -1 **e.** -3 ▒ -8 **f.** $\dfrac{1}{2}$ ▒ $\dfrac{1}{6}$

Absolute Value

The **absolute value** of a number represents the distance the number is from zero on the number line. For example, $+9$ and -9 are both 9 units from 0 but in opposite directions, so they have the same absolute value, 9. Symbolically, you write $|9| = 9$ and $|-9| = 9$ and $|0| = 0$. Notice that opposites have the same absolute value. More generally, the absolute value of a number represents the size or magnitude of the number. *The absolute value of a nonzero number is always positive.*

3. Evaluate each of the following.

   a. $|17|$ **b.** $\left|-\dfrac{3}{4}\right|$ **c.** $|0|$ **d.** $|-3.75|$

4. **a.** What number has the same absolute value as -57?

b. What number has the same absolute value as 38.5?

c. What number has an absolute value of zero?

d. Can the absolute value of a number be negative? Explain.

Adding Signed Numbers

In this activity, a thermometer model illustrates addition and subtraction of signed numbers. On a Celsius thermometer, 0° represents the temperature at which water freezes and 100° represents the temperature at which water boils. The thermometers in the following problems show temperatures from −10°C to +10°C.

5. a. What is the practical meaning of positive numbers in the Celsius thermometer model?

b. What is the practical meaning of negative numbers in the Celsius thermometer model?

You can use signed numbers to represent a change in temperature. A rise in temperature is indicated by a positive number, and a drop in temperature is indicated by a negative number. As shown on the thermometer below, a rise of 6° from −2° results in a temperature of 4°, symbolically written as −2° + 6° = 4°.

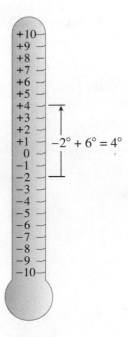

6. Answer the following questions using the thermometer models.

a. What is the result if a temperature starts at 5° and rises 3°? Symbolically, you are calculating 5° + 3°. Use the thermometer below to demonstrate your calculation.

b. If a temperature starts at −5° and rises 3°, what is the result? Write the calculation symbolically. Use the thermometer below to demonstrate your calculation.

c. If a temperature starts at 5° and drops 3°, the resulting temperature is 2°. Symbolically, you are calculating 5° + (−3°). Use the thermometer below to demonstrate your calculation.

d. What is the result when a temperature drops 3° from −5°? Write the calculation symbolically. Use the thermometer below to demonstrate your calculation.

7. a. In what direction do you move on the thermometer when you add positive degrees to a starting temperature in Problem 6?

b. In each case, is the result greater or less than the starting number?

8. a. When you add negative degrees to a starting temperature in Problem 6, in what direction do you move on the thermometer?

b. In each case, is the result greater or less than the starting number?

9. a. Evaluate each of the following mentally. Use your calculator to check your result.

    i. $4 + 6$ **ii.** $6 + 8$ **iii.** $-7 + (-2)$

    iv. $-16 + (-10)$ **v.** $(-0.5) + (-1.4)$ **vi.** $-\dfrac{5}{2} + \left(-\dfrac{3}{2}\right)$

b. In each calculation in part a, what do you notice about the signs of the two numbers being added?

c. How do the signs of the numbers being added determine the sign of the result?

d. How do you calculate the numerical part of the result from the numerical parts of the numbers being added?

10. a. Evaluate each of the following mentally. Use your calculator to check your result.

    i. $4 + (-6)$ **ii.** $-6 + 8$ **iii.** $7 + (-2)$

    iv. $-16 + (10)$ **v.** $-\dfrac{6}{2} + \dfrac{7}{2}$ **vi.** $0.5 + (-2.8)$

b. In each calculation in part a, what do you notice about the signs of the numbers being added?

c. How do the signs of the numbers being added determine the sign of the result?

d. How do you calculate the numerical part of the result from the numerical parts of the numbers being added?

11. Evaluate each of the following. Use your calculator to check your result.

a. $-8 + (-6)$ **b.** $9 + (-12)$ **c.** $6 + (-8)$ **d.** $-7 + (-9)$

e. $16 + 10$ **f.** $16 + (-10)$ **g.** $-9 + 8$ **h.** $-2 + (-3)$

i. $9 + (-5)$ **j.** $3 + (-6)$ **k.** $-\dfrac{5}{8} + \dfrac{1}{8}$ **l.** $-\dfrac{5}{6} + \left(-\dfrac{1}{6}\right)$

m. $\dfrac{3}{4} + \left(-\dfrac{1}{4}\right)$ **n.** $\dfrac{1}{2} + \dfrac{1}{3}$ **o.** $-5.9 + (-4.7)$ **p.** $0.50 + 0.06$

q. $(-5.75) + 1.25$ **r.** $-12.1 + 8.3$ **s.** $-6 + \left(-\dfrac{2}{3}\right)$ **t.** $5 + \left(-1\dfrac{3}{4}\right)$

Subtracting Signed Numbers

In context, subtracting a signed number often reflects a change in value. The ending value is written first, and the starting value is subtracted from it.

Procedure

Calculating Change in Value

Change in value is calculated by subtracting the starting value from the ending value. That is,

change in value (difference) = ending value − starting value.

Example 1

The change in value from 2° to 9° is 9° − 2° = 7°. Here, 9° is the final temperature, 2° is the original (or initial) temperature, and 7° is the change in temperature. Notice that the temperature has risen; therefore, the change in temperature of 7° is positive, as shown on the thermometer on the left.

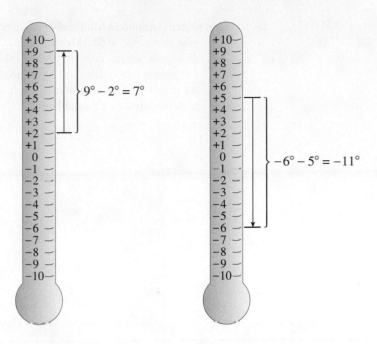

Suppose that − 6° *is the final temperature and* 5° *is the original (or initial) temperature. Symbolically, this is written* − 6° − 5°, *which produces a result of* − 11°, *as indicated on the thermometer on the right. The significance of a negative change is that the temperature has fallen.*

12. a. What is the change in temperature from 3° to 5°? That is, what is the difference between the final temperature, 5°, and the initial temperature, 3°? Symbolically, you are calculating +5° − (+3°). Use the thermometer below to demonstrate this calculation. Has the temperature risen or fallen?

b. The change in temperature from 3° to −5° is −8°. Symbolically, you write −5° − (+3°) = −8°. Use the thermometer below to demonstrate this calculation. Has the temperature risen or fallen?

c. A temperature in Montreal last winter rose from $-3°$ to $5°$. What was the change in temperature? Write the calculation symbolically, and determine the result. Use the thermometer below to demonstrate your calculation.

d. The temperature on a March day in Montana was $-3°$ at noon and $-5°$ P.M. What was the change in temperature? Write the calculation symbolically, and determine the result. Use the thermometer below to demonstrate your calculation. Has the temperature risen or fallen?

13. When calculating a change in value, in which position in the subtraction calculation must the initial value be placed—to the left or to the right of the subtraction symbol?

Your calculations in Problem 12 demonstrated that subtraction is used to calculate the *change* or *difference* in values.

Opposites

You can use the concept of opposites to relate the operations of addition and subtraction of integers.

Opposites are numbers that when added together give a sum of zero.

Example 2   *The numbers 10 and -10 are opposites because $10 + (-10) = 0$.*

14. a. What is the opposite of 5?

b. What is the opposite of −8?

c. $0 - 5 =$
$0 + (-5) =$

d. $0 - 12 =$
$0 + (-12) =$

e. $0 - (-8) =$
$0 + 8 =$

f. $0 - (-6) =$
$0 + 6 =$

g. From your experience in parts c–f, what can you conclude about the result of subtracting a number from zero? What about adding a number to zero?

h. From your experience with parts c–f, is it reasonable to believe that subtracting a number gives the same result as adding its opposite? Explain.

Subtracting a signed number is equivalent to *adding its opposite*. Therefore, when subtracting a signed number, *two* changes must be made. The subtraction symbol is changed to an addition symbol, *and* the number following the subtraction symbol is replaced by its opposite. The transformed problem is evaluated as an addition problem.

Example 3 *The expression* $(-4) - 6$ *becomes* $(-4) + (-6)$, *which equals* -10.

opposite

$$-4 - 6 = -4 + (-6) = -10$$

change
subtraction
to addition

15. Convert each of the following to an equivalent addition problem and evaluate. Use a calculator to check your results.

a. $4 - 11$

b. $-8 - (-6)$

c. $9 - (-2)$

d. $-7 - 1$

e. $10 - 15$

f. $6 - (-4)$

g. $-8 - 10$

h. $-2 - (-5)$

i. $-\dfrac{5}{8} - \dfrac{1}{8}$

j. $\dfrac{5}{6} - \left(-\dfrac{1}{6}\right)$ **k.** $-\dfrac{3}{4} - \left(-\dfrac{1}{4}\right)$ **l.** $\dfrac{1}{2} - \dfrac{1}{3}$

m. $5.9 - (-4.7)$ **n.** $-3.75 - 1.25$

o. $-6 - \left(-\dfrac{2}{3}\right)$ **p.** $5 - \left(-\dfrac{3}{4}\right)$

General Method for Adding and Subtracting Signed Numbers

Number lines and models such as thermometers provide a good visual approach to adding and subtracting signed numbers. However, you will find it more efficient to use a general method when dealing with signed numbers in applications. A convenient way to add and subtract signed numbers is to use the concept of absolute value.

The procedures you have developed in this activity for adding and subtracting signed numbers can be restated using absolute value.

Procedure

Addition and Subtraction of Signed Numbers

When *adding* two numbers with the *same* sign, add the absolute values of the numbers. The sign of the sum is the same as the sign of the original numbers.

When *adding* two numbers with *opposite* signs, determine their absolute values and then subtract the smaller from the larger. The sign of the sum is the same as the sign of the number with the larger absolute value.

When *subtracting* two numbers, rewrite the problem as an equivalent addition problem.

Example 4   *Perform the indicated operations on the following signed numbers.*

a. $-6 + (-13)$ **b.** $-29 + 14$ **c.** $5 - (-2)$

SOLUTION

a. The numbers being added have the same sign, so first add their absolute values.

$$|-6| + |-13| = 6 + 13 = 19$$

Because the signs of given numbers are the same, the answer will have that sign. Therefore, $-6 + (-13) = -19$.

b. The numbers being added have different signs, so determine their absolute values and subtract the smaller from the larger.

$$|-29| - |14| = 29 - 14 = 15$$

The sign of the answer is the sign of the number with the greater absolute value. Therefore, $-29 + 14 = -15$

c. $5 - (-2) = 5 + |-2| = 5 + 2 = 7.$

16. Perform the indicated operations.

a. $-31 + (-23)$ **b.** $-31 + 23$

c. $-52 + (-12)$ **d.** $-14 + (-23)$

e. $0 + (-23)$ **f.** $23 + (-15)$

g. $-6 - 8$ **h.** $7 - (-4)$

17. The temperature at noon is 58°F and is projected to decrease 12°F by midnight. Express the calculation to determine the temperature at midnight as both an addition and a subtraction. Determine the predicted temperature at midnight.

18. Your checking account is overdrawn with a balance of $-\$100$. You make a deposit of \$37. Express the calculation to determine your new balance as an addition and as a subtraction. Determine your resulting balance.

19. The temperature at 6 P.M. is 70°F and is projected to fall to 53°F by midnight. Express the calculation to determine the predicted change in temperature as a subtraction and as an addition. Determine the change in temperature.

20. Your checking account balance at the beginning of the month was $-\$65$. By the end of the month, it was \$55. Express the calculation to determine the change in your balance as a subtraction and as an addition. Determine the change in your balance.

SUMMARY: ACTIVITY 1.12

1. The collection of positive counting numbers, negative counting numbers, and zero is known as the set of **integers**.

2. The following is a number line scaled with the integers.

3. **Change in value** is calculated by subtracting the starting value from the ending value.

4. Two numbers whose sum is zero are called **opposites**. On a number line, opposites are equidistant from 0, one lying to the left of 0 and the other to the right of 0.

5. **Subtracting a signed number** is equivalent to *adding the opposite* of the signed number.

6. The **absolute value** of a number, also called its *magnitude*, represents the distance the number is from zero on the number line.

7. The absolute value of a nonzero number is always positive.

8. When *adding* **two numbers with the** *same* **sign**, add the absolute values of the numbers. The sign of the sum is the same as the sign of the numbers being added.

9. When *adding* **two numbers with** *opposite* **signs**, determine their absolute values and then subtract the smaller from the larger. The sign of the sum is the same as the sign of the number with the larger absolute value.

EXERCISES: ACTIVITY 1.12

1. In France, the number zero is used in hotel elevators to represent the ground floor.

   a. Describe the floor numbered -2.

   b. If you are on the floor numbered -3, would you take the elevator up or down to get to the floor labeled -1?

2. Another illustration of adding signed numbers can be found on the gridiron—that is, on the football field. On the number line, a gain is represented by a move to the right, whereas a loss results in a move to the left. For example, a 7-yard loss and a 3-yard gain result in a 4-yard loss. This can be written symbolically as $-7 + 3 = -4$.

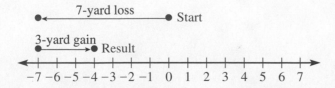

Use the number line to complete the following.

    a. A 6-yard gain and a 3-yard gain result in a _____

    b. A 5-yard gain and a 2-yard loss result in a _____.

    c. A 4-yard loss and a 3-yard loss result in a _____.

    d. A 1-yard gain and an 8-yard loss result in a _____.

3. Write each of the situations in Exercise 2 as an addition problem.

    a.   **b.**

    c.   **d.**

4. The San Francisco 49ers are famous for deciding their first 10 plays before the game begins. If the yardage gained or lost from each play is totaled, the sum represents the total yards gained (or lost) for those plays. One play sequence proceeded as follows.

Play	1	2	3	4	5	6	7	8	9	10
Yards	Lost 5	Gained 7	Gained 9	Lost 5	Gained 15	Gained 4	Lost 6	Gained 20	Lost 1	Gained 5

Determine the total yards gained (or lost) for these 10 plays.

5. Evaluate each of the following. Use a calculator to verify your result.

    a. $-3 + (-7)$ **b.** $-6 + 2$ **c.** $4 + (-10)$

    d. $16 - 5$ **e.** $7 - 12$ **f.** $-2 - 8$

    g. $-5 + (-10)$ **h.** $(-6) + 9$ **i.** $-8 - 10$

    j. $-4 - (-5)$ **k.** $0 - (-6)$ **l.** $-3 + 0$

    m. $9 - 9$ **n.** $9 - (-9)$ **o.** $0 + (-7)$

    p. $-4 + (-5)$ **q.** $-8 - (-3)$ **r.** $(-9) - (-12)$

    s. $5 - 8$ **t.** $(-7) - 4$ **u.** $-\dfrac{5}{9} + \dfrac{4}{9}$

    v. $2.5 - 3.1$ **w.** $-75 - 20$ **x.** $33 - 66$

6. The current temperature is $-12°C$. If it is forecast to be $5°$ warmer tomorrow, what is tomorrow's predicted temperature?

7. The current temperature is $-4°C$. If it is forecast to be $7°$ colder tomorrow, what is tomorrow's predicted temperature?

8. On a cold day in April, the temperature rose from $-5°C$ to $7°C$. What is the change in temperature?

9. A heat wave hits New York City, increasing the temperature an average of $3°C$ per day for 5 days, starting with a high of $20°C$ on June 6. What is the approximate high temperature on June 11?

10. Insert the appropriate inequality symbol between the two numbers in each of the following number pairs.

    a. 5   3   **b.** -5 -3 **c.** -8 -1

    d. -4 -7 **e.** -1 -2 **f.** $\dfrac{1}{5}$ $\dfrac{1}{3}$

11. Based on the 2010 U.S. Census, New York and Ohio each lost 2 seats in the U.S. House of Representatives. In contrast, Texas gained 4 seats and Florida gained 2. Write a signed number that represents the total change in the number of legislature seats for these states.

12. The highest recorded temperature in the United States was $134°F$ at Death Valley, California, in 1913. The record low temperature was $-80°F$ at Prospect Creek, Alaska, in 1971. Determine the difference between the highest and lowest temperatures. Note that the phrase *difference between* means to subtract the numbers in the order stated.

13. The following table contains the median sale prices for existing single-family homes in the United States for 2007–2013. Complete the table determining the change from one year to the next.

YEAR	MEDIAN SALE PRICE ($)	CHANGE FROM PREVIOUS YEAR
2007	217,000	
2008	198,100	
2009	172,100	
2010	173,100	
2011	166,200	
2012	180,200	
2013	171,100	

14. On September 29, 2008 the Dow Jones index (DJIA) dropped a record 777.68 points falling to a close of 10,365.45.

    a. What was the previous daily closing value?

    b. By what percent did the Dow Jones index decrease that day?

15. During a one-month period, the price of a gallon of 87-octane gasoline dropped 7 cents, then dropped another 5 cents, rose 6 cents, and finally dropped 3 cents. Write a signed number that represents the total change in gasoline price per gallon and interpret the result in a sentence.

16. In one class, during the first two weeks of the semester, 4 students withdrew, 9 students were added, and 5 students were dropped as no-shows. Write a signed number that represents by how many students the class size changed.

17. Although careful steps are taken by retail stores and banks to ensure internal control of cash, errors do occur. A cash over message on the register tape means that there is too much money. Cash short means that there is a shortage. The cash short and over results for a certain teller for one week are as follows:

MONDAY	TUESDAY	WEDNESDAY	THURSDAY	FRIDAY
$10 Short	$5 Over	$15 Short	Balanced	$20 Over

    a. Representing a cash short amount as a negative quantity, determine the total of the cash short and over amounts for the week.

    b. Determine the total cash short and over for the week by adding the absolute values of the amounts.

    c. Compare the results in parts a and b. Why might the total in part b be a better indicator of the teller's work performance?

18. The following table gives the elevation (in feet) of several locations relative to sea level.

LOCATION	ELEVATION RELATIVE TO SEA LEVEL
Mount Everest	29,035
Mount McKinley	20,320
Pikes Peak	14,110
Mauna Kea	13,710
Sea Level	0
Death Valley	−282
Dead Sea	−1,312
Ocean floor near Hawaii	−16,400
Mariana Trench	−35,827

Determine the difference in elevation between each of the following.

a. Mount McKinley in Alaska and Mauna Kea in Hawaii

b. Death Valley in California and Dead Sea between Israel and Jordan

c. Pikes Peak in Colorado and the ocean floor near Hawaii

19. Data on exports and imports by countries around the world are often shown in graphs for comparison purposes. The following graph is one example. The number line in the graph represents net exports in billions of dollars for eight countries. Net exports are obtained by subtracting imports from exports; a negative net export means that the country imported more goods than it exported.

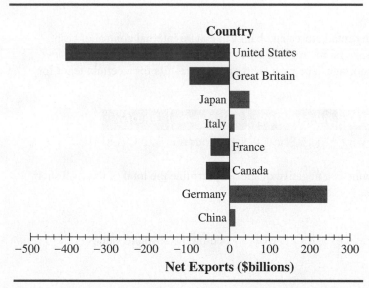

Country

Net Exports ($billions)

Source: *The Economist*, Nov. 9, 2013, p. 96, "Economic and Financial Indicators."
Data are for mid-2013.

a. Estimate the net amount of exports for Japan. Is your answer a positive or negative integer? Explain what this number tells you about the imports and exports of Japan.

b. Estimate the net amount of exports for the United States. Is your answer a positive or negative integer? Explain what this number tells you about the imports and exports of the United States.

c. Determine the absolute value for the net exports of Great Britain.

d. What is the opposite of the net amount of exports of Germany?

20. Profits and losses in millions of dollars per quarter over a 2-year period for a telecommunications company are shown in the bar graph. Determine the total profit or loss for the company over the 2-year period.

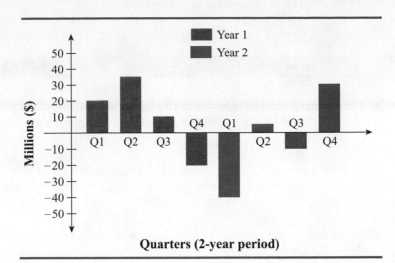

Activity 1.13

Shedding the Extra Pounds

Objective

I. Multiply and divide signed numbers.

You have joined a health and fitness club to lose some weight. The club's registered dietitian and your personal trainer helped you develop a special diet and exercise program. Your weekly weight gain (positive value) or weekly weight loss (negative value) over the first 7 weeks of the program is recorded in the following table.

Week Number	1	2	3	4	5	6	7
Change In Weight (Lb)	−2	−2	3	−2	−2	3	−2

I. a. Identify the 2 weeks during which you gained weight. What was your total weight gain over those 2 weeks? Write your answer as a signed number and in words.

b. Identify the 5 weeks during which you lost weight. What was your total weight loss over those 5 weeks? Write your answer as a signed number and in words.

c. At the end of the first 7 weeks, what is the total change in your weight?

In the solution of Problem 1b, you may have used the fact that the repeated loss of 2 pounds in weeks 1, 2, 4, 5 and 7 can be represented as a product:

$$-2 + (-2) + (-2) + (-2) + (-2) = 5(-2) = -10.$$

Because the multiplication of integers is commutative, it follows that

$$5(-2) = -2(5) = -10.$$

In the remainder of this activity, you will explore the multiplication and division of signed numbers.

Multiplication of Signed Numbers

Recall that opposites are numbers with the same absolute value that lie at equal distances from 0 on the number line, on opposite sides of 0. Just as multiplication by 1 leaves any number unchanged, multiplication by −1 changes the sign of a number, resulting in its opposite. For example, $-1(3) = -3$ and $-1(-4) = 4$. As a consequence, every negative number can be written as the product of its opposite (a positive number) and −1. The only exception is 0. The number 0 is neither positive nor negative, and multiplication by −1 still results in 0.

Therefore, multiplying any number by −1 can be considered a sort of toggle operation that changes the sign of the number. Each time a nonzero number is multiplied by −1, the given number changes sign. If a number is multiplied by several factors of −1, the sign of the resulting product depends on the number of factors of −1.

2. a. Determine the product $-1(-1)(-1)(5)$.

b. Determine the product of $-1(-1)(-1)(-1)(-1)(-1)(5)$.

c. How does the number of factors of −1 affect the sign of the product?

Because all negative numbers contain a factor of -1, it follows from the results of Problem 2 that the product of 2 signed numbers with opposite signs is negative. The product of 2 negative numbers is positive. If the product contains more than 2 signed numbers, the associative and commutative properties of multiplication allow you to collect and group the factors of -1. Therefore,

Sign of a Product

i. an even number of negative factors will result in a positive product

ii. an odd number of negative factors will result in a negative product

For example, the product $(-3)(2)(-4)$ can be viewed as $(-1)(3)(2)(-1)(4)$ or $(-1)(-1)(3)(2)(4) = 24$.

3. Multiply each of the following. Use your calculator to check your answers.

 a. $7(-3)$ **b.** $-5(12)$ **c.** $6(-8)$ **d.** $-9(7)$

 e. $-12(4)$ **f.** $10(-5)$ **g.** $-2.34(10)$ **h.** $7.5(-4.2)$

 i. $-\dfrac{1}{3} \cdot \left(\dfrac{7}{8}\right)$ **j.** $3\dfrac{2}{3}\left(-1\dfrac{1}{4}\right)$

4. Multiply each of the following. Use your calculator to check your answers.

 a. $(-11)(-4)$ **b.** $(5)(13)$ **c.** $(-6)(-8)$

 d. $(-12)(-3)$ **e.** $(-9)(-8)$ **f.** $(15)(4)$

 g. $(-4.567)(-100)$ **h.** $(-7)(-4.50)$ **i.** $\left(-\dfrac{3}{8}\right)\left(-\dfrac{2}{21}\right)$

5. Multiply the following. Use your calculator to check your answers.

 a. $(-4)(-5)(-3)$ **b.** $(-6)(-2)(-1)$

 c. $2(-3)(-5)$ **d.** $4(-2)(-3)(-1)$

 e. $3(-1)(-1)(-1)(-1)$ **f.** $(-1)(-1)(1)(-1)(-1)(-1)$

 g. $(-1)(-1)(-1)(-1)(-1)(-1)$ **h.** $(-2)(-3)(-4)(5)(-6)$

Division of Signed Numbers

Suppose your friend is trying to put on weight and gains 8 pounds over a 4-week period. This is an average gain of $\dfrac{8 \text{ pounds}}{4 \text{ weeks}} = 8 \div 4 = 2$ pounds per week.

Suppose you lose 12 pounds during the same 4-week period. Your average loss per week can be represented by the quotient $\dfrac{-12 \text{ pounds}}{4 \text{ weeks}}$ or $-12 \div 4$. Division is the inverse operation to multiplication. Therefore,

$$\frac{-12}{4} = -12 \div 4 = -3 \text{ because } 4 \cdot (-3) = -12.$$

A weight loss of 12 pounds over 4 weeks means that you lost an average of 3 pounds per week.

6. Use the multiplication check for division to obtain the only reasonable answer for each of the following. Then use your calculator to check your answers.

   a. $33 \div 3$

   b. $-24 \div 6$

   c. $15 \div (-5)$

   d. $\dfrac{-36}{4}$

   e. $\dfrac{-56}{-7}$

   f. $\dfrac{42}{-2}$

   g. $(-64) \div (-8)$

   h. $\dfrac{-63}{-9}$

   i. $\dfrac{81}{-9}$

   j. $\dfrac{-30.45}{2.1}$

7. Use the patterns observed in Problem 6 to answer parts a and b.

   a. The sign of the quotient of two numbers with the same sign is _____.

   b. The sign of the quotient of two numbers with opposite signs is _____.

8. Determine the quotient of each of the following.

   a. $\dfrac{-24}{8}$

   b. $\dfrac{24}{-8}$

   c. $-\dfrac{24}{8}$

   d. $-\dfrac{18}{-6}$

   e. $-\dfrac{-28}{-7}$

   f. $-\dfrac{-8}{2}$

   g. Explain the effect of having one negative sign in fractions of the form in parts a–c. Does the result depend on the location of the negative sign?

A negative fraction can be written equivalently in three ways with the negative sign in three possible positions: in front of the fraction, in the numerator, or in the denominator.

That is, $-\dfrac{6}{3}$, $\dfrac{-6}{3}$ and $\dfrac{6}{-3}$ are all equivalent fractions whose numerical value is -2.

Division Involving Zero

Recall that division is defined to be the inverse of multiplication. That is, given a product of 2 numbers (called factors), dividing the product by one of the factors produces the other factor.

For example, because $12 = 3 \cdot 4$,

i. dividing 12 by the factor 4 yields the other factor, 3 ($12 \div 4 = 3$ or equivalently $\dfrac{12}{4} = 3$)

ii. dividing 12 by the factor 3 yields the other factor, 4 ($12 \div 3 = 4$ or equivalently $\dfrac{12}{3} = 4$).

One important exception occurs when one of the factors is 0, because 0 has the unique property that $0 \cdot$ any number $= 0$.

For example, consider the product $0 = 0 \cdot 9$.

i. Dividing the 0 on the left by the factor 9 does yield the other factor, 0 ($0 \div 9 = 0$ or equivalently $\dfrac{0}{9} = 0$).

ii. However, if you attempt to divide the 0 on the left by the zero factor on the right, the value $0 \div 0$ would seem to yield the other factor, 9.

Simarily, consider the product $0 = 0 \cdot 5$.

i. Dividing the 0 on the left by the factor 5 does yield the other factor, 0 ($0 \div 5 = 0$ or equivalently $\dfrac{0}{5} = 0$).

ii. However, if you attempt to divide the 0 on the left by the zero factor on the right, the value $0 \div 0$ would **now** seem to yield the other factor, 5. Of course, 5 does not equal 9!

These last two examples demonstrate two very important facts about divisions that involve zero:

1. When 0 is divided by a nonzero number, the result is always 0.

$$\frac{0}{nonzero\ number} = 0 \text{ because } 0 \cdot nonzero\ number = 0.$$

2. When 0 is divided by 0, the quotient can be any number at all. Because there is no specific value of this quotient, its value is **undefined**.

$\dfrac{0}{0}$ has no specific value because $0 \cdot$ every number $= 0$; its value is undefined.

The quotient $\dfrac{0}{0}$ is also called **indeterminate** because a specific value cannot be determined.

One last quotient involving 0 is of the form $\dfrac{nonzero\ number}{0}$. Can a value be assigned to this quotient? If so, you could rewrite this quotient as an equivalent product as shown below.

$$\dfrac{nonzero\ number}{0} = value \quad \text{would mean that} \quad 0 \cdot value \text{ would equal a nonzero number}$$

But because $0 \cdot \text{every number} = 0$, there is no possible value that can ever make this true. This demonstrates a third important fact about divisions that involve zero:

> **3.** When a nonzero number is divided by 0, the quotient has no possible value. Because there is no such value, the quotient is **undefined**.
>
> $\dfrac{nonzero\ number}{0}$ has no possible value because $0 \cdot \text{every number} = 0$;   its value is undefined.

One consequence of attempting to divide by 0 occurs in programming, where it can creep in if the programmer has not written code that anticipates and prevents this calculation. The program will experience a "fatal error" and may either crash or enter an endless loop. A calculator may display an error message or the word FALSE.

9. Evaluate each of the following, if possible.

    a. $0 \div 9$ **b.** $0 \div (-3)$ **c.** $\dfrac{0}{-5}$

    d. $9 \div 0$ **e.** $(-3) \div 0$ **f.** $\dfrac{-5}{0}$

10. Evaluate the following expressions, if possible.

    a. $\dfrac{20 - 5 \cdot 2^2}{8}$ **b.** $\dfrac{7 \cdot 4 - 10}{6 - 2 \cdot 3}$

    c. $\dfrac{3^2 - (-3)^2}{8 + (-2)^3}$ **d.** $\dfrac{-4^2 + (-2)^4}{8 - (-2)^3}$

SUMMARY: ACTIVITY 1.13

I. When you multiply or divide two numbers with **opposite** signs, the result is always negative.

$$(-3) \cdot 4 = -12; \quad 5 \cdot (-6) = -30; \quad 15 \div (-3) = -5; \quad (-21) \div 3 = -7$$

2. When you multiply or divide two numbers with the **same** sign, the result is always positive.

$$3 \cdot 4 = 12; \quad (-5) \cdot (-6) = 30; \quad 15 \div 3 = 5; \quad (-21) \div (-3) = 7$$

3. When you multiply a string of signed numbers, the product is positive when the number of negative factors is even and negative when the number of negative factors is odd.

$5 \cdot (-2) \cdot (-3) \cdot (-1) = -30$ because there are three (odd number) negative factors in the expression.

$5 \cdot (-2) \cdot (3) \cdot (-1) = 30$ because there are two (even number) negative factors in the expression.

4. When zero is divided by a nonzero number, the quotient is always zero. When any number is divided by zero, the result has no numerical value—it is undefined.

EXERCISES: ACTIVITY 1.13

I. Determine the product or quotient mentally. Use your calculator to check your answers.

a. $-5(-7)$

b. $(-3)(-6)$

c. $-4 \cdot 6$

d. $(-8)(-2)$

e. $-1(2.718)$

f. $-3.14(-1)$

g. $-6(0)$

h. $4(-7)$

i. $(-3)(-8)$

j. $32 \div (-8)$

k. $-25 \div (-5)$

l. $0 \div (-9)$

m. $-5 \div 0$

n. $(-12) \div 6$

o. $16 \div (-2)$

p. $-48 \div 6$

q. $-63 \div (-1)$

r. $-6 \div 12$

2. Evaluate each of the following. Use a calculator to verify your answers.

   a. $\dfrac{1}{2} \div (-2)$ **b.** $-2 \div \left(\dfrac{1}{2}\right)$ **c.** $(2.1)(-8)$

   d. $11(-3)(-4)$ **e.** $0.2(-0.3)$ **f.** $-1.6 \div (-4)$

   g. $\left(-\dfrac{3}{4}\right)\left(-\dfrac{2}{9}\right)$ **h.** $-4\dfrac{2}{3} \div \left(-\dfrac{7}{9}\right)$ **i.** $(-5)(-6)(-1)(3)(-2)$

3. Without actually performing the following calculations, determine the sign of the product.

   a. $(-6)(-3)(2)(-4)$ **b.** $(2)(-5)(2)(-5)(2)$ **c.** $(-2)(-2)(-2)(-2)$

4. Summarize the results of Exercise 3 in your own words.

5. A cold front moves through your hometown in January, dropping the temperature an average of 3° per day for 5 days. What is the change in temperature over a 5-day period? Write your answer as a signed number.

6. Yesterday, you wrote three checks, each for $21.86. How did these checks affect your balance? State your answer in words and as a signed number.

7. Your friend is very happy that he lost $21\frac{1}{2}$ pounds in 6 weeks. On average, how many pounds did he lose per week?

8. During the last year, your friend lost $9876.34 in stocks. Express her average monthly loss as a signed number.

9. Stock values are reported daily in some newspapers. Reports usually state the values at both the opening and closing of a business day. Suppose you own 220 shares of a technology stock.

   a. At the highest point today, your stock is worth $26.03 per share, but by the close of the New York Stock Exchange, it had dropped to $24.25 per share. What is the total change in the value of your 220 shares from the highest point of the day to the close of business? State your answer in words and as a signed number.

b. Your stock opened today at $23.71 and gained $0.54 per share by the end of the day. What was the total value of your shares after the stock exchange closed?

10. You are one of three equal partners in a company.

a. In 2014, your company made a profit of $300,000. How much money did you and each of your partners make? Write the answer in words and as a signed number.

b. In 2013, your company experienced a loss of $150,000. What was your share of the loss? Write the answer in words and as a signed number.

c. Over the 2-year period from January 1, 2013, to December 31, 2014, what was the net profit for each partner? Express your answer in words and as a signed number.

d. Suppose that in 2015, your corporation suffered a further loss of $180,000. Over the 3-year period from January 1, 2013, to December 31, 2015, what was the net change for each partner? Express your answer in words and as a signed number.

11. Each individual account in your small business has a current dollar balance, which can be either positive (a credit) or negative (a debit). To determine your current total account balance, you add all the individual balances.

Suppose your records show the following individual balances.

$$-220 \quad -220 \quad 350 \quad 350 \quad -220 \quad -220 \quad -220 \quad 350$$

a. What is the total of the individual positive balances?

b. What is the total of the individual negative balances?

c. In part b, did you sum the five negative balances? Or did you multiply -220 by 5? Which method is more efficient? What is the sign of the result of your calculation?

d. What is the total balance of these accounts?

12. You own 180 shares of an energy stock. The following table lists the changes (in dollars) during a given week for the price of a single share. Did you earn a profit or suffer a loss on the value of your shares during the given week? How much profit or loss?

Day	Mon.	Tues.	Wed.	Thurs.	Fri.
Daily Change	+0.38	−0.75	−1.25	−1.13	2.55

13. You plan to dive to a shipwreck located 168 feet below sea level. You can dive at the rate of approximately 2 feet per second. Use negative integers to represent the distance below sea level in the following calculations.

a. Calculate the depth you dove in 1 minute.

b. Calculate where you are in relationship to the shipwreck after 1 minute.

14. You are interning at the weather station. This week the low temperatures were 4°C, −8°C, 4°C, −1°C, −2°C, −2°C, and −2°C. Determine the average low temperature for the week.

Activity 1.14

Order of Operations Revisited

Objectives

1. Use the order of operations convention to evaluate expressions involving signed numbers.

2. Evaluate expressions that involve negative exponents.

3. Distinguish between such expressions as -5^4 and $(-5)^4$.

4. Write very small numbers in scientific notation.

You can combine the procedures for adding and subtracting signed numbers with those for multiplying and dividing signed numbers by using the order of operations convention you learned in Activity 1.3, Properties of Arithmetic. The order of operations convention is valid for all numbers, positive and negative, including decimals, fractions, and numerical expressions with exponents. Your ability to follow these procedures correctly with all numbers will help you use the formulas that you will encounter in applications.

1. Calculate the following expressions by hand. Then use your calculator to check your answers.

a. $6 + 4(-2)$ b. $-6 + 2 - 3$ c. $(6 - 10) \div 2$

d. $-3 + (2 - 5)$ e. $(2 - 7)5$ f. $-3(-3 + 4)$

g. $-2 \cdot 3^2 - 15$ h. $(2 + 3)^2 - 10$

i. $-7 + 8 \div (5 - 7)$ j. $2.5 - (5.2 - 2.2)^2 + 8$

k. $4.2 \div 0.7 - (-5.6 + 8.7)$ l. $\dfrac{1}{4} - \left(\dfrac{2}{3} \cdot \dfrac{9}{8}\right)$

m. $\dfrac{5}{6} \div (-10) + \dfrac{7}{12}$

Negation and Exponentiation

2. Evaluate -3^2 by hand, and then use your calculator to check your answer.

In Problem 2, two operations are to be performed on the number 3, exponentiation and negation. Negation can be considered as multiplication by -1. Therefore, by the order of operations convention, 3 is first squared to obtain 9 and then 9 is multiplied by -1 to determine the result, -9.

3. Evaluate $(-3)^2$ by hand, and then check with your calculator.

In Problem 3, parentheses are used to indicate that the number 3 is negated *first* and then -3 is squared to produce the result, 9.

Example 1   *Evaluate* $(-2)^4$ *and* -2^4.

SOLUTION

In $(-2)^4$, the parentheses indicate that 2 is first negated and then -2 is written as a factor 4 times.

$$(-2)^4 = (-2)(-2)(-2)(-2) = 16$$

In -2^4, the negative indicates multiplication by -1. Because exponentiation is performed before multiplication, the 2 is raised to the fourth power first and then the result is multiplied by -1 (negated).

$$-2^4 = -1 \cdot 2^4 = -1 \cdot 2 \cdot 2 \cdot 2 \cdot 2 = (-1) \cdot 16 = -16$$

4. Evaluate the following expressions by hand. Then use your calculator to check your answers.

   a. -5^2 **b.** $(-5)^2$ **c.** $(-3)^3$

   d. -1^4 **e.** $2 - 4^2$ **f.** $(2 - 4)^2$

   g. $-5^2 - (-5)^2$ **h.** $(-1)^8$ **i.** $-5^2 + (-5)^2$

   j. $-\left(\dfrac{1}{2}\right)^2 + \left(\dfrac{1}{2}\right)^2$

Negative Exponents

So far in this book, you have encountered only zero or positive integers as exponents. Are numbers with negative exponents meaningful? How would you calculate an expression such as 10^{-5}? In the following problems, you will discover answers to those questions.

5. **a.** Complete the following table.

EXPONENTIAL FORM	EXPANDED FORM	VALUE
10^5	$10 \cdot 10 \cdot 10 \cdot 10 \cdot 10$	100,000
10^4	$10 \cdot 10 \cdot 10 \cdot 10$	10,000
10^3	$10 \cdot 10 \cdot 10$	1000
10^2		
10^1		
10^0	1	1
10^{-1}	$\dfrac{1}{10}$	$\dfrac{1}{10}$ or 0.1
10^{-2}		
10^{-3}		
10^{-4}		

b. Use the pattern developed in the preceding table to determine the relationship between each negative exponent in column 1 and the number of zeros in the denominator of the corresponding fraction and the decimal result in column 3. List and explain everything you observe.

The pattern you observed in the table in Problem 5 that involves raising base 10 to a negative exponent can be generalized to any nonzero number as base.

A negative exponent always signifies a reciprocal. That is, $5^{-1} = \dfrac{1}{5}$ and $2^{-3} = \dfrac{1}{2^3} = \dfrac{1}{8}$.

Symbolically,

$$b^{-n} = \frac{1}{b^n}, \text{ where } b \neq 0.$$

Example 2 *Write each of the following without a negative exponent. Then write each answer as a simplified fraction and as a decimal.*

a. $10^{-5} = \dfrac{1}{10^5} = \dfrac{1}{100,000} = 0.00001$

b. $3^{-2} = \dfrac{1}{3^2} = \dfrac{1}{9} = 0.\overline{1}$

c. $(-4)^{-3} = \dfrac{1}{(-4)^3} = \dfrac{1}{-64} = -0.015625$

d. $6 \cdot 10^{-3} = \dfrac{6}{10^3} = \dfrac{6}{1000} = \dfrac{3}{500} = 0.006$

6. Write each of the following expressions without a negative exponent. Then write each result as a fraction and as a decimal.

a. 10^{-4} **b.** $3.25 \cdot 10^{-3}$ **c.** 5^{-2}

d. $(-5)^{-2}$ **e.** -5^{-2} **f.** 4^{-3}

g. $-3^{-2} + 3^{-2}$ **h.** $-2^{-3} + 2^3$

Scientific Notation Revisited

In Activity 1.3, you learned that scientific notation provides a compact way of writing very large numbers. Recall that any number can be written in scientific notation by expressing the number as the product of a decimal number between 1 and 10 times an integer power of 10. For example, 53,000,000 can be written in scientific notation as 5.3×10^7.

$$\underbrace{N}_{\substack{\text{number from} \\ \text{1 to 10}}} \times \underbrace{10^n}_{\substack{\text{power of 10,} \\ n \text{ an integer}}}$$

7. The star nearest Earth (excluding the Sun) is Proxima Centauri, which is 24,800,000,000,000 miles away. Write this number in scientific notation.

Very small numbers can also be written in scientific notation. For example, a molecule of DNA is about 0.000000002 meter wide. To write this number in scientific notation, you need to move the decimal point to the right to obtain a number between 1 and 10.

0.000000002.   Move the decimal point 9 places to the right.

Therefore, 0.000000002 can be written as 2.0×10^n. You will determine the value of n in Problem 8.

8. **a.** If you multiply 2.0 by 10,000, or 10^4, you move the decimal point _____ places to the _____

   b. If you multiply 2.0 by 0.00001, you move the decimal point _____ places to the _____

   c. What is the relationship between each negative exponent in column 1 in the table in Problem 5 and the number of placeholding zeros in the decimal form of the number in column 3?

   d. Write 0.00001 as a power of 10.

   e. Use the results from parts b–d to determine the value of n in the following: $0.000000002 = 2.0 \times 10^n$.

   f. How does your calculator display the number 0.000000002?

9. Write each of the following in scientific notation.

   a. 0.0000876   **b.** 0.00000018

   c. 0.00000000000000781

10. The thickness of human hair ranges from fine to coarse, with coarse hair about 0.0071 inch in diameter. Describe how you would convert this measurement into scientific notation.

Calculations Involving Numbers Written in Scientific Notation

11. a. Recall from Activity 1.3 that you used the $\boxed{\text{EE}}$ key on your calculator to enter numbers written in scientific notation. For example, $10^2 = 1 \times 10^2$ is entered as $\boxed{1}\,\boxed{\text{EE}}\,\boxed{2}$. Try it.

b. Now enter $10^{-2} = 1 \times 10^{-2}$ as $\boxed{1}\,\boxed{\text{EE}}\,\boxed{(-)}\,\boxed{2}$. What is the result written as a decimal? As a fraction?

c. Use the $\boxed{\text{EE}}$ key to enter your answers from Problem 9 into your calculator. Record your results, and compare them with the numbers given in Problem 9.

12. The wavelength of X-rays is 1×10^{-8} centimeter, and that of ultraviolet light is 2×10^{-5} centimeter.

a. Which wavelength is shorter?

b. How many times shorter? Write your answer in standard notation and in scientific notation.

13. a. The mass of an electron is about $\dfrac{1}{2000}$ that of a proton. If the mass of a proton is 1.7×10^{-24} gram, determine the mass of an electron.

b. The mass of one oxygen molecule is 5.3×10^{-23} gram. Determine the mass of 20,000 molecules of oxygen.

14. Use your calculator to evaluate the following expressions.

a. $\dfrac{16}{(4 \times 10^{-2}) - (2 \times 10^{-3})}$
b. $\dfrac{3.2 \times 10^3}{(8.2 \times 10^{-2})(3.0 \times 10^2)}$

SUMMARY: ACTIVITY 1.14

1. To indicate that a negative number is raised to a power, the number must be enclosed in parentheses.

   Example: "Square the number -4" is written as $(-4)^2$, which means $(-4)(-4)$ and equals 16.

2. An expression of the form $-a^2$ always signifies the negation (opposite) of a^2 and is *always* a negative quantity (for nonzero a).

   Example: -4^2 represents the negation (opposite) of 4^2 and, therefore, has the value -16.

3. A negative exponent always signifies a reciprocal. Symbolically, this is written as $b^{-n} = \dfrac{1}{b^n}$,

   where $b \neq 0$.

   Examples: $6^{-1} = \dfrac{1}{6}$ and $5^{-3} = \dfrac{1}{5^3} = \dfrac{1}{125}$

4. A positive number less than 1 can be written in scientific notation using negative powers of 10 as follows:

   i. Count the number of digits from the decimal point to the right (past) of the first nonzero digit. Make a mental note of this count.

   ii. Rewrite the original number by placing the decimal point directly after its first nonzero digit and then multiplying by 10 raised to the negative of your count.

   Example: Write 0.00425 in scientific notation. Count 3 digits from the decimal point to the right of 4. Therefore, $0.00425 = 4.25 \cdot 10^{-3}$

EXERCISES: ACTIVITY 1.14

1. To discourage random guessing on a multiple-choice exam, a professor assigns 5 points for a correct response, deducts 2 points for an incorrect answer, and gives 0 points for leaving the question blank.

   a. What is your score on a 35-question exam if you have 22 correct and 7 incorrect answers?

   b. Express your score as a percentage of the total points possible.

2. You are remodeling your bathroom. To finish the bathroom floor, you decide to install molding around the base of each wall. The molding needs to go all the way around the room except for a 39-inch gap at the doorway. The dimensions of the room are $8\frac{2}{3}$ feet by $12\frac{3}{4}$ feet. You determine the number of feet of molding you need by evaluating the following expression: $2\left(8\frac{2}{3} + 12\frac{3}{4}\right) - 3\frac{1}{4}$. How much molding do you need?

3. **a.** Complete the following table.

EXPONENTIAL FORM	EXPANDED FORM	VALUE
2^5	$2 \cdot 2 \cdot 2 \cdot 2 \cdot 2$	
2^4	$2 \cdot 2 \cdot 2 \cdot 2$	
2^3		
2^2		
2^1		
2^0		
2^{-1}	$\dfrac{1}{2}$	
2^{-2}	$\dfrac{1}{2} \cdot \dfrac{1}{2}$	
2^{-3}		
2^{-4}		

b. Use the pattern in the table to determine the simplified fraction form of 2^{-7}. What is the connection between this fraction and the value of 2^7?

4. Evaluate the following expressions by hand. Verify your answers with a calculator.

a. $(8 - 17) \div 3 + 6$ **b.** $-7 + 3(1 - 5)$ **c.** $-3^2 \cdot 2^2 + 25$

d. $1.6 - (1.2 + 2.8)^2$ **e.** $\dfrac{5}{16} - 3\left(\dfrac{3}{16} - \dfrac{7}{16}\right)$ **f.** $\dfrac{3}{4} \div \left(\dfrac{1}{2} - \dfrac{5}{12}\right)$

g. -1^{-2} **h.** $4^3 - (-4)^3$ **i.** $(6)^2 - (-6)^2$

5. Use your calculator to evaluate the following expressions.

a. $\dfrac{-3 \times 10^2}{(3 \times 10^{-2}) - (2 \times 10^{-2})}$ **b.** $\dfrac{1.5 \times 10^3}{(-5.0 \times 10^{-1})(2.6 \times 10^2)}$

6. Evaluate the following expressions:

a. 6^{-2} **b.** $(-6)^2$ **c.** -6^2 **d.** $(-6)^{-2}$

7. Write the following numbers in scientific notation.

a. 0.0056   **b.** 0.0000001234   **c.** 5,305,000,000   **d.** 105,000

8. Write the following numbers in standard notation.

a. 3.57×10^4 **b.** 2.61×10^{-4} **c.** 1.025×10^8

9. The radius of a hydrogen atom is approximately 0.000000005 centimeter. Express this number in scientific notation.

10. You have found 0.4 gram of gold while panning for gold on your vacation.

a. How many pounds of gold do you have ($1 \text{ g} = 2.2 \times 10^{-3}$ lb.)? Write your result in decimal form and in scientific notation.

b. If you were to tell a friend how much gold you have, would you state the quantity in grams or in pounds? Explain.

Cluster 4   What Have I Learned?

1. Explain how to add two signed numbers and how to subtract two signed numbers. Use examples to illustrate each.

2. Explain how to multiply two signed numbers and how to divide two signed numbers. Use examples to illustrate each.

3. What would you suggest to your classmates to avoid confusing addition and subtraction procedures with multiplication and division procedures?

4. **a.** Without actually performing the calculation, determine the *sign* of the product $(-0.1)(+3.4)(6.87)(-0.5)(+4.01)(3.9)$. Explain.

   b. Determine the *sign* of the product $(-0.2)(-6.5)(+9.42)(-0.8)(1.73)(-6.72)$. Explain.

   c. Determine the *sign* of the product $(-1)(-5.37)(-3.45)$. Explain.

   d. Determine the *sign* of the product $(-4.3)(+7.89)(-69.8)(-12.5)(+4.01)(-3.9)(-78.03)$. Explain.

   e. What rule is suggested by the results you obtained in parts a–d?

5. a. If −2 is raised to the power 4, what is the sign of the result? What is the sign if −2 is raised to the sixth power?

b. Raise −2 to the eighth power, tenth power, and twelfth power. What are the signs of the numbers that result?

c. What general rule involving signs is suggested by the preceding results?

6. a. Raise −2 to the third power, fifth power, seventh power, and ninth power.

b. What general rule involving signs is suggested by the preceding results?

7. The following table contains the daily midnight temperatures (in °F) in Batavia, New York, for a week in January.

Day	Sun.	Mon.	Tues.	Wed.	Thurs.	Fri.	Sat.
Temperature (°F)	−7	11	11	11	−7	−7	11

a. Determine the average daily temperature for the week.

b. Did you use the most efficient way to do the calculation? Explain.

8. a. What are the values of 3^2, 3^0, and 3^{-2}?

b. What is the value of any nonzero number raised to the zero power?

c. What is the meaning of any nonzero number raised to a negative power? Give two examples using −2 as the exponent.

9. a. The diameter of raindrops in drizzle near sea level is reported to be approximately 30×10^{-2} millimeter. Write this number in standard decimal notation and in scientific notation.

b. What is the measurement of these raindrops in centimeters (10 mm = 1 cm)? Write your answer in standard notation and in scientific notation.

c. From your experience with scientific notation, explain how to convert numbers from standard notation to scientific notation and vice versa.

Cluster 4 How Can I Practice?

Calculate by hand. Use your calculator to check your answers.

1. $15 + (-39)$

2. $-43 + (-28)$

3. $-0.52 + 0.84$

4. $-7.8 + 2.9$

5. $-32 + (-45) + 68$

6. $-46 - 63$

7. $53 - (-64)$

8. $8.9 - (-12.3)$

9. $-75 - 47$

10. $-34 - (-19)$

11. $-4.9 - (-2.4) + (-5.6) + 3.2$

12. $16 - (-28) - 82 + (-57)$

13. $-1.7 + (-0.56) + 0.92 - (-2.8)$

14. $\dfrac{2}{3} + \left(-\dfrac{3}{5}\right)$

15. $-\dfrac{3}{7} + \left(\dfrac{4}{5}\right) + \left(-\dfrac{2}{7}\right)$

16. $-\dfrac{5}{9} - \left(-\dfrac{8}{9}\right)$

17. $2 - \left(\dfrac{3}{2}\right)$

18. $1 - \left(\dfrac{3}{4}\right) + 1 - \dfrac{3}{4}$

19. $0 - \left(-\dfrac{7}{10}\right) + \left(-\dfrac{1}{2}\right) - \dfrac{1}{5}$

20. $\dfrac{-48}{12}$

21. $\dfrac{63}{-9}$

22. $\dfrac{121}{-11}$

23. $\dfrac{-84}{-21}$

24. $-125 \div -25$

25. $-2.4 \div 6$

26. $\dfrac{24}{-6}$

27. $-\dfrac{24}{6}$

28. $-0.8(12)$

29. $4(-0.06)$

30. $-9(-11)$

31. $-4(-0.6)(-5)(-0.01)$

32. $0.5(-7)(-2)(-3)$

33. $-4(-4)$

34. $(-4)(-4)$

35. $4(-4)$

36. You have $85.30 in your bank account. You write checks for $23.70 and $35.63. You then deposit $325.33. Later, you withdraw $130.00. What is your final balance?

37. You lose 4 pounds, then gain 3 pounds back. Later you lose 5 more pounds. If your aim is to lose 12 pounds, how many more pounds do you need to lose? State your answer in words and as a signed number.

38. You and your friend Patrick go on vacation. Patrick decides to go scuba diving, but you prefer to do some mountain climbing. You decide to separate for part of the day and do your individual activities. While you climb to an altitude of 2567 feet above sea level, Patrick is 49 feet below sea level. What is the vertical distance between you and Patrick?

39. In parts a–e, state your answer as a signed number and in words.

   a. The temperature is 2°C in the morning but drops to −5°C by night. What is the change in temperature?

   b. The temperature is −12°C in the morning but is expected to drop 7°C during the day. What will be the evening temperature?

   c. The temperature is −17°C this morning but is expected to rise 9°C by noon. What will be the noon temperature?

   d. The temperature is −8°C in the morning but drops to −17°C by night. What is the change in temperature?

   e. The temperature is −14°C in the morning and −6°C at noon. What is the change in temperature?

40. At the end of a hockey season, a defenseman summarized his +/− total in the following chart. For example: In the first column, in three games, his +/− total was +6. His total +/− for the 3 games was +18.

No. of Games	3	4	12	8	6	11	3	15	9	6	3
+/− Total	+6	+4	+3	+2	+1	0	−1	−2	−3	−4	−5

Write an expression that will determine the +/− total for the season. What is his +/− total for the season?

41. The following table contains the daily midnight temperatures in Buffalo for a week in January.

Day	Sun.	Mon.	Tues.	Wed.	Thurs.	Fri.	Sat.
Temp. (°F)	−3°	6°	−3°	−12°	6°	−3°	−12°

    a. Write an expression to determine the average daily temperature for the week.

    b. What is the average daily temperature for the week?

42. Evaluate the following expressions. Use your calculator to check your result.

    a. $-6 \div 2 \cdot 3$

    b. $(-3)^2 + (-7) \div 2$

    c. $(-2)^3 + (-9) \div (-3)$

    d. $(-14 - 4) \div 3 \cdot 2$

    e. $(-4)^2 - [-8 \div (2 + 6)]$

    f. $-75 \div (-5)^2 + (-8)$

    g. $(-3)^3 \div 9 - 6$

43. **a.** On average, fog droplets measure 20×10^{-3} millimeter in diameter at sea level. How many inches is this ($1 \text{ mm} = 0.03937$ in.)?

    b. At sea level, the average diameter of raindrops is approximately 1 millimeter. How many inches is this?

    c. How many times larger are raindrops than fog droplets?

C
Appendix

Skills Check 2 in Appendix C

Summary

The bracketed numbers following each concept indicate the activity in which the concept is discussed.

CONCEPT/SKILL	DESCRIPTION	EXAMPLE
Problem solving [1.2]	Problem solving strategies include: • discussing the problem • organizing information • drawing a picture • recognizing patterns • doing a simpler problem	You drive for 2 hours at an average speed of 47 mph. How far do you travel? $d = rt$ $d = \dfrac{47 \text{ mi.}}{\text{hr.}} \cdot 2 \text{ hr.} = 94 \text{ mi.}$
The commutative property of addition [1.3]	Changing the order of the terms yields the same sum.	$9 + 8 = 17; \quad 8 + 9 = 17$
The commutative property of multiplication [1.3]	Changing the order of the factors yields the same product.	$13 \cdot 4 = 52; \quad 4 \cdot 13 = 52$
Subtraction is not commutative [1.3]	The result is different when the order is reversed, so subtraction is not commutative. The two results are opposites.	$13 - 7 = 6; \quad 7 - 13 = -6$
Division is not commutative [1.3]	The result is different when the order is reversed, so division is not commutative. The two results are reciprocals.	$8 \div 10 = 0.8;$ $10 \div 8 = 1.25$
Associative property of addition [1.3]	Given three terms, it makes no difference whether the first two numbers or the last two numbers are added first. This property is written symbolically as $(a + b) + c = a + (b + c)$	$5 + (6 + 7)$ $= (5 + 6) + 7 = 18$
Associative property of multiplication [1.3]	When multiplying three factors, it makes no difference whether the first two numbers or the last two numbers are multiplied first. The same product results. This property is written symbolically as $(a \cdot b) \cdot c = a \cdot (b \cdot c)$	$(5 \cdot 7) \cdot 12$ $= 5 \cdot (7 \cdot 12) = 420$
Square Root [1.3]	When a number is squared (raised to the 2nd power), taking the square root reverses the process and returns the absolute value of the original number. The symbol $\sqrt{}$ indicates the square root.	Because $6^2 = 36$, $\sqrt{36} = 6$
Order of operations [1.3] and [1.14]	Use the order of operations convention for evaluating arithmetic expressions. Perform operations inside all grouping symbols first. This includes expressions in the numerator and denominator of a fraction, inside a radical, and within a pair of parentheses. All operations are performed in the following order.	$2 \cdot (3 + 4^2 \cdot 5) - 9$ $= 2 \cdot (3 + 16 \cdot 5) - 9$ $= 2 \cdot (3 + 80) - 9$ $= 2 \cdot (83) - 9$ $= 166 - 9$ $= 157$

CONCEPT/SKILL	DESCRIPTION	EXAMPLE
	1. Evaluate all exponential expressions as you read the expression from left to right. 2. Do all multiplication and division as you read the expression from left to right. 3. Do all addition and subtraction as you read the expression from left to right.	$-4^2 - 18 \div 3 \cdot (-2)$ $= -16 - 18 \div 3 \cdot (-2)$ $= -16 - 6 \cdot (-2)$ $= -16 - (-12)$ $= -16 + 12$ $= -4$
Exponential notation [1.3]	A number written as 10^4 is in exponential form; 10 is the base, and 4 is the exponent. The exponent indicates how many times the base appears as a factor.	$10^4 = \underbrace{10 \cdot 10 \cdot 10 \cdot 10}_{\text{10 is used as a factor 4 times}}$
A number raised to the zero power [1.3]	Any nonzero number raised to the zero power equals 1.	$2^0 = 1; \quad 50^0 = 1; \quad 200^0 = 1;$ $(-2)^0 = 1; \quad (5 \cdot 6)^0 = 1$
Scientific notation [1.3] and [1.14]	When a number is written as the product of a decimal number between 1 and 10 and a power of 10, it is expressed in scientific notation.	$0.0072 = 7.2 \times 10^{-3}$ $43{,}100 = 4.31 \times 10^4$
Simple average (mean) [1.5]	Add all the values, and divide the sum by the number of values.	$\dfrac{71 + 66 + 82 + 86}{4} = 76.25$
Weighted average [1.5]	To compute a weighted average of several components: 1. Multiply each data value by its respective weight. 2. Sum these weighted data values.	Weights of tests 1, 2, and 3: $\dfrac{1}{5}$ Weight of final exam: $\dfrac{2}{5}$ Weighted average: $\dfrac{1}{5} \cdot 71 + \dfrac{1}{5} \cdot 66 +$ $\dfrac{1}{5} \cdot 82 + \dfrac{2}{5} \cdot 86 = 78.2$
Sum of the weights [1.5]	The sum of the weights used to compute a weighted average will always equal 1.	$\dfrac{1}{5} + \dfrac{1}{5} + \dfrac{1}{5} + \dfrac{2}{5} = 1$
Ratio [1.6]	A ratio is a quotient that compares two similar numerical quantities, such as "part" to "total." Percent always indicates a ratio out of 100.	4 out of 5 is a ratio. It can be expressed as a fraction, $\dfrac{4}{5}$; a decimal, 0.8; or a percent, 80%.
Converting from decimal to percent format [1.6]	Move the decimal point 2 places to the right, and then attach the % symbol.	0.125 becomes 12.5%. 0.04 becomes 4%. 2.50 becomes 250%.
Converting from percent to decimal format [1.6]	Drop the percent symbol, and move the decimal point 2 places to the left.	35% becomes 0.35. 6% becomes 0.06. 200% becomes 2.00.

CONCEPT/SKILL	DESCRIPTION	EXAMPLE
Proportional reasoning: Applying a known ratio to a given piece of information [1.7]	known total · known ratio = unknown part $\dfrac{\text{known part}}{\text{known ratio}} = \text{unknown total}$	Forty percent of the 350 children play an instrument; 40% is the known ratio, and 350 is the total. $350 \cdot 0.40 = 140$. A total of 140 children play an instrument. Twenty-four children, constituting 30% of the marching band, play the saxophone; 30% is the known ratio, and 24 is the part. $24 \div 0.30 = 80$. A total of 80 children are in the marching band.
Relative change [1.8]	$\text{relative change} = \dfrac{\text{actual change}}{\text{original value}}$	A quantity changes from 25 to 35. The actual change is 10; the relative change is $\dfrac{10}{25}$, or 40%.
Growth factor [1.9]	When a quantity increases by a specified percent, the ratio of its new value to its original value is called the *growth factor* associated with the specified percent increase. The growth factor is determined by adding the specified percent increase to 100% and then changing this percent to its decimal form.	To determine the growth factor associated with a 25% increase, add 25% to 100% to obtain 125%. Change 125% to a decimal to obtain the growth factor 1.25.
Applying a growth factor to an original value [1.9]	original value · growth factor = new value	A $120 item increases by 25%. Its new value is $120 \cdot 1.25 = \$150$.
Applying a growth factor to a new value [1.9]	new value ÷ growth factor = original value	An item has already increased by 25% and is now worth $200. Its original value was $200 \div 1.25 = \$160$.
Decay factor [1.9]	When a quantity decreases by a specified percent, the ratio of its new value to its original value is called the *decay factor* associated with the specified percent decrease. The decay factor is determined by subtracting the specified percent decrease from 100% and then changing this percent to its decimal form.	To determine the decay factor associated with a 25% decrease, subtract 25% from 100%, to obtain 75%. Change 75% into a decimal to obtain the decay factor 0.75.

CONCEPT/SKILL	DESCRIPTION	EXAMPLE
Applying a decay factor to an original value [1.9]	original value · decay factor = new value	A $120 item decreases by 25%. Its new value is $120 · 0.75 = $90.
Applying a decay factor to a new value [1.9]	new value ÷ decay factor = original value	An item has already decreased by 25% and is now worth $180. Its original value was $180 ÷ 0.75 = $240.
Applying a sequence of percent changes [1.10]	The cumulative effect of a sequence of percent changes is the product of the associated growth or decay factors.	The effect of a 25% increase followed by a 25% decrease is 1.25 · 0.75 = 0.9375. That is, the item is now worth only 93.75% of its original value. The item's value has decreased by 6.25%.
Using unit (or dimensional) analysis to solve rate and conversion problems [1.11]	1. Identify the measurement unit of the result. 2. Set up the sequence of multiplications so that the appropriate units divide out, leaving the appropriate measurement unit of the result. 3. Multiply and divide the numbers as indicated to obtain the numerical part of the result. 4. Check that the appropriate units divide out, leaving the expected unit for the result.	To convert your height of 70 inches to centimeters: $70 \text{ in.} \cdot \dfrac{2.54 \text{ cm}}{1 \text{ in.}} = 177.8 \text{ cm}$ To estimate how far a 16-gallon tank of fuel will last with mileage of 31 miles per gallon: $16 \text{ gal.} \cdot \dfrac{31 \text{ mi.}}{\text{gal.}} = 16 \cdot 31 \text{ mi.} = 496 \text{ mi.}$
Metric units [1.11]	The base units for length, mass and volume in the metric system are the meter (m), gram (g) and liter (L).	A tree 20 m in height. 2.4 g of a prescription. 14 L of gasoline.
Metric prefixes [1.11]	The base units can be multiplied by powers of ten using prefixes: giga (G) = 10^9 nano (n) = 10^{-9} mega (M) = 10^6 micro (μ) = 10^{-6} kilo (k) = 10^3 milli (m) = 10^{-3} hecto (h) = 10^2 centi (c) = 10^{-2} deka (da) = 10^1 deci (d) = 10^{-1}	1 millimeter = 1 mm = 10^{-3} m 1 kilometer = 1 km = 1000 m 1 kilogram = 1 kg = 1000 g
Integers [1.12]	Integers consist of the positive and negative counting numbers and zero.	$\ldots -3, -2, -1, 0, 1, 2, 3, \ldots$
Number line [1.12]	Use a number line scaled with integers to visualize relationships between positive and negative numbers.	

CONCEPT/SKILL	DESCRIPTION	EXAMPLE
Order on the number line [1.12]	If $a < b$, then a is left of b on a number line. If $a > b$, then a is right of b on a number line.	$-3 < -2;$ $4 < 7;$ $8 > -2$
Absolute value [1.12]	The absolute value of a number, a, written $|a|$, is the distance of a from zero on a number line. It is always positive or zero.	$|-2| = 2;$ $|2| = 2$
Change in value [1.12]	Calculate a change in value by subtracting the (original) value from the final value.	If the temperature changed from 8° at noon to 10° in the evening, the change in temperature is $10 - 8 = 2$. The temperature increased 2°.
Opposites [1.12]	Different numbers that have the same absolute value are called *opposites*. The sum of two opposite numbers is zero.	$-9 + 9 = 0$
Adding signed numbers [1.12]	When adding two numbers with the same sign, add the absolute values of the numbers. The sign of the sum is the same as the sign of the original numbers.	$11 + 6 = 17$ $-7 + (-8) = -15$
	When adding two signed numbers with opposite signs, find their absolute values and then subtract the smaller from the larger. The sign of the sum is the sign of the number with the larger absolute value.	$-9 + 12 = 3$ $14 + (-18) = -4$
Subtracting signed numbers [1.12]	To subtract two signed numbers, change the operation of subtraction to addition, and change the number being subtracted to its opposite; then follow the rules for adding signed numbers.	$4 - (-9) = 4 + 9 = 13$ $-6 - 9 = -6 + (-9) = -15$ $-8 - (-2) = -8 + 2 = -6$
Multiplying or dividing two numbers with opposite signs [1.13]	When you multiply or divide two numbers with *opposite* signs, the result is always negative.	$-9 \cdot 5 = -45$ $56 \div (-7) = -8$
Multiplying or dividing two numbers with the same signs [1.13]	When you multiply or divide two numbers with the *same* sign, the result is always positive.	$(-3)(-8) = 24$ $(-72) \div (-6) = 12$
Multiplying or dividing a string of numbers [1.13]	When you multiply a string of signed numbers, the product is positive when the number of negative factors is even and negative when the number of negative factors is odd.	$(-5)(-4)(-3) = -60$ $(-5)(-4)(-3)(-2) = 120$
Dividing 0 by a nonzero number [1.13]	When 0 is divided by a nonzero number, the result is always 0.	Because $0 = 0 \cdot 5, 0 \div 5 = 0$ or equivalently $\dfrac{0}{5} = 0,$

145

CONCEPT/SKILL	DESCRIPTION	EXAMPLE
Dividing 0 by 0 [1.13]	When 0 is divided by 0, the quotient has no unique value. Therefore, you assign no value to this quotient and you say it is not defined.	Because $0 = 0 \cdot$ any number, $0 \div 0$ would appear to have many possible values (any number). Therefore, we cannot assign a specific value to this quotient and we say that $\dfrac{0}{0}$ is undefined.
Dividing a nonzero number by 0 [1.13]	When a nonzero number is divided by 0, the quotient has no possible value. Therefore, we assign no value to this quotient and we say it is not defined.	$\dfrac{\textit{nonzero number}}{0} = \textit{value}$ would mean that $0 \cdot \textit{value}$ would equal a nonzero number. But because $0 \cdot$ every number $= 0$, there is no possible value for which $0 \cdot \textit{value} = $ a nonzero number. Therefore, there is no possible value of the quotient and we say that $\dfrac{\textit{nonzero number}}{0}$ is undefined.
Powers of signed numbers [1.14]	To indicate that a negative number is raised to a power, the number *must* be enclosed in parentheses. An expression of the form $-a^2$ *always* signifies the negation (opposite) of a^2 and is *always* a negative quantity (for nonzero a).	$(-4)^2 = (-4)(-4) = 16$ $-4^2 = -(4)(4) = -16$
Whole numbers raised to negative exponents [1.14]	A negative exponent always signifies a reciprocal. Symbolically, this is written as $b^{-n} = \dfrac{1}{b^n}$, where $b \neq 0$.	$7^{-2} = \dfrac{1}{7^2} = \dfrac{1}{49}$

Gateway Review

1. Write the number 2.0202 in words.

2. Write the number fourteen and three thousandths in standard notation.

3. What is the place value of the 4 in the number 3.06704?

4. Evaluate: $3.02 + 0.5 + 7 + 0.004$

5. Subtract 9.04 from 21.2.

6. Evaluate: $6.003 \cdot 0.05$

7. Divide 0.0063 by 0.9.

8. Round 2.045 to the nearest hundredth.

9. Change 4.5 to a percent.

10. Change 0.3% to a decimal.

11. Change 7.3 to an improper fraction.

12. Change $\dfrac{3}{5}$ to a percent.

13. Write the following numbers in order from largest to smallest: 1.001, 1.1, 1.01, $1\dfrac{1}{8}$.

14. Evaluate: 3^3

15. Evaluate: 6^0

16. Evaluate: 4^{-2}

17. Evaluate: $(-4)^3$

18. Evaluate: $|16|$

19. What number has the same absolute value as 25?

20. Evaluate: $|-12|$

21. Write 0.0000543 in scientific notation.

22. Write 3.7×10^4 in standard notation.

23. Determine the average of your exam scores: 25 out of 30, 85 out of 100, and 60 out of 70. Assume that each exam counts equally.

Answers to all Gateway exercises are included in the Selected Answers appendix.

24. Change $\dfrac{9}{2}$ to a mixed number.

25. Change $5\dfrac{3}{4}$ to an improper fraction.

26. Reduce $\dfrac{15}{25}$ to lowest terms.

27. Evaluate: $\dfrac{1}{6} + \dfrac{5}{8}$

28. Calculate: $5\dfrac{1}{4} - 3\dfrac{3}{4}$

29. Evaluate: $\dfrac{2}{9} \cdot 3\dfrac{3}{8}$

30. Divide: $\dfrac{6}{11}$ by $\dfrac{8}{22}$

31. Evaluate: $2\dfrac{5}{8} \cdot 2\dfrac{2}{7}$

32. What is 20% of 80?

33. On one exam, you answered 12 out of 20 questions correctly. On a second exam, you answered 16 out of 25 correctly. On which exam did you do better?

34. This year the number of customer service complaints to a cable company increased by 40% over last year's number. If there were 3500 complaints this year, how many were made last year?

35. Twenty percent of the students in your math class scored between 80 and 90 on the last test. In a class of 30, how many students does this include?

36. In one summer, the number of ice-cream stands in your town decreased from 10 to 7. What is the percent decrease in the number of stands?

37. A campus survey reveals that 30% of the students are in favor of the proposal for no plastic take-out boxes in the cafeteria. If 540 students responded to the survey, how many students are in favor of the proposal?

38. You read 15 pages of your psychology text in 90 minutes. At this rate, how many pages can you read in 4 hours?

39. The scale at a grocery store registered 0.62 kilogram. Convert this weight to pounds (1 kg = 2.2 lb.).

40. A turtle walks at the rate of 2 meters per minute. What is the turtle's rate in inches per second (1 in. = 0.0254 mm)?

41. $-5 + 4 =$

42. $2(-7) =$

43. $-3 - 4 - 6 =$

44. $-12 \div (-4) =$

45. $-5^2 =$

46. $(-1)(-1)(-1) =$

47. $-5 - (-7) =$

48. $-3 + 2 - (-3) - 4 - 9 =$

49. $3 - 10 + 7 =$

50. $\left(-\dfrac{1}{6}\right)\left(-\dfrac{3}{5}\right) =$

51. $7 \cdot 3 - \dfrac{4}{2} =$

52. $4(6 + 7 \cdot 2) =$

53. $3 \cdot 4^3 - \dfrac{6}{2} \cdot 3 =$

54. $5(7 - 3) =$

55. A deep-sea diver dives from the surface to 133 feet below the surface. If the diver swims down another 27 feet, determine his depth.

56. You lose $200 in the stock market on each of 3 consecutive days. Represent your total loss as a product of signed numbers, and write the result.

57. You have $85 in your checking account. You write a check for $53, make a deposit of $25, and then write another check for $120. How much is left in your account?

58. The mass of the hydrogen atom is about 0.000 000 000 000 000 000 000 002 gram. If 1 gram is equal to 0.0022 pound, determine the mass of a hydrogen atom in pounds. Express the result in scientific notation.

59. Evaluate the following expressions by hand. Then check your results using a calculator.

a. $\sqrt{15^2}$

b. $\sqrt{5^2 - 4^2}$

c. $\sqrt{\dfrac{64}{8 + 1}}$

d. $\dfrac{\sqrt{100}}{8 + 2}$

e. $\dfrac{6 + 9}{\sqrt{25}}$

Variable Sense

Arithmetic is the branch of mathematics that deals with counting, measuring, and calculating. Algebra is the branch that deals with variables and the relationships between and among variables. Variables and their relationships, expressed in tabular, graphical, verbal, and symbolic forms, will be the central focus of this chapter.

Cluster 1   Symbolic Rules and Expressions

Activity 2.1

Symbolizing Arithmetic

Objectives

1. Generalize from an arithmetic calculation to a symbolic representation by utilizing variables.

2. Evaluate algebraic expressions.

To communicate in a foreign language, you must learn the language's alphabet, grammar, and vocabulary. You must also learn to translate between your native language and the foreign language you are learning. The same is true for algebra, which, with its symbols and grammar, is the language of mathematics. To become confident and comfortable with algebra, you must practice speaking and writing it, as well as translating between it and the natural language you are using (English, in this book).

In this activity, you will note the close connections between algebra and arithmetic. Arithmetic involves determining a numerical value using a given sequence of operations. Algebra focuses on this sequence of operations, not on the numerical result. The following problems will highlight this important connection.

Formulas

Formulas are found everywhere in our modern society, including science, economics, finance, medicine, engineering, and computer applications. Formulas can help explain the forces of planetary motion as well as the secrets of DNA. You can build computers and send information around the world. Most apps on modern devices use formulas to edit photos or to create realistic games.

Formulas are essentially algebraic statements that describe the arithmetic relationship among specific quantities. For example, a rectangle can be described completely by its base and height.

- The area of a rectangle is calculated as the product of its base and height.

- The perimeter is calculated as twice the sum of its base and height.

Because different rectangles have different dimensions, it is convenient to use symbols to represent the quantities associated with an arbitrary rectangle. For convenience, the symbol used is often the first letter of the word for the quantity. Therefore, the area and perimeter of a rectangle can be described, respectively, by the following familiar geometric formulas.

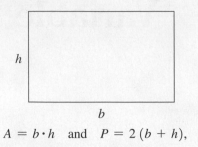

$$A = b \cdot h \quad \text{and} \quad P = 2\,(b + h),$$

where A represents area, b represents base, h represents height, and P represents perimeter.

The symbols b, h, A, and P are called **variables** because their values vary from one situation to another.

Definition

A **variable**, usually represented by a letter, is a quantity that may change, or vary, in value from one particular instance to another.

Although the area and perimeter formulas contain variables that represent specific quantities, the formula is *not* about the variables. The formulas describe the calculations:

Multiplication to calculate area and addition followed by multiplication by 2 to calculate perimeter.

The symbolic representations, $b \cdot h$ and $2(b + h)$, are called **algebraic expressions.** Such expressions present in a compact form the sequence of arithmetic operations to be performed on these variables. Because the multiplication symbol $\times$ can be confused with the variable x, a dot is used to indicate the operation of multiplication. Generally, the product of two or more variables is written with no symbol between them. The operation is understood to be multiplication. Therefore, $b \cdot h$ is written as bh.

The variables contained in an algebraic expression are often called **input variables**, or simply **inputs**. When the input variables in an algebraic expression are replaced with specific numerical values, these values are called **replacement values.** The entire calculation process is called **evaluating the algebraic expression**. The result obtained at the end of the evaluation process is called the **output.** Often, but not always, the output is denoted by its own symbol. The output variables in the algebraic expressions bh denoted by A. The output variable for $2(b + h)$ is denoted by P.

1. Determine the area and perimeter of rectangles with the following dimensions. Include appropriate units of measure.

   a. base 10 inches, height 4 inches

   b. base 2 inches, height 18 inches

   c. base $3\frac{1}{4}$ inches, height $2\frac{3}{8}$ inches

2. Two familiar formulas associated with circles are $A = \pi r^2$ and $C = 2\pi r$.

   a. Identify by name the quantities represented by the variables in these two formulas. Describe what variable quantity that each symbol represents.

   b. Identify which variable is the input and which is the output.

   c. Describe in words the sequence of operations indicated by the algebraic expression in each formula.

   d. Determine the area and circumference for a circle having radius $r = 2.5$ cm. Include the appropriate units of measurement.

3. Several weight-loss programs assign points to prepared or packaged foods that take into account the food's fat F, carbohydrate C, protein P, and fiber B content in grams. The point value for a specific item of food can be determined by evaluating the algebraic expression

$$\frac{F}{4} + \frac{C}{9} + \frac{P}{10} - \frac{B}{12}.$$

   a. How many input variables are contained in the expression? What quantity does the output represent? What would be an appropriate symbol to represent this output?

   b. Determine the point value of one serving of the item having the nutrition facts given in the label to the left.

Nutrition Facts

8 servings per container

Serving size	2/3 cup (55g)

Amount per 2/3 cup

Calories   **230**

% DV*		
12%	**Total Fat** 8g	
5%	Saturated Fat 1g	
	Trans Fat 0g	
0%	**Cholesterol** 0mg	
7%	**Sodium** 160mg	
12%	**Total Carbs** 37g	
14%	Dietary Fiber 4g	
	Sugars 1g	
	Added Sugars 0g	
	Protein 3g	
10%	**Vitamin D** 2mcg	
20%	**Calcium** 260mg	
45%	**Iron** 8mg	
5%	**Potassium** 235mg	

* Footnote on Daily Values (DV) and calories reference to be inserted here.

Constructing Algebraic Expressions in Context

In Problems 1–3, the variable quantities and formulas were provided. However, often the challenge is to identify the variable quantities involved in a situation and then to construct the corresponding algebraic expression.

4. Suppose you use a wire cutter to snip a 24-inch piece of wire into two parts. One of the parts measures 8 inches.

   a. What is the length of the second part?

   b. Write down and describe in words the arithmetic operation you used to determine your answer.

5. Now you use a wire cutter to snip another 24-inch piece of wire into two parts. This time one of the parts measures 10 inches.

   a. What is the length of the second part?

b. Write down and describe in words the arithmetic operation you used to determine your answer.

6. Consider the situation in which a 24-inch piece of wire is cut into two parts. Suppose you have no measuring device available; so you will denote the length of the first part by the symbol x, representing the variable length.

   a. Using Problems 4 and 5 as a guide, represent the length of the second part symbolically in terms of x.

   b. What are the reasonable replacement values for the variable x? What are their units of measurement?

Notice that the algebraic expression in Problem 6a provides a general method for calculating the length of the second part, given the length x of the first part. In each case, you subtract the length of the first part from 24.

Now suppose you take the two parts of the snipped wire and bend each to form a square as shown in the following figure.

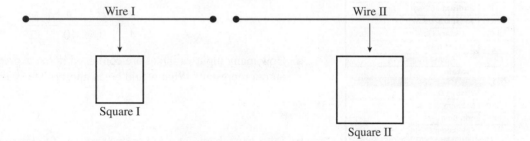

Remember, the area of a square is calculated by squaring the length of its side. The perimeter of a square is calculated as 4 times the length of one side.

7. Refer back to Problem 5, in which one piece of wire measured 10 inches.

   a. Explain in words how you can use the known length of the wire to determine the area of the first square. What is this area?

   b. Write down the sequence of arithmetic operation(s) you used to determine your answer in part a.

   c. Determine the area of the second square.

   d. Write down the sequence of arithmetic operation(s) you used to determine your answer in part c.

8. Refer back to Problem 6, in which one piece of wire cut from a 24-inch piece measured x inches.

   a. Using Problem 7b as a guide, represent the area of the first square as an algebraic expression in terms of x.

   b. Similarly, represent the area of the second square as an algebraic expression in terms of x.

   c. Replace x with 10 in your algebraic expressions in parts a and b to calculate the areas of the two squares formed when a length of 10 inches is cut from a 24-inch piece of wire. Compare your results with the answers in Problem 7a and c.

Here is an additional situation to explore arithmetically and algebraically.

9. Suppose you have 8 dimes and 6 quarters in a desk drawer.

   a. Explain how you can use this information to determine the total value of these coins. What is the value?

   b. Write down and describe in words the sequence of arithmetic operations you used to determine your answer.

   c. Suppose you have d dimes and q quarters in your pocket. Represent the total number of coins as an algebraic expression using d and q.

   d. Now represent the total value of these coins as an algebraic expression using d and q.

   e. Evaluate the algebraic expressions in parts c and d when there are 8 dimes and 6 quarters.

In the algebraic expression from Problem 9d each variable is multiplied by a number. In such a situation, it is not necessary to include a multiplication symbol. When no arithmetic symbol appears between a number and variable or between two variables, the operation is always understood to be multiplication.

Some special terminology is often used to describe important features of an algebraic expression. The **terms** of an algebraic expression are the parts of the expression that are added or subtracted. For example, the expression in Problem 9d has two terms—namely, $0.10d$ and $0.25q$.

A number or numerical constant that multiplies a variable is called the **coefficient** of the variable. For example, in the expression in Problem 9d, the coefficient of the variable d is 0.10 and the coefficient of the variable q is 0.25.

The two contexts in this section, the wire and the coins, illustrate a general four-step process for representing the relationship between variable quantities by an algebraic expression.

Step 1. Identify the variable quantity or quantities within the context.
In the first case, the variable quantity was the length of the snipped wire, measured in inches.
In the second case, the variable quantity was the number of each type of coin.

Step 2. Choose a letter or symbol with which to represent each quantity.
In the first case, the length of the wire was represented by x.
In the second case, the number of dimes and quarters was represented by d and q, respectively.

Step 3. Use several reasonable replacement values to help determine the appropriate sequence of arithmetic operations required by the problem.

Step 4. Generalize the arithmetic calculations by replacing the numerical values with the letters or symbols identified in step 2. The sequence of operations from the numerical examples should all be the same as those in the algebraic expression.

Interpreting Algebraic Expressions in Context

In the previous problems in this activity, you identified the variable quantities and formed algebraic expressions to represent the relationship between the variable quantities. In this section, the variables will be described and the algebraic expression(s) in the variables will be given. Your task will be to interpret in context the quantity represented by each algebraic expression.

Note: You should draw a sketch of any geometric figure referenced in the problem.

10. The area of a rectangle is 24 square inches.

$$A = 24 \text{ in.}^2$$

b

a. If b represents the length of the base in inches, then what does $\dfrac{24}{b}$ represent? What are its units of measure? Hint: You need to recall the relationship between the base and the area of a rectangle.

b. What quantity does the expression $2\left(b + \dfrac{24}{b}\right)$ represent?

11. You have two student loans to help you pay for college. You obtained the first loan at a local credit union at 6.50% interest. The second is a Stafford Loan with an interest rate of 3.86%.

a. Let c represent the amount borrowed at the local credit union. What does $0.065c$ represent?

b. Let s represent the amount of the Stafford Loan. What does $0.0386s$ represent?

c. What does $c + s$ represent?

d. What does $0.065c + 0.0386s$ represent?

Constructing and Evaluating Formulas

12. For many items such as large equipment, machinery, and automobiles, their monetary value decreases over time due to use, wear and tear, or obsolescence. In this context, the decrease is called **depreciation**. For accounting purposes, depreciation may be calculated by a number of methods. The most common method of depreciation is a constant annual monetary decrease over the useful life of the item. The annual depreciation of an item depends on its original cost (dollars), its salvage value (dollars), and its useful life (years). This decrease in value is measured in dollars per year.

a. The annual depreciation of an item is calculated by subtracting its salvage value from its original cost and then dividing this difference by the useful life. Identify the three quantities involved in calculating depreciation, and represent each with a symbol or letter.

b. Translate the sequence of operations among the variable quantities into a formula for calculating annual depreciation.

c. A new car costs $25,000, has an estimated life of 10 years, and has a salvage value of $2,000. Use the formula in part b to determine the annual depreciation of the car.

13. You discover a formula that gives the number of calories K needed each day by a moderately active male who weighs w pounds, is h inches tall, and is a years old.

$$K = 9.65w + 19.69h - 10.54a + 102.3$$

If a 20-year old male is moderately active, weighs 165 lbs, is 5 ft 11 in. tall, determine the number of calories needed per day.

Generic Algebraic Expressions

You will also frequently encounter algebraic expressions that are generic in nature. Their variables have no context, and the expression has no meaning except as a compact way of describing the sequence of operations to be performed. The most common default symbol for the input variable in a generic expression is x.

14. Consider the two algebraic expressions $9 + x$ and $9 - x$.

    a. Evaluate $9 + x$ for $x = 3$, and then evaluate $9 - x$ for $x = -3$.

    b. Evaluate $9 + x$ for $x = -4$ and then evaluate $9 - x$ for $x = 4$.

    c. Explain what parts a and b illustrate regarding addition and subtraction of opposites.

Universal Language

It is unlikely that any individual is literate in all the world's languages. However, we all have the ability to be literate in the language of algebra. Formulas and algebraic expressions provide a powerful tool for representing and studying relationships between related variable quantities. The result and consequences of these investigations affect almost every aspect of the world around us—from entertainment to our very survival. The remainder of Chapter 2 will further your fluency in this language of algebra.

SUMMARY: ACTIVITY 2.1

1. A **variable**, usually represented by a letter, is a quantity that may change, or vary, in value from one particular instance to another.

2. A **constant** is a quantity whose value will not change (remains constant). A constant can be a specific number or may be represented by a letter or symbol whose value does not change.

3. An **algebraic expression** is a compact way of describing the sequence of arithmetic operations that need to be performed on a variable or variables.

4. A **replacement value** for a variable is a numerical value that is assigned to that variable and replaces the variable within an algebraic expression.

5. An algebraic expression is **evaluated** when a replacement value is substituted for the variable (or variables) and the sequence of arithmetic operations is performed to produce a single output value.

6. The **terms** of an algebraic expression are the parts of the expression that are added or subtracted.

7. The **coefficient** of a variable is a number or constant that multiplies the variable.

8. A general four-step process for representing the relationship between variable quantities by an algebraic expression.

    Step 1. Identify the variable quantity or quantities within the context.

    Step 2. Choose a letter or symbol with which to represent each quantity.

    Step 3. Use several reasonable replacement values to help determine the appropriate sequence of arithmetic operations required by the problem.

    Step 4. Generalize the arithmetic calculations by replacing the numerical values with the letters or symbols identified in Step 2.

EXERCISES: ACTIVITY 2.1

1. You currently have $125 in your debit card account. If *d* represents the cost of an item you purchased with your card, write an expression in terms of *d* that represents the new balance of the debit card.

2. Let *n* represent the number of songs you purchased on iTunes for $0.99 each and *m* represent the number of songs you purchased for $1.29 each.

   a. Write an algebraic expression that represents the total number of songs you purchased.

   b. Write an expression that represents the total amount of money you spent on the songs.

3. **a.** Suppose you have *p* pennies, *n* nickels, and *d* dimes in your pocket. Represent the total number of the coins as an algebraic expression in terms of *p*, *n*, and *d*.

   b. Represent the total value of the coins as an algebraic expression in terms of *p*, *n*, and *d*.

4. You and your friend are standing facing each other 100 yards apart.

   a. Suppose your friend walks *y* yards toward you and stops. Represent his distance from you symbolically in terms of *y*.

   b. Suppose, instead, he walks *y* yards away from you and stops. Represent his distance from you symbolically in terms of *y*.

5. An equilateral triangle has sides of length *s*.

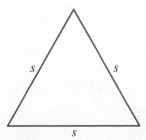

   a. If each side is increased by 5, represent the new perimeter in terms of *s*.

   b. If each original side is doubled, represent the new perimeter in terms of *s*.

6. **a.** The markdown of every piece of clothing in the college bookstore is 20% of the original price of the item. If *p* represents the original price, write an algebraic expression that can be used to determine the sale price.

Exercise numbers appearing in color are answered in the Selected Answers appendix.

b. Complete the following table by evaluating the expression in part a with the given input values.

Input Price, p ($)	50	100	150	200
Output Sale Price ($)				

7. **a.** The perimeter of a rectangle is twice the length plus twice the width. If l represents the length and w represents the width, write an algebraic expression that represents the perimeter.

b. Calculate the perimeter of a rectangle with length 10 inches and width 17 inches.

c. The area of a rectangle is the product of its length and width. Write an algebraic expression in l and w that represents its area.

d. Calculate the area of a rectangle with length 10 inches and width 17 inches.

8. A fruit vendor at the local farmer's market sells apples for $0.90 each and pears for $1.20 each. A customer buys A apples and P pears. Explain what the following expressions represent in this context using appropriate units of measure.

a. $A + P$

b. $0.90\,A$

c. $0.90A + 1.20P$

d. $10 - (0.90A + 1.20P)$

9. A running track has the shape of a square capped with two semicircles, each on opposite sides as shown in the figure below. Because the size of the track depends on the length of the side of the square, represent this variable side length with the symbol s.

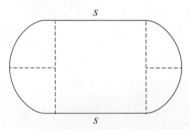

a. Construct an algebraic expression representing the *perimeter* of the track in terms of s.

b. Construct an algebraic expression representing the *area* of the track in terms of s.

c. Evaluate the algebraic expressions from parts a and b for a track whose side length is 50 yards.

10. Evaluate each of the following algebraic expressions for the given value(s).

a. $3x - 10$, for $x = -2.5$

b. $5 - 2.7x$, for $x = 10.25$

c. $\frac{1}{2}bh$, for $b = 10$ and $h = 6.5$

d. $5x - 7$, for $x = -2.5$

e. $55 - 6.7x$, for $x = 10$

f. $\frac{1}{2}bh$, for $b = 5$ and $h = 5$

g. $\frac{1}{2}h(b + B)$, for $h = 10$, $b = 5$, $B = 7$

11. A cardboard box used for flat rate shipping by a postal service has length 18 inches, width 13 inches, and height 5 inches.

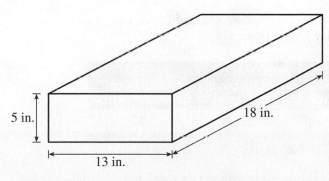

a. The volume of a rectangular box is the product of the length times the width times the height. Determine the volume of the box.

b. Suppose each dimension is increased by x inches. Represent the length, width, and height of the new box symbolically in terms of x.

c. Write a single algebraic expression that symbolically expresses the volume of the new box in terms of x.

d. Now evaluate your algebraic expression for the volume when $x = 3$ to determine the volume when each side of the box is increased by three inches.

12. Your mathematics instructor calculates a student's average A for her course using the formula.

$$A = 0.3\,q + 0.4\,t + 0.2\,f + 0.1\,h,$$

where q = quiz average, t = test average, f = final exam score, and h = homework average.

Determine the course average for a student having the following grades. Round your answer to the nearest tenth.

quizzes: 75, 80, 91, 85; tests: 88, 90, 93; final exam: 92; homework average: 85

13. The formula used by the NCAA to calculate the quarterback passing efficiency rating is

$$R = \frac{8.4Y + 330T + 100C - 200I}{A},$$

where R = quarterback rating
A = passes attempted
C = passes completed
Y = passing yardage
T = touchdown passes
I = number of interceptions

In the 2012–2013 regular season, Johnny Manziel, quarterback for Texas A&M, and David Ash, quarterback for the University of Texas, had the following player statistics:

PLAYER	PASSES ATTEMPTED	PASSES COMPLETED	PASSING YARDAGE	NUMBER OF TOUCHDOWN PASSES	NUMBER OF INTERCEPTIONS
Johnny Manziel	434	295	3706	26	9
David Ash	318	214	2699	19	8

a. Determine the quarterback rating for Johnny Manziel for the 2012–2013 NCAA football season. Round to the nearest tenth.

b. Determine the quarterback rating for David Ash for the same season. Round to the nearest tenth.

c. Visit the NCAA website, and select stats to obtain the rating of your favorite quarterback.

Blood Alcohol Concentration—Symbolic Representation

Activity 2.2

Blood Alcohol Levels

Objectives

1. Identify input and output in situations involving two variable quantities.

2. Determine the replacement values for a variable within a given situation.

3. Represent and interpret an input/output relationship algebraically by formula, numerically as data pairs in a table and graphically as points on a graph.

As of 2014, all states and the District of Columbia have adopted the 0.08% blood alcohol concentration as the maximum legal limit for drunk driving. Several factors may influence blood alcohol concentration, including the rate at which the individual's body processes alcohol, the amount of food eaten prior to drinking, and the concentration of alcohol in the drink. In this scenario, you are only considering the effect on blood alcohol concentration of the number of beers consumed in an hour and the weight of the individual.

1. The number of 12-ounce beers someone consumes in an hour is a variable (that is, the number consumed varies from person to person). Let this variable be represented by the letter n.

   a. Can n be reasonably replaced with a negative value, such as –2? Explain.

   b. Can n be reasonably replaced with zero? Explain.

   c. What is the possible collection of replacement values for n in this situation?

2. Weight varies from person to person. Let the letter w represent this variable. What is the possible collection of reasonable replacement values for w, measured in pounds, in this situation?

3. The concentration of alcohol in the blood (expressed as a percent) also varies. Use the letter B to represent this variable.

   a. Can B be reasonably replaced with zero? Explain.

   b. Can B be reasonably replaced with 100? Explain.

The algebraic relationship between the variables n, number of 12-ounce beers consumed in one hour, w, drinker's weight in pounds, and B, their blood alcohol level in percent, can be described by the formula

$$B = \frac{600n}{w(169 + 0.6n)}$$

The variables n and w in the algebraic expression on the right are called the **input** variables because when you *input* replacement values for these variables the algebraic expression can be evaluated. The *output,* or result, of this calculation represents the corresponding value for B. As a consequence, B is designated the **output** variable. Typically, the output value is associated in some way with, or is a consequence of, the input value(s).

4. **a.** Two friends, one weighing 200 pounds, and the other weighing 130 pounds, each consume 3 beers in one hour. Determine the blood alcohol – concentration of each person.

b. Do either of the persons exceed the legal limit for driving while intoxicated? Does this result seem reasonable? Explain.

5. Rewrite the formula so that it describes the algebraic relationship between the two variables n and B for an individual weighing 200 pounds. Identify the input and output variables in this new formula

Numerical Representation

The relationship between n and B can also be described numerically using a table of paired data values. The **input variable** is the one that is listed first, either in the top row of a horizontal table (as seen below) or in the left column of a vertical one. The **output variable** is placed below or to the right, depending on the table layout.

Number of Beers in an Hour, n	1	2	3	5	6	7	8	10
Blood Alcohol Concentration (%),* B	0.018	0.035	0.053	0.087	0.104	0.121	0.138	0.171

*Based on body weight of 200 pounds

6. Why is it reasonable in this situation to designate the number of beers consumed in an hour as the input and the blood alcohol concentration as the output?

7. Use the formula in Problem 5 to verify the table entries and then calculate the blood alcohol concentration for a 200-pound person who has consumed 4 beers in one hour and 9 beers in one hour.

Notice that a table can reveal *numerical* patterns and relationships between the input and output variables. You can also describe these patterns and relationships *verbally* using such terms as *increases*, *decreases*, or *remains the same*.

8. According to the preceding table, as the number of beers consumed in 1 hour increases, what happens to the blood alcohol concentration?

9. From the preceding table, determine the number of beers a 200-pound person can consume in 1 hour without exceeding the recommended legal measure of drunk driving.

Graphical Representation

The relationship between n and B can also be represented graphically (visually) as plotted points on a rectangular coordinate system. The resulting graph is often helpful in detecting trends or other information not apparent in a table or a symbolic rule.

On a graph, the input variable is referenced on the **horizontal axis** and the output variable is referenced on the **vertical axis**.

The following graph provides a visual representation of the beer consumption and blood alcohol level data from the preceding table, based on a body weight of 200 pounds. Note that each of the 10 input/output data pairs in that table corresponds to a plotted point on the graph. The resulting graph of data points is called a **scatterplot**. For example, drinking seven beers in an hour is associated with a blood alcohol concentration of 0.121%. If you read across to 7 along the input (horizontal) axis and move up to 0.121 on the output (vertical) axis, you locate the point that represents the **ordered pair** of numbers $(7, 0.121)$. Similarly, you would label the other points on the graph by ordered pairs of the form (n, B), where n is the input value and B is the output value. Such ordered pairs are called the **coordinates** of the point.

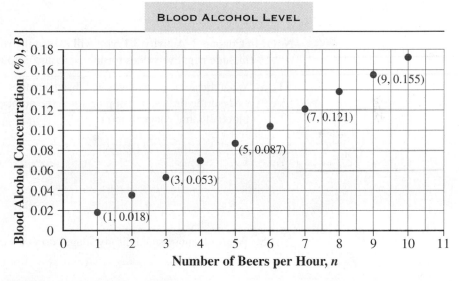

BLOOD ALCOHOL LEVEL

Based on body weight of 200 pounds

Observations

- The distance represented by space between adjacent tick marks on an axis is determined by the range of particular replacement values for the variable represented on that axis. The process of determining and labeling an appropriate distance between tick marks is called **scaling**.

- The axes are often labeled and scaled differently. The way axes are labeled and scaled depends on the context of the problem. Note that the tick marks to the left of zero on the horizontal axis are negative; similarly, the tick marks below zero on the vertical axis are negative.

- On each axis, equal distance between adjacent pairs of tick marks must be maintained. For example, if the distance between the first two tick marks is 10 units, then all adjacent tick marks on that axis must also differ by 10 units.

10. a. From the graph, approximate the blood alcohol concentration of a 200-pound person who consumes 2 beers in an hour.

Write this input/output correspondence using ordered-pair notation.

    b. Estimate the number of beers that a 200-pound person must consume in 1 hour to have a blood alcohol concentration of 0.104.

    Write this input/output correspondence using ordered-pair notation.

    c. When $n = 4$, what is the approximate value of B from the graph?

    Write this input/output correspondence using ordered-pair notation.

    d. As the number of beers consumed in an hour increases, what happens to the blood alcohol concentration?

Note: In Chapter 3, Activity 3.1, you will extend the rectangular coordinate system to include negative values for the input and output variables.

11. a. What are some advantages and disadvantages of using a graph when you are trying to describe the relationship between the number of beers consumed in an hour and blood alcohol concentration?

    b. What advantages and disadvantages do you see in using a table?

Medicare Spending

12. Medicare is a government program that helps senior citizens pay for medical expenses. As the U.S. population ages and greater numbers of senior citizens join the Medicare rolls each year, the expense and quality of health service becomes an increasing concern. The following graph presents Medicare expenditures from 1970 through 2015 (projected), where the data points have been connected to emphasize the overall pattern or trend over time. Use the graph to answer the following questions.

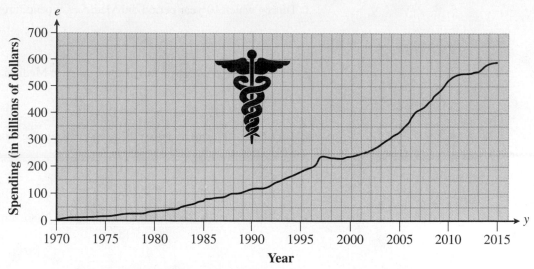

Source: Centers for Medicare and Medicaid Services

a. Identify the input variable and the output variable in this situation.

b. Use the graph to estimate the Medicare expenditures for the years in the following table.

YEAR, y	MEDICARE EXPENDITURES, e (IN BILLIONS OF DOLLARS)
1970	
1975	
1980	
1985	
1990	
1995	
2000	
2005	
2010	
2015	

c. What letters in the table in part b are used to represent the input variable and the output variable?

d. Estimate the year in which the expenditures reached $100 billion.

e. Estimate the year in which the expenditures reached $200 billion.

    f. During which 10-year period did Medicare expenditures change the least?

    g. During which 5-year period was the change in Medicare expenditures the greatest?

    h. For which 1-year period does the graph indicate the most rapid increase in Medicare expenditures?

    i. In what period was there almost no change in Medicare expenditures?

SUMMARY: ACTIVITY 2.2

1. The set of **replacement values** for a variable is the collection of numbers that make sense within the given context.

2. An input/output relationship can be represented **numerically** by a table of paired data values.

3. An input/output relationship can be depicted **graphically** as plotted points on a rectangular coordinate system. This set of points is called a scatterplot. The input variable is referenced on the **horizontal axis** and the output variable on the **vertical axis** so that each plotted point has an ordered pair (input value, output value).

EXERCISES: ACTIVITY 2.2

1. On December 3, 1992, the first text message was sent in the United Kingdom from a computer to a mobile device, using the Short Message Service (SMS). The message was "Merry Christmas." In 1999, 100 billion text messages were sent worldwide. In 2005, 1 trillion text messages were sent. In 2012, the number of text messages sent worldwide surpassed 8.5 trillion.

A recent study revealed that people who text while driving are 23 times more likely to have an accident. Another study concluded that driving while texting is more dangerous than driving while under the influence of alcohol. It is estimated that people in the age group of 18 to 24 send an average of 71 texts daily.

The following table gives estimates of the number of text messages (in billions) sent each month in the United States.

YEAR, y	2005	2006	2007	2008	2009	2010	2011	2012
Number of monthly text messages (in billions), n	7	8	30	86	153	188	212	220

    a. Identify the input and output variables.

b. Construct an appropriately scaled and labeled scatterplot of the given data.

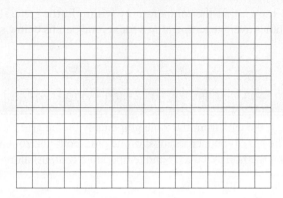

c. Describe any patterns or trends that you observe.

2. While most medications do not affect driving ability, some prescription and over-the-counter medicines can cause reactions that may make it unsafe to drive. Driving under the influence of certain medications can be as serious as driving under the influence of alcohol.

a. When you take medicine, your body metabolizes and eliminates the medication so that only an undetectable trace will ultimately remain. The *half-life* of a medication is the time it takes for your body to eliminate one-half the amount present; we say the amount of the medication decays. If the half-life of a medication is 1 hour, the decay factor each hour is 0.5. Therefore, if you start with a 20-milligram dose, the amount remaining is $20 \cdot (0.5)$ mg after one hour, $20 \cdot (0.5)^2$ mg after 2 hours, $20 \cdot (0.5)^n$ mg after n hours. Use the formula, $A = 20(0.5)^n$, to complete the following table.

Number of Hours, n	0	1	2	3	4
Amount (mg) of Drug in Your Body, A					

b. Why is it reasonable in this situation to designate n, the number of hours after taking the medication, as the input and A, the amount of medication in your body, as the output?

c. Construct a scatterplot of the data pairs on a appropriately scaled and labeled coordinate axes.

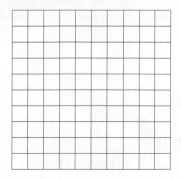

d. As the number of hours increase, what happens to the amount of medication in your body?

3. Measurements in wells and mines have shown that the temperatures within Earth generally increase with depth. The following table shows average temperatures for several depths below sea level.

D, Depth (km) Below Sea Level	0	25	50	75	100	150	200
T, Temperature (°C)	20	600	1000	1250	1400	1700	1800

a. Choose an appropriate scale to represent the data from the table graphically on the grid following part d. The data points should be spread out across most of the grid, both horizontally and vertically. Place depth (input) along the horizontal axis and temperature (output) along the vertical axis.

b. How many units does each tick mark on the horizontal axis represent?

c. How many units does each tick mark on the vertical axis represent?

d. Explain your reasons for selecting the particular scales that you used.

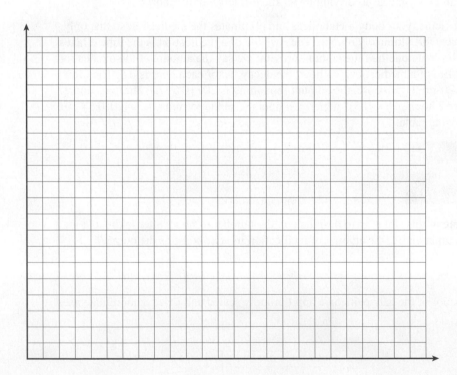

e. Which representation (table or graph) presents the information and trends in this data more clearly? Explain your choice.

4. Suppose the variable n represents the number of DVDs a student owns.

a. Can n be reasonably replaced with a negative value, such as -2? Explain.

b. Can n be reasonably replaced with the number 0? Explain.

c. What is a possible collection of replacement values for the variable n in this situation?

5. Suppose the variable t represents the average daily Fahrenheit temperature in Oswego, a city in upstate New York, during the month of February in any given year.

  a. Can t be reasonably replaced with a temperature of $-3°F$? Explain.

  b. Can t be reasonably replaced with a temperature of $-70°F$? Explain.

  c. Can t be reasonably replaced with a temperature of $98°F$? Explain.

  d. What is a possible collection of replacement values for the variable t that would make this situation realistic?

6. Belgian statistician Adolphe Quetelet developed the body mass index (BMI) formula in the 1800s. BMI is an internationally used measure to determine one's weight status.

BMI	Weight Status
Below 18.5	Underweight
18.5–24.9	Normal
25–29.9	Overweight
30 or greater	Obese

The formula for BMI, B, is

$$B = \frac{703w}{h^2},$$

where w is weight in pounds and h is height in inches.

  a. What is the body mass index of a person who weighs 145 pounds and is 68 inches tall.

  b. Suppose your friend weighs 180 pounds. Substitute this value for w to obtain a formula for B in terms of height h.

  c. Complete the following table using the formula for body mass index of a 180-pound person.

h, Height in Inches	60	64	68	72	76	80
B, Body Mass Index						

  d. Produce a scatterplot of the data points in your table on an appropriately scaled and labeled coordinate axis. Enhance the graph by connecting the points with line segments to more clearly see the pattern between the height and BMI.

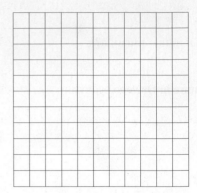

e. What happens to the body mass index as height increases? Does this make sense in the context of the situation? Explain.

f. Use the table or the graph to estimate the values of *h* for which a 180-pound person has a normal weight status.

7. Since the National Highway Traffic Safety Administration began recording alcohol-related fatalities in 1982, drunk driving fatalities have decreased considerably from the 21,113 deaths in 1982. To investigate the effect of our nation's effort to fight drunk driving, go online to find the number of alcohol-impaired deaths each year since 2001.

a. Identify the two variable quantities in this situation. Represent each quantity with a letter.

b. Determine which variable in part *a* is the *input* and which is the *output*.

c. Record the results of your online search in a table. Construct an appropriately scaled and labeled scatterplot of the data.

d. Describe any patterns or trends that you observe in the graph.

Activity 2.3

College Expenses

Objectives

1. Write verbal rules that represent relationships between input and output variables.

2. Translate verbal rules into symbolic rules.

3. Determine input/output values from a graph.

You are considering taking some courses at your local community college on a part-time basis for the upcoming semester. For a student who carries fewer than 12 credits (the full-time minimum), the tuition is $173 for each credit hour taken. You have a limited budget, so you need to calculate the total cost based on the number of credit hours.

1. **a.** Describe in words the input variable and the output variable in the college tuition situation.

   b. What are reasonable replacement values for the input?

   c. What is the tuition bill if you only register for a 3-credit-hour accounting course?

   d. Use the replacement values you determined in part b to complete the following table.

Number of Credit Hours	1	2	3	4	5	6	7	8	9	10	11
Tuition ($)											

   e. Write a rule in words (called a **verbal rule**) that describes how to calculate the total tuition bill for a student carrying fewer than 12 credit hours.

   f. What is the change in the tuition bill as the number of credit hours increases from 2 to 3? Is the change in the tuition bill the same amount for each unit increase in credit hours?

   g. Translate the verbal rule in part e into a symbolic rule.

The relationship between the number of credit hours taken and the total cost of tuition (output) in Problem 1 was described verbally in words as well as symbolically.

> **Definition**
>
> A **symbolic rule** is a mathematical statement that defines an **output** variable as an algebraic expression in terms of the **input** variable. The symbolic rule is essentially a recipe that describes how to determine the output value corresponding to a given input value.

2. Sketch a graph of the tuition cost data in the table in Problem 1d. Choose a scale for each axis that will enable you to plot all 11 points. Label the horizontal axis to represent the number of credit hours taken and the vertical axis to represent the corresponding tuition costs.

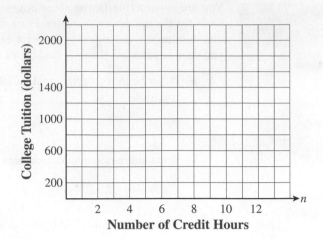

Number of Credit Hours

Note that the graph of the tuition cost data consists of 11 distinct points that are not connected. The input variable number of credit hours is defined only for whole numbers. This relationship is not defined for input values between these particular values.

Caution.   In order to use the graph for a relationship such as the tuition cost situation to make predictions or to recognize patterns, it is convenient to connect the points with line segments. This changes the replacement values for the input variable shown in the graph from "some values" to "all values." Therefore, you need to be cautious. Connecting data points may cause confusion when working with real-world situations.

Additional College Expenses

3. At the beginning of the semester, you buy a lunch debit card worth $150. The daily lunch special costs $6 in the college cafeteria.

   a. Fill in the following table. Use the number of lunch specials you purchase as the input variable.

NUMBER OF LUNCH SPECIALS PURCHASED	REMAINING BALANCE ON YOUR MEAL CARD
0	
5	
10	
15	
20	

   b. Write a verbal rule describing the relationship between the input variable, number of lunch specials, and the output variable, remaining balance.

   c. Translate the verbal rule in part b into a symbolic rule.

d. What are possible replacement values for the input variable? Explain.

e. Because the input/output relationship in the debit card situation can be defined symbolically, the table in part a can be constructed using technology. Note that the input variable *n* in the table increases by 5 units. In such a case, you say that the input increases by an increment of 5. The steps to build an input/output table using the TI-84 Plus C are given in Appendix E. The TI screens for the debit card situation should appear as follows:

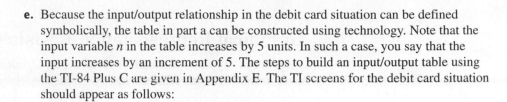

4. You are a graphic arts major and need to purchase a camera and lenses for your photography class in the fall semester. There is a special promotion for students in which you can purchase an $850 complete camera outfit by making monthly payments for 2 years at 0% interest. You are required to make a 10% down payment, with the balance to be paid in 24 equal monthly payments.

a. Determine the amount of down payment required.

b. Determine the amount owed after the down payment.

c. What are the monthly payments?

d. Let *A* represent the total amount paid after *n* payments are made. What is the amount *A* when *n* = 0? What does this value of *A* represent in this situation?

e. Complete the following table.

PAYMENT NUMBER, *n*	AMOUNT PAID, *A* ($)
0	
1	
2	
3	

f. Express the relationship between the output *A*, total amount you have already paid, and the input *n*, number of months that have elapsed, as a verbal rule and then as a symbolic rule.

g. What is the amount due after 3 payments?

h. Express the relationship between the output B, balance remaining, and the input n, number of months that have elapsed, as a verbal rule and then as a symbolic rule.

Graphical Representation Revisited

There are two types of calculations involving an input/output relationship that are important in the problem-solving process.

• Given an input value, determine the corresponding output value.

• Given an output value, determine the corresponding input value(s).

If the situation can be defined symbolically, both of these calculations can be done using algebraic methods. However, if the input/output relationship can only be represented graphically, an estimate of both of these calculations can be done by the following visual methods using the graph.

> **1.** Given an input value, graphically determine the corresponding output value:
>   **a.** Locate the given *input value* on the *horizontal axis*.
>   **b.** Move vertically (up or down) until you reach the point on the graph.
>   **c.** The *y-coordinate* of the point represents the corresponding *output value*.
>
> **2.** Given an output value, graphically determine the corresponding input value:
>
>   **a.** Locate the given *output value* on the *vertical axis*.
>   **b.** Move horizontally to the point(s) on the graph; there may be more than one.
>   **c.** The *x-coordinate* of each point represents the *input value*.

In addition, as you have already observed in Activity 2.2, trends and patterns in the input/output relationship can often be more clearly understood when the relationship is presented in visual form. Reading and interpreting graphs is an important skill and is reinforced throughout this textbook.

5. When you were born, your grandparents invested $1000 for you in a local bank. The following graph shows how your investment grows.

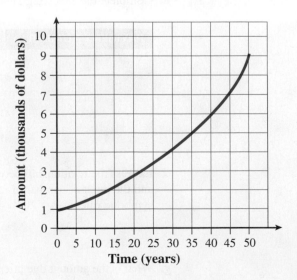

   a. Which variable is the input variable?

   b. How much money did you have when you were 10 years old?

   c. Estimate in what year your original investment would have doubled.

   d. If your college bill is estimated to be $3000 in the first year of college, will you have enough to pay the bill with these funds? (Assume that you attend when you are 18.) Explain.

   e. Assume that you expect to be married when you are 30 years old. You figure that you will need about $5000 for your share of the wedding and honeymoon expenses. Also assume that you left the money in the bank and did not use it for your education. Will you have enough money to pay your share of the wedding and honeymoon expenses? Explain.

SUMMARY: ACTIVITY 2.3

1. A **symbolic rule** is a mathematical statement that defines an **output** variable as an algebraic expression in terms of the **input** variable. The symbolic rule is essentially a recipe that describes how to determine the output value corresponding to a given input value.

2. Given an input value, the corresponding output value can be determined graphically as follows:

   a. Locate the given *input value* on the *horizontal axis*.

   b. Move vertically (up or down) until you reach the point on the graph.

   c. The *y-coordinate* of the point represents the corresponding *output value*.

3. Given an output value, the corresponding input value can be determined graphically as follows:

   a. Locate the given *output value* on the *vertical axis*.

   b. Move horizontally to the point(s) on the graph; there may be more than one.

   c. The *x-coordinate* of each point represents the *input value*.

EXERCISES: ACTIVITY 2.3

1. **a.** Determine the relationship common to all input and output pairs in each of the following tables. For each table, write a verbal rule that describes how to calculate the output *y* from its corresponding input *x*. Then complete the tables.

Exercise numbers appearing in color are answered in the Selected Answers appendix.

i.

INPUT, x	OUTPUT, y
2	4
4	8
6	12
8	16
10	
20	
25	

ii.

INPUT, x	OUTPUT, y
2	4
3	9
4	16
5	
6	
7	
8	

iii.

INPUT, x	OUTPUT, y
0	10
2	8
4	6
6	
8	
10	

b. Translate each verbal rule in part a into a symbolic rule.

   i.   **ii.**   **iii.**

c. Graph each input/output pair from part a on an appropriately scaled and labeled set of coordinate axes.

   i. 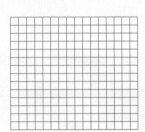   **ii.** 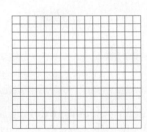   **iii.**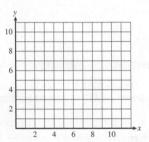

2. a. Complete the following tables using the verbal rule in column 2.

i.

INPUT	OUTPUT IS 10 MORE THAN THE INPUT
1	
3	
5	
7	
9	
12	
14	

ii.

INPUT	OUTPUT IS 3 TIMES THE INPUT, PLUS 2
−5	
−4	
−3	
−2	
−1	
0	
1	
2	

b. Translate each verbal rule in part a into a symbolic rule.

   i. $y = x + 10$ **ii.** $y = 3x + 2$

3. You are considering taking a part-time job that pays $9.50 per hour.

   a. What would be your gross pay if you work 22 hours one week?

b. Write a verbal rule that describes how to determine your gross weekly pay based on the number of hours worked.

c. Translate the verbal rule in part b into a symbolic rule. Let *h* represent the number of hours worked and *P* represent the gross weekly pay.

d. Complete the following table using the rule in part b.

NUMBER OF HOURS WORKED (INPUT), h	WEEKLY PAY (OUTPUT), p
6	
12	
18	
20	
22	
28	
30	

e. What are realistic replacement values for the input variable, the number of hours worked?

f. Graph the information given in the table in part d. Remember to use properly scaled and labeled axes.

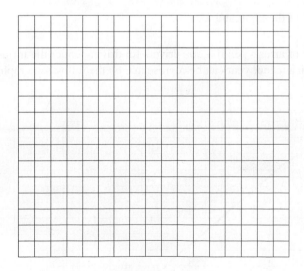

g. Explain your reasons for selecting the particular scales that you used.

h. Which representation (table or graph) presents the information and trends more clearly? Explain your choice.

4. Let *x* represent the input variable. Translate each of the phrases into an algebraic expression.

  a. The input decreased by 7

  b. Five times the difference between the input and 6

  c. Seven increased by the quotient of the input and 5

  d. Twelve more than one-half of the square of the input

  e. Twenty less than the product of the input and −2

  f. The sum of three-eighths of the input and five

5. Let *x* represent the input variable and *y* represent the output variable. Translate each of the following verbal rules into a symbolic rule.

  a. The output is the input decreased by ten.

  b. The output is three times the difference between the input and four.

  c. The output is nine increased by the quotient of the input and six.

  d. The output is seven more than one-half of the square of the input.

  e. The output is fifteen less than the product of the input and −4.

  f. The output is the sum of one-third of the input and ten.

6. The relationship between the number of arrests per 100,000 drivers for driving under the influence of alcohol and the driver's age in a recent year can be represented by the following graph.

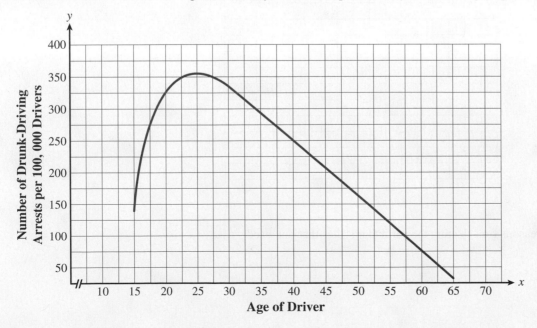

a. What is the input variable in this situation? What is the output variable?

b. Determine the age that corresponds to the greatest number of arrests. Estimate the number of arrests per 100,000 drivers for this age group.

c. Which age group has 200 arrests per 100,000 drivers?

d. Describe in words what the graph is telling you about this situation. Indicate whether the graph rises, falls, or is constant and whether the graph reaches a minimum (smallest) or maximum (largest) output value.

7. The following graph shows the height, h, in meters, of an eagle above the ground in terms of its time, t, in seconds, in flight.

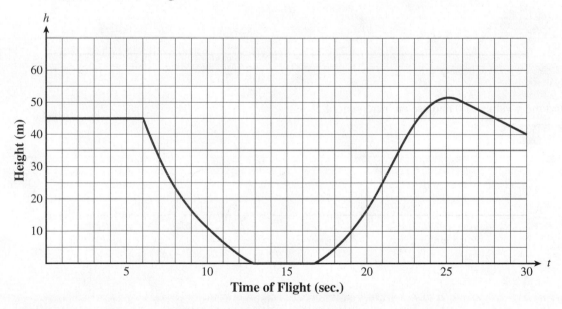

a. Identify the input and output variables.

b. What is the initial height of the eagle (when $t = 0$) of the eagle?

c. Determine the height of the eagle after 8 seconds of flight.

d. Determine at what time of flight the eagle is 20 meters from the ground.

e. What is the eagle's maximum height during the 30-second flight?

f. Determine the value of t for which $h = 0$. Interpret what this means in the context of the situation.

g. Describe in words what the graph is telling you about this situation. Indicate whether the graph rises, falls, or is constant and whether the graph reaches a minimum (smallest) or maximum (largest) output value.

In Exercises 8 and 9, construct a table of values of four ordered pairs for the given equation. Check your results using the table feature of your grapher.

8. $y = x^2$. Start the input variable x at 3, and use an increment of 2.

x	y

9. $y = 3.5x + 6$. Start the x-values at 0 and use an increment of 5.

x	y

Activity 2.4

Are They the Same?

Objectives

1. Identify equivalent algebraic expressions by examining their outputs.

2. Identify equivalent algebraic expressions by comparing their graphs.

3. Write algebraic expressions that involve grouping symbols.

In the problem-solving process using algebra, it is important to identify equivalent forms of algebraic expressions. Algebraic expressions are said to be **equivalent** if identical inputs always produce the same output.

Equivalent Expressions

A major road construction project in your neighborhood is forcing you to take a 3-mile detour each way when you leave and return home. To compute the round-trip mileage for a routine trip, you must double the usual one-way mileage, adding in the 3-mile detour.

Does it matter in which order you perform these operations? That is, to determine the round-trip mileage (output), do you

Rule 1:   double the usual one-way mileage (input) and then add 3, or

Rule 2:   add 3 to the usual one-way mileage (input) and then double the result?

1. Complete the following table to determine whether there is a difference in the results from using these two methods.

USUAL MILEAGE (INPUT)	RULE 1: TO OBTAIN THE OUTPUT, DOUBLE THE INPUT, THEN ADD 3.	RULE 2: TO OBTAIN THE OUTPUT, ADD 3 TO THE INPUT, THEN DOUBLE.
8		
10		
15		

2. Do rules 1 and 2 generate the same output values (round-trip mileage)?

3. For each of the rules given in the preceding table, let x represent the input and y represent the output. Translate each verbal rule into a symbolic rule.

   Rule 1:

   Rule 2:

4. From which sequence of operations do you obtain the correct round-trip mileage? Explain why.

Because algebraic expressions that are **equivalent** always produce the same outputs for identical inputs, the expressions $2x + 3$ (from rule 1) and $(x + 3) \cdot 2$ (from rule 2) are not equivalent. The table in Problem 1 shows that an input such as $x = 8$ produces different outputs.

5. The graphs of rules 1 and 2 are given on the same axes. Explain how the graphs demonstrate that the expressions $2x + 3$ and $(x + 3) \cdot 2$ are not equivalent.

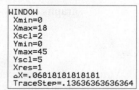 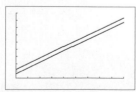

6. You are starting a new job to earn money to pay for college. The pay rate is $12 per hour for the first 40 hours. For any hours worked over 40, you are paid an overtime pay rate that is $1\frac{1}{2}$ times the regular rate. Assume that you will work more than 40 hours each week.

a. Identify the input and output variables in this situation.

b. Determine the overtime pay rate.

c. One method to calculate your weekly wages is as follows:

Rule 1: To obtain the weekly wages, multiply overtime hours by 1.5, add 40, and then multiply the sum by 12.

If n represents the number of overtime hours and w represents the weekly wage, translate the verbal rule into a symbolic rule.

d. A coworker suggests an alternative method to calculate the weekly wage.

Rule 2: To obtain weekly wages, multiply the number of overtime hours by $18 and add that result to 12 times the number of regular hours.

Translate the verbal rule into a symbolic value for w in terms of n.

e. Use the symbolic rules in parts b and c to complete the following table.

n, NUMBER OF OVERTIME HOURS	w, WEEKLY WAGE USING RULE 1	w, WEEKLY WAGE USING RULE 2
0		
3		
5		
10		

f. What do you notice about the outputs generated by rules 1 and 2?

g. Is the expression $12(1.5x + 40)$ equivalent to the expression $18x + 480$? Explain.

h. Use technology to sketch a graph of each rule. What is the relationship between the graphs? Is it what you expected? Explain.

You may have noticed that the result in Problem 6g, the equivalence of the expressions $12(1.5x + 40)$ and $18x + 480$, can also be verified using the Distributive Property. This important property of algebra will be discussed in detail in Activity 2.9 and will play a major role in allowing you to simplify algebraic expressions.

7. a. For each of the following rules, translate the verbal rule into a symbolic rule. Let x represent the input and y the output. Then complete the tables.

Rule 1: To obtain the output, divide the sum of the input and 10 by 2.

x	0	2	4	6	8
y					

Rule 2: To obtain the output, take $\frac{1}{2}$ of the input and then add 10.

x	0	2	4	6	8
y					

b. Is the expression $\dfrac{x + 10}{2}$ equivalent to $\dfrac{x}{2} + 10$? Explain.

8. a. For each of the following rules, translate the given verbal rule into a symbolic rule. Let x represent the input and y the output. Then complete the tables.

Rule 1: To obtain y, square the sum of x and 3.

x	y
-1	
0	
2	
5	
10	

Rule 2: To obtain y, add 9 to the square of x.

x	y
1	
0	
2	
5	
10	

b. What do you notice about the outputs generated by rules 1 and 2?

c. Is the expression $(x + 3)^2$ equivalent to the expression $x^2 + 9$? Explain.

d. If you were to graph rule 1 and rule 2, how would you expect the graphs to compare?

Using the commutative and associative properties for addition and multiplication is an effective way of generating equivalent algebraic expressions. For example:

EQUIVALENT ALGEBRAIC EXPRESSIONS	PROPERTY
$x4$ and $4x$	Commutative Property of Multiplication
$5(x + 2)$ and $(x + 2)5$	Commutative Property of Multiplication
$2(3x)$ and $(2 \cdot 3)x = 6x$	Associative Property of Multiplication
$(x + 1) + 3$ and $x + (1 + 3) = x + 4$	Associative Property of Addition
$4 + (9 + x)$ and $x + 13$	Associative Property of Addition followed by Commutative Property of Addition

9. The following expressions are equivalent. Identify which property of addition or multiplication justifies the equivalence.

   a. $(2x + 5) + 1$ and $2x + 6$

   b. $2(5x)$ and $10x$

   c. $x(3)(5)$ and $15x$

SUMMARY: ACTIVITY 2.4

1. Two algebraic expressions are said to be **equivalent** if outputs obtained from the same replacement values are always the same.

2. The graphs of rules that define equivalent expressions are identical.

EXERCISES: ACTIVITY 2.4

1. a. Complete the following table.

x	$x \cdot 3$	$3x$
-4		
-1		
0		
3		

   b. Do the expressions $x \cdot 3$ and $3x$ produce the same output value when given the same input value?

   c. In part b, the input value times 3 gives the same result as 3 times the input value. What property of multiplication does this demonstrate?

d. Use technology to sketch a graph of $y_1 = x \cdot 3$ and $y_2 = 3x$ on the same coordinate axes. How do the graphs compare? Your screens should appear as follows.

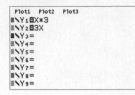 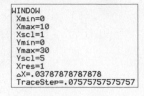 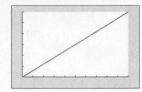

2. a. Is the expression $x - 3$ equivalent to $3 - x$? Complete the following table to help justify your answer.

x	x − 3	3 − x
−5		
−3		
0		
1		
3		

b. What correspondence do you observe between the output values in columns 2 and 3? How is the expression $3 - x$ related to the expression $x - 3$?

c. Is the operation of subtraction commutative?

d. Use technology to sketch a graph of $y_1 = x - 3$ and $y_2 = 3 - x$. How do the graphs compare? Why does this show that $x - 3$ and $3 - x$ are not equivalent expressions? Your screens should appear as follows.

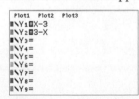 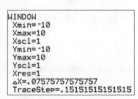 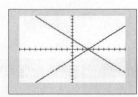

3. a. Write a verbal rule that describes the sequence of operations indicated by the given symbolic rule. Then complete the tables.

i. $y = x^2$

x	y
−2	
−1	
0	
2	
3	

ii. $y = -x^2$

x	y
−2	
−1	
0	
2	
3	

iii. $y = (-x)^2$

x	y
−2	
−1	
0	
2	
3	

b. Which, if any, of the expressions x^2, $-x^2$, and $(-x)^2$ are equivalent? Explain.

The symbolic rules in the following exercises each contain two or more arithmetic operations. In Exercises 4 and 5, write a verbal rule, complete the table, and answer the question about your results.

4.

x	$y = 2x + 1$	$y = 2(x + 1)$
−1		
0		
2		
5		

Are the expressions $2x + 1$ and $2(x + 1)$ equivalent? Why or why not?

5.

x	$y = 3x^2 + 1$	$y = (3x)^2 + 1$
−1		
0		
2		
5		

Are the expressions $3x^2 + 1$ and $(3x)^2 + 1$ equivalent? Why or why not?

6. You need to estimate the amount of carpeting for a rectangular commercial space measuring approximately 45 feet by 60 feet. To ensure that you order enough carpeting to account for inevitable waste, a colleague suggests that you increase each measurement by 10% and then calculate the area. Another colleague suggests that you first calculate the area and then increase it by 10%.

   a. Are these suggestions equivalent?

   b. If the carpeting costs $22 per square yard, how much more will you pay if you use the larger measurement estimate?

Cluster 1 What Have I Learned?

1. Describe several ways that relationships between variables may be represented. What are some advantages and disadvantages of each?

2. Obtain a graph from a local newspaper, magazine, or textbook in your major field. Identify the input and output variables. The input variable is referenced on which axis? Describe any trends in the graph.

3. You will be graphing some data from a table in which the input values range from 0 to 150 and the output values range from 0 to 2000. Assume that your grid is a square with 16 tick marks across and up.

   a. How many units does the distance between tick marks on the horizontal axis represent?

   b. How many units does the distance between tick marks on the vertical axis represent?

Exercise numbers appearing in color are answered in the Selected Answers appendix.

Cluster 1 How Can I Practice?

1. You are a scuba diver and plan a dive in the St. Lawrence River. The water depth in the diving area does not exceed 150 feet. Let x represent your depth in feet *below* the surface of the water.

    a. What are the possible replacement values to represent your depth from the surface?

    b. Would a replacement value of 0 feet be reasonable? Explain.

    c. Would a replacement value of -200 be reasonable? Explain.

    d. Would a replacement value of 12 feet be reasonable? Explain.

2. Fish need oxygen to live, just as you do. The amount of dissolved oxygen (D.O.) in water is measured in parts per million (ppm). Trout need a minimum of 6 ppm to live.

    The data in the table show the relationship between the temperature of some water samples and the amount of dissolved oxygen present.

Temp (°C)	11	16	21	26	31
D.O. (in ppm)	10.2	8.6	7.7	7.0	6.4

    a. Represent the data in the table graphically. Place temperature (input) along the horizontal axis and dissolved oxygen (output) along the vertical axis.

    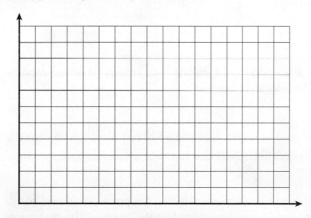

    b. What general trend do you notice in the data?

    c. In which of the 5° temperature intervals given in the table does the dissolved oxygen content change the most?

   d. Which representation (table or graph) presents the information and trends more clearly?

3. A square has perimeter 80 inches.

   a. What is the length of each side?

   b. If the length of each side is increased by 5 inches, what will the new perimeter equal?

   c. If each side of the original square is increased by x inches, write the new perimeter in terms of x.

4. Let x represent the input variable. Translate each of the following phrases into an algebraic expression.

   a. 20 less than twice the input   **b.** the sum of half the input and 6

5. Evaluate each algebraic expression for the given value(s).

   a. $2x + 7$, for $x = -3.5$ **b.** $8x - 3$, for $x = 1.5$

   c. $2k + 3h$, for $k = -2.5, h = 9$

6. a. A concert ticket costs p dollars for an adult ticket and q dollars for a student. Write an expression that represents the total cost for 4 adults and 6 students.

   b. You purchase an item for x dollars and give the sales associate a $20 bill. Write an algebraic expression that represents the change you will receive.

7. The cost to own and operate a car depends on many factors, including gas prices, insurance costs, size of the car, and finance charges. Using a cost calculator found on the Internet, you determine that the average cost to own and operate a small sedan in the United States is $0.449 per mile.

   a. Determine the total weekly cost of owning and operating your small sedan if you anticipate driving 200 miles per week.

   b. Identify the input and output variables in this situation.

c. Let C represent the total cost of owning and operating a small sedan and n represent the number of miles driven. Write an equation for C in terms of n.

d. What are the practical replacements for the input?

e. If you drive an average of 15,000 miles per year, determine the total cost for a year.

f. Use technology to create a table of input/output values beginning with an input of zero. Use increments of 15,000 miles.

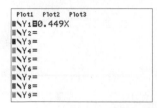 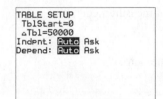 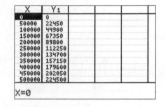

8. The following graph shows the unemployment rate in a large metropolitan area since 2000. The variable n represents the number of years since 2000.

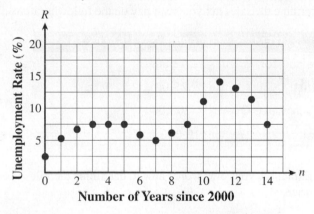

a. Identify the input and output variables.

b. In what year was the unemployment rate the highest? Estimate the rate in that year.

c. In what year was the unemployment rate the lowest? Estimate the rate in that year.

d. Estimate during which year the unemployment rate was 10%.

Cluster 2   Solving Equations

Activity 2.5

Let's Go Shopping

Objectives

1. Solve an equation of the form $ax = b$, $a \neq 0$, for x using an algebraic approach.

2. Solve an equation of the form $x + a = b$ for x using an algebraic approach.

3. Use the four-step process to solve problems.

The sales tax collected on taxable items in Allegany County in western New York is 8.5%. The tax you must pay depends on the price of the item you are purchasing.

1. What is the sales tax on a dress shirt that costs $20?

2. a. Because you are interested in determining the sales tax given the price of an item, which variable is the input?

  b. What are its units of measurement?

3. a. Which variable is the output?

  b. What are its units of measurement?

4. Determine the sales tax you must pay on the following items.

ITEM	PRICE ($)	SALES TAX ($)
Calculator	12.00	
Shirt	25.00	
Microwave Oven	200.00	
Car	15,000.00	

5. a. Explain in your own words how to determine the sales tax on an item of a given price.

  b. Translate the verbal rule in part a into a symbolic rule, with x representing the input (price) and y representing the output (sales tax).

Solving Equations of the Form $ax = b$, $a \neq 0$, Using an Algebraic Approach

You can diagram the symbolic rule $y = 0.085x$ in the following way.

Start with x (price) $\longrightarrow$ multiply by 0.085 $\longrightarrow$ to obtain y (sales tax).

How might you determine the price of a fax machine for which you paid a sales tax of $21.25? In this situation you know the sales tax (output) and want to determine the price (input). To accomplish this, "reverse the direction" in the preceding diagram and replace the operation (multiplication) with its inverse (division).

Start with y (sales tax) $\longrightarrow$ divide by 0.085 $\longrightarrow$ to obtain x (price).
Start with 21.25 $\longrightarrow$ 21.25 ÷ 0.085 $\longrightarrow$ to obtain 250.

Therefore, the price of a fax machine for which you paid $21.25 sales tax is $250.

6. Use this reverse process to determine the price of a scanner for which you paid a sales tax of $61.20.

The reverse process illustrated above can also be done in a more common and formal way, as demonstrated in Example 1.

> **Example 1** *The sales tax for the fax machine was $21.25. In the symbolic rule* $y = 0.085x$, *replace* y *with* 21.25 *and proceed as follows:*
>
> $21.25 = 0.085x$ This is the equation to be solved.
>
> $\dfrac{21.25}{0.085} = \dfrac{0.085x}{0.085}$ Divide both sides of the equation by 0.085, or equivalently, multiply by $\dfrac{1}{0.085}$.
>
> $250 = x$

The approach demonstrated in Example 1 is the **algebraic method** of solving the equation $21.25 = 0.085x$. In this situation, you needed to undo the multiplication of x and its coefficient 0.085 by dividing each side of the equation by the coefficient 0.085. The value obtained for x (here, 250) is called the **solution** of the equation. Note that the solution of the equation is complete when the variable x is isolated on one side of the equation, with coefficient 1.

7. Use the symbolic rule $y = 0.085x$ to determine the price of a Blu-ray DVD player for which you paid a sales tax of $17.85.

Solving Equations of the Form $x + a = b$ Using an Algebraic Approach

8. Suppose you have a $15 coupon that can be used on any purchase over $100 at your favorite clothing store.

   a. Determine the discounted price of a sports jacket having a retail price of $116.

   b. Identify the input variable and the output variable.

   c. Let x represent the input and y represent the output. Write a symbolic rule that describes the relationship between the input and output. Assume that your total purchase will be greater than $100.

d. Complete the following table using the symbolic rule you determined in part c.

x, RETAIL PRICE ($)	y, DISCOUNT PRICE ($)
105	
135	
184	
205	

You can also diagram this symbolic rule in the following way.

Start with x (retail price) $\longrightarrow$ subtract 15 $\longrightarrow$ to obtain y (discount price).

9. Suppose you are asked to determine the retail price if the discount price of a suit is $187.

  a. Is 187 an input value or an output value?

  b. Reverse the direction in the diagram above, replacing the operation (subtraction) with its inverse (addition).

  Start with $y = 187$ (output) _____ to obtain x (input).

  c. Use this reverse process to determine the retail price of the suit.

The equation in Problem 9 can be solved using an algebraic approach.

Example 2   *Solve the equation* $187 = x - 15$ *for x.*

$187 = x - 15$	**Equation to be solved.**
$\underline{+15 \qquad + 15}$	**Add the term 15 to each side to isolate the variable.**
$202 = x$	

---

10. Use the symbolic rule $y = x - 15$ to determine the retail price, x, of a trench coat whose discounted price is $203.

In this activity, you have practiced solving equations involving a single operation—addition, subtraction, multiplication, or division—by performing the appropriate inverse operation. Recall that addition and subtraction are inverse operations, as are multiplication and division. Use this algebraic method to solve the equations arising in the following problems.

11. Set up and solve the appropriate equation in each case.

  a. $y = 7.5x$ **b.** $z = -5x$

  Determine x when $y = 90$.   Determine x when $z = -115$.

c. $y = \dfrac{x}{4}$

Determine x when $y = 2$.

d. $p = \dfrac{2}{3}x$

Determine x when $p = -18$.

12. Use an algebraic approach to determine the input x for the given output value.

a. $y = x - 10$

Determine x when $y = -13$.

b. $y = 13 + x$

Determine x when $y = 7$.

c. $p = x + 4.5$

Determine x when $p = -10$.

d. $x - \dfrac{1}{3} = s$

Determine x when $s = 8$.

Recall George Polya's four-step process for solving problems that you encountered in Activity 1.2 on page 4. Solve each of the following by applying the four steps of problem solving. Use the strategy of solving an algebraic equation in each problem.

13. In preparing for a trip with friends during spring break, your assignment is to make the travel arrangements. You can rent a car for $75 per day with unlimited mileage. If you have budgeted $600 for car rental, how many days can you drive?

14. You need to drive 440 miles to get to your best friend's wedding. How fast must you drive to get there in 8 hours?

SUMMARY: ACTIVITY 2.5

1. A **solution** of an equation containing one variable is a replacement value for the variable that produces equal values on both sides of the equation.

2. An **algebraic approach** to solving an equation for a given variable is complete when the variable (such as x) is isolated on one side of the equation with coefficient 1. To isolate the variable, you apply the appropriate inverse operation, as follows:

   a. Multiplication and division are inverse operations. Therefore,

      i. to undo multiplication of the input by a nonzero number, divide each side of the equation by that number

      ii. to undo division of the input by a number, multiply each side of the equation by that number

   b. Addition and subtraction are inverse operations. Therefore,

      i. to undo addition of a number to the input, subtract that number from each side of the equation

      ii. to undo subtraction of a number from the input, add that number to each side of the equation

EXERCISES: ACTIVITY 2.5

1. You just graduated from your local community college and have decided to enroll as a part-time student (taking fewer than 12 credit hours) at the university. Full-time college work does not fit into your present financial or personal situation. The cost per credit hour at the university is $275 per credit hour.

   a. What is the cost of a 3-credit-hour literature course?

   b. Write a symbolic rule to determine the total tuition for a given number of credit hours. Let n represent the number of credit hours taken (input) and T represent the total tuition paid (output).

   c. What are the possible replacement values for n?

   d. Complete the following table.

CREDIT HOURS	TUITION PAID ($)
1	
2	
3	
4	

Exercise numbers appearing in color are answered in the Selected Answers appendix.

e. Suppose you have enrolled in a psychology course and a computer course, each of which is 3 credit hours. Use the symbolic rule in part b to determine the total tuition paid.

f. Use the symbolic rule from part b to determine algebraically the number of credit hours carried by a student with a tuition bill of $1925.

2. a. The average amount, A, of precipitation in Boston during March is 4 times the average amount, P, of precipitation in Phoenix. Translate the verbal rule into a symbolic rule.

b. What is the amount of precipitation in Boston if there are 2 inches in Phoenix?

c. What is the amount of precipitation in Phoenix if there are 24 inches in Boston?

3. a. The depth (inches) of water that accumulates in the spring soil from melted snow can be determined by dividing the cumulative winter snowfall (inches) by 12. Translate the verbal statement into a symbolic rule, using I for the accumulated inches of water and n for the inches of fallen snow.

b. Determine the amount of water that accumulates in the soil if 25 inches of snow falls.

c. Determine the total amount of winter snow that produces 6 inches of water in the soil.

4. a. Profit is calculated as revenue minus expenses. If a company's expenses were $10 million, write a symbolic rule that expresses profit, P (in millions of dollars), in terms of revenue, R (in millions of dollars).

    b. Determine the profit when revenue is $25 million.

    c. Determine the revenue that will produce a profit of $5 million.

For part a of Exercises 5–10, determine the output when you are given the input. For part b, use an algebraic approach to solve the equation for the input when you are given the output.

5. $3.5x = y$

    a. Determine y when $x = 15$.   **b.** Determine x when $y = 144$.

6. $z = -12x$

    a. Determine z when $x = -7$.   **b.** Determine x when $z = 108$.

7. $y = 15.3x$

    a. Determine y when $x = -13$.   **b.** Determine x when $y = 351.9$.

8. $y = x + 5$

    a. Determine y when $x = -11$.   **b.** Determine x when $y = 17$.

9. $y = x + 5.5$

    a. Determine y when $x = -3.7$.   **b.** Determine x when $y = 13.7$.

10. $z = x - 11$

    a. Determine z when $x = -5$.   **b.** Determine x when $z = -4$.

In Exercises 11 and 12, solve each problem using the four-step process for solving problems.

11. Your goal is to save \$1200 to pay for next year's books and fees. How much must you save each month if you have 5 months to accomplish your goal?

12. You have enough wallpaper to cover 240 square feet. If your walls are 8 feet high, what wall length can you paper?

Although the emphasis in this activity is solving problems using an algebraic approach, there are many situations in which the relationship between the variables cannot be represented by a symbolic rule. Exercises 13 and 14 illustrate the important role that numerical and graphical methods play in solving input / output problems (even if a symbolic rule exists). These methods will continue to be explored and developed throughout the textbook.

13. An object is dropped from the top of the Willis Tower (formerly called the Sears Tower) in Chicago. The table below gives the object's distance from the ground (the output) at a given time after it is dropped (the input).

Time (sec.)	1	2	3	4	5	6	7
Distance (ft.)	1435	1387	1307	1195	1051	875	667

    a. Estimate the number of seconds it takes the object to be 1000 feet above the ground.

    b. The Willis Tower is 1451 feet tall. Approximately how many seconds will it take the object to be halfway to the ground?

    c. If you knew the symbolic rule for this input/output relation, how might your answers be improved?

14. You wake up in the morning feeling lousy. You take your temperature at 8 A.M. and continue to monitor it throughout the day. The following graph shows how your temperature changes over a 7-hour period.

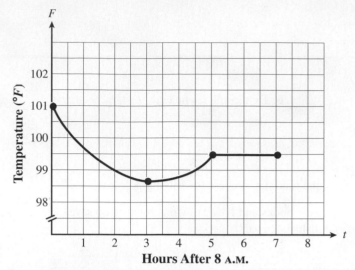

a. Identify the input and output in this situation.

b. What was your temperature at 8 A.M.?

c. Estimate your minimum temperature during the indicated time period. At what time did this occur?

d. During which period of time was your temperature decreasing? When was it increasing?

e. Estimate your temperature at noon.

f. How was your temperature changing from 1 P.M. to 3 P.M.? What was the temperature?

Activity 2.6

Leasing a Copier

Objectives

1. Model contextual situations with symbolic rules of the form $y = ax + b, a \neq 0$.

2. Solve equations of the form $ax + b = c$, $a \neq 0$.

As part of your college program, you are a summer intern in a law office. The office manager asks you to gather some information about leasing a copy machine for the office. The sales representative at Eastern Supply Company recommends a 50-copy/minute copier to satisfy the office's copying needs. The copier leases for $300 per month, plus 1.5 cents a copy. Maintenance fees are covered in the monthly charge. The lawyers would own the copier after 39 months.

1. **a.** The total monthly cost depends on the number of copies made. Identify the input and output variables.

   b. Write a statement in words to determine the total monthly cost in terms of the number of copies made during the month.

   c. Complete the following table.

Number of Copies	5000	10,000	15,000	20,000
Monthly Cost ($)				

   d. Translate the statement in part b into an equation. Let n represent the number of copies (input) made during the month and c represent the total monthly cost (output).

2. **a.** Use the result obtained in Problem 1d to determine the monthly cost if 12,000 copies are made.

   b. Is 12,000 a replacement value for the input or the output variable?

In Problem 2, you determined the cost by evaluating the expression $0.015n + 300$ for $n = 12,000$. This was accomplished by performing the sequence of operations shown in the diagram.

Start with n $\longrightarrow$ multiply by 0.015 $\longrightarrow$ add 300 $\longrightarrow$ to obtain c.

3. Suppose the monthly budget for leasing the copier is $675.

   a. Is 675 the input value for n or the output value for c?

   b. To determine the input, n, for a given output value of c, reverse the sequence of operations in the preceding diagram and replace each operation with its inverse. Recall that addition and subtraction are inverse operations and that multiplication and division are inverse operations. Complete the following.

   Start with c $\longrightarrow$ _subtract 300_ $\longrightarrow$ _divide by 0.015_ $\longrightarrow$ to obtain n.

   c. Use the sequence of operations in part b to determine the number of copies that can be made for a monthly budget of $c = 675$.

Solving Equations Algebraically

Once you understand the concept of using the reverse process to solve equations (see Problem 3), you will want a more systematic algebraic procedure to follow.

4. Use the equation in Problem 1 to write an equation that can be used to determine the number, n, of copies that can be made with a monthly budget of $c = 675$.

In Problem 1 you constructed a symbolic rule involving *two* operations.

In Problem 2 you evaluated this rule using the indicated sequence of operations.

In Problem 3 you solved an equation by performing the sequence in reverse order, replacing each original operation with its inverse. This algebraic solution is commonly written in a vertical format in which each operation is performed on a separate line. Example 1 illustrates this format.

Example 1   *The following example illustrates a systematic algebraic procedure by undoing two operations to solve an equation.*

Solve for n:

$$675 = 0.015n + 300$$
$$\underline{-300 \qquad\qquad -\ 300}$$
$$375 = 0.015n$$

Step 1: To undo the addition of the term 300, subtract 300 from each side of the equation.

$$\frac{375}{0.015} = \frac{0.015n}{0.015}$$
$$25{,}000 = n$$

Step 2: To undo the multiplication by the coefficient 0.015, divide each side of the equation by 0.015.

Check:

$$675 \overset{?}{=} 0.015(25{,}000) + 300$$
$$675 \overset{?}{=} 375 + 300$$
$$675 = 675$$

Note that the algebraic procedure emphasizes that an equation can be thought of as a scale whose arms are in balance. The equal sign can be thought of as the balancing point.

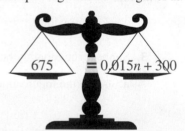

As you perform the appropriate inverse operation to solve the equation $675 = 0.015n + 300$ for n, you must maintain the balance as you perform each step in the process. If you subtract the term 300 from one side of the equal sign, you must subtract 300 from the other side. Similarly, if you divide one side of the equation by the coefficient 0.015, you must divide the other side by 0.015.

5. Use the algebraic procedure outlined in Example 1 to solve the equation $120 = 3x + 90$.

6. In each of the following, replace y with the given value and solve the resulting equation for x.

a. If $y = 3x - 5$ and $y = 10$, determine x.

b. If $y = 30 - 2x$ and $y = 24$, determine x.

c. If $y = \frac{3}{4}x - 21$ and $y = -9$, determine x.

d. If $-2x + 15 = y$ and $y = -3$, determine x.

7. Complete the following tables.

a. $y = 4x - 11$

x	y
6	
	53

b. $y = -5x - 80$

x	y
12	
	35

SUMMARY: ACTIVITY 2.6

1. To solve an equation containing two or more operations, perform the sequence of operations in *reverse* order, replacing each original operation with its *inverse*.

EXERCISES: ACTIVITY 2.6

1. In 2014, the number of full-time students at your local community college was 1850. The admissions office anticipated that the enrollment will increase at a constant rate of 90 students per year for the next several years.

   a. Write a verbal rule to determine the enrollment of full-time students in a given number of years after 2014.

   b. Translate the verbal rule in part a into a symbolic rule. Use N to represent the total enrollment of full-time students and t to represent the number of years after 2014.

   c. Determine the total enrollment of full-time students in 2017.

   d. In what year is the total enrollment of full-time students expected to be 2300?

2. The social psychology class is organizing a campus-entertainment night to benefit charities in the community. You are a member of the budget committee for the class project. The committee suggests a $10 per person admission donation for food, nonalcoholic beverages, and entertainment. The committee determines that the fixed costs for the event (food, drinks, posters, tickets, etc.) will total $2100. The college is donating the use of the gymnasium for the evening.

   a. The total revenue (gross income before expenses are deducted) depends on the number, n, of students who attend. Write an expression in terms of n that represents the total revenue if n students attend.

   b. Profit is the net income after expenses are deducted. Write a symbolic rule expressing the profit, p, in terms of the number, n, of students who attend.

   c. If the gymnasium holds a maximum of 700 people, what is the maximum amount of money that can be donated to charity?

   d. Suppose that the members of the committee want to be able to donate at least $1500 to community charities. How many students must attend to have a profit of $1500?

3. You are planning an extended trip to Mexico. You have a new smartphone so that you can stay in touch with your friends, but are concerned about roaming charges because you know you will be outside the home network service area. Your plan costs $35 per month plus $0.25 per minute for roaming charges.

   a. Write a statement to determine your total monthly cost for your smartphone per month during your extended stay in Mexico.

   b. Translate the statement in part a into an equation, using c to represent the total monthly cost and n to represent the total roaming minutes for the month.

   c. Determine your monthly cost if 250 minutes of roaming charges are added to your bill.

   d. If you budget $75 per month for your smartphone, how many minutes of roaming charges can you incur?

4. In the summer of 2009, during a period of economic woes for the United States, President Barack Obama's "Cash for Clunkers" program was wildly popular. The program was intended to aid the auto industry, thereby boosting the economy. It was also a green initiative because consumers replaced their cars with cars getting better fuel mileage. Suppose you had a 1995 Jeep Grand Cherokee 4WD with a combined city/highway fuel economy of 15 miles per gallon. In return for it, you would receive a $4500 credit toward the purchase of a new car provided that you chose one with fuel economy that was at least 10 miles per gallon more than that of your Jeep. This incentive, in addition to the 0% interest loans that were being offered by the troubled auto industry, was too much for you to resist. Your new car payment was $420 per month.

   a. Write an equation to determine the total amount, p, paid on the car after n months, including the $4500 credit.

   b. Use the equation you wrote in part a to complete the following table.

Number of Months, n	1	2	3	4	5	6
Total Amount Paid, p ($)						

   c. Determine the amount, p, paid on the car after 9 months.

   d. The car you chose was a 2009 Ford Escape Hybrid, with combined fuel economy of 30 miles per gallon, and the price of your car after negotiation, including taxes and all fees, was $29,700. How many years did it take you to pay off the loan?

5. Archaeologists and forensic scientists use the length of human bones to estimate the height of individuals. A person's height, h, in centimeters, can be determined from the length of the femur, f (the bone from the knee to the hip socket), in centimeters, using the following formulas:

Man: $h = 69.089 + 2.238f$
Woman: $h = 61.412 + 2.317f$

a. A partial skeleton of a man is found. The femur measures 50 centimeters. How tall was the man?

b. What is the length of the femur for a woman who is 150 centimeters tall?

6. Let p represent the perimeter of an isosceles triangle that has two equal sides of length a and a third side of length b. The formula for the perimeter is $p = 2a + b$. Determine the length of the equal side of an isosceles triangle having a perimeter of $\frac{3}{4}$ yard and a third side measuring $\frac{1}{3}$ yard.

7. The recommended weight of an adult male is given by the formula $w = \frac{11}{2}h - 220$, where w represents his recommended weight in pounds and h represents his height in inches. Determine the height of a man whose recommended weight is 165 pounds.

8. Even though housing prices have been declining in many parts of the country, in your neighborhood, they have been increasing steadily since you bought your home in 2006. You purchased your home in 2006 for $130,000, and it steadily increased its value by $3500 a year. Let V represent the market value (in dollars) of your home; x, the length of time you have lived there (in years).

a. Complete the following table.

YEAR	x	V, MARKET VALUE
2006		
2011		
2014		

b. Write a symbolic rule that expresses V in terms of x.

c. Determine the expected value of your home in 2016.

d. In which year is the value of your home expected to reach $186,000?

9. Your local utility company offers a free energy audit of your home. The report is based on a temperature difference of 68°F inside and 28°F outside. The resulting report indicates that a wall of the house containing 40 square feet of glass (single-pane) and 120 square feet of standard insulated (R-11) wall lost 2035 BTU per hour. If the heat loss per square foot per hour for a standard wall is 3.7 BTU/hr., determine the heat lost per square foot per hour for glass (single-pane).

10. According to a 2012 U.S. Census Bureau Report, a person with a bachelor's degree could expect to have an annual salary that is $21,528 more than that of someone who has only a high school diploma.

a. Let x represent the annual salary of a person with a high school diploma in 2012. Write an algebraic expression in terms of x that represents the average annual salary of a person with a bachelor's degree in 2012.

b. Two individuals, one with a high school diploma and one with a bachelor's degree, would have a total annual salary of $89,336. Write an equation that represents the relationship between the two average salaries in part a.

c. Solve the equation in part b to determine the average annual salary of a person at each level of education in 2012.

11. The total cost, c, of purchasing tickets for a college football game through a ticket service depends on the number of tickets, t, you purchase. For a particular game, the tickets cost \$41.50 each plus a \$6.00 service charge for the order.

    a. Write an equation that relates the total cost, c, to the number of tickets purchased, t.

    b. If you purchase 6 tickets, what is the total cost of your order?

    c. If your total cost was \$296.50, how many tickets did you purchase?

12. Solve each of the following equations for x.

    a. $10 = 2x + 12$ **b.** $-27 = -5x - 7$

    c. $3x - 26 = -14$ **d.** $24 - 2x = 38$

    e. $5x - 12 = 15$ **f.** $-4x + 8 = 8$

    g. $12 + \dfrac{1}{5}x = 9$ **h.** $\dfrac{2}{3}x - 12 = 0$

    i. $0.25x - 14.5 = 10$

Complete the tables in Exercises 13–16 using algebraic methods. Indicate the equation that results from replacing x or y with its assigned value.

13. $y = 2x - 10$

x	y
4	
	14

14. $y = 20 + 0.5x$

x	y
3.5	
	-10

15. $y = -3x + 15$

x	y
$\dfrac{2}{3}$	
	-3

16. $y = 12 - \dfrac{3}{4}x$

x	y
-8	
	-6

The Algebra of Weather

Objectives

1. Solve a formula for a specified variable.

Windchill

On Monday morning, you listen to the news and weather before going to class. The meteorologist reports that the temperature is 25°F, a balmy February day on the campus of SUNY Oswego in New York State, but he adds that a 30-mile-per-hour wind makes it feel like 8°F.

Curious, you do some research and learn that windchill is the term commonly used to describe how cold your skin feels when it is exposed to the wind. You also discover a formula that allows you to calculate windchill as a temperature relative to air temperature and a 30-mile-per-hour wind.

1. Windchill temperature, w (°F), produced by a 30-mile-per-hour wind at various air temperatures, t (°F), can be modeled by the formula

$$w = 1.36t - 26.$$

Complete the following table using the given formula.

Air temperature, t (°F)	−15	5	30
Windchill temperature, w (°F) (30-mph wind)			

2. **a.** Use the formula $w = 1.36t - 26$ to determine the windchill temperature if the air temperature is 7°F.

   b. On a cold day in New York City, the wind is blowing at 30 miles per hour. If the windchill temperature is reported to be −18°F, then what is the air temperature on that day? Use the formula $w = 1.36t - 26$.

The formula $w = 1.36t - 26$ is said to be solved for w in terms of t because the variable w is isolated on one side of the equation. To determine the windchill temperature, w, for air temperature $t = 7°F$ (Problem 2a), you substitute 7 for t in $1.36t - 26$ and do the arithmetic:

$$w = 1.36(7) - 26 = -16.48°F$$

In Problem 2b, you were asked to determine the air temperature, t, for a windchill temperature of −18°F. In this case, you replaced w with −18 and solved the resulting equation

$$-18 = 1.36t - 16$$

for t.

Each time you are given a windchill temperature and asked to determine the air temperature, you need to set up and solve a similar equation. Often it is more convenient and efficient to solve the original formula $w = 1.36t - 26$ for t symbolically and then evaluate the new rule to determine values of t.

Example 1 *Solving the formula w = 1.36t − 26 for t is similar to solving the equation −18 = 1.36t − 26 for t.*

$$-18 = 1.36t - 26 \qquad\qquad w = 1.36t - 26$$
$$\underline{+26 = \qquad\quad +26} \qquad\qquad \underline{+26 = \qquad\quad +26}$$
$$8 = 1.36t \qquad\qquad w + 26 = 1.36t$$
$$\frac{8}{1.36} = \frac{1.36t}{1.36} \qquad\qquad \frac{w + 26}{1.36} = \frac{1.36t}{1.36}$$
$$6 \approx t \qquad\qquad \frac{w + 26}{1.36} = t$$

The new formula is $t = \dfrac{w + 26}{1.36}$ or $t = \dfrac{1}{1.36}w + \dfrac{26}{1.36}$, which is approximately equivalent to $t = 0.735w + 19.12$, after rounding.

> To solve the equation $w = 1.36t - 26$ for t means to rewrite the equation so that t becomes the output variable and w the input variable.

3. Redo Problem 2b using the new formula derived in Example 1 that expresses t in terms of w.

4. a. If the wind speed is 15 miles per hour, the windchill temperature, w, can be approximated by the formula $w = 1.28t - 19$, where t is the air temperature in degrees Fahrenheit. Solve the formula for t.

b. Use the new formula from part a to determine the air temperature, t, if the windchill temperature is $-10°$F.

Weather Balloon

A weather balloon is launched at sea level. The balloon is carrying instruments that measure temperature during the balloon's trip. After the balloon is released, the data collected shows that the temperature dropped 0.0117°F for each meter the balloon rose.

5. a. If the temperature at sea level is 50°F, determine the temperature at a height of 600 meters above sea level.

b. Write a verbal rule to determine the temperature at a given height above sea level on a 50°F day.

c. If t represents the temperature (°F) a height of m meters above sea level, translate the verbal rule in part b into a symbolic rule.

d. Complete the following table.

Meters Above Sea Level, m	500	750	1000
Temperature, t (50°F Day)			

Recall that subtracting a number (or symbol) can be understood as adding its opposite. It is helpful to think of the symbolic rule $t = 50 - 0.0117m$ equivalently as

$$t = 50 + (-0.0117)m$$

6. a. Solve the formula $t = 50 - 0.0117m$ for m.

b. Water freezes at 32°F. Determine the height above sea level at which water will freeze on a 50°F day. Use the formula from part a.

Temperature to What Degree?

Since the 1970s, for almost all countries in the world, the Celsius scale is the choice for measuring temperature. The Fahrenheit scale is still the preference for nonscientific work in just a few countries, among them Belize, Jamaica, and the United States.

The scales are each based on the freezing point and boiling point properties of water and a straightforward relationship exists between the two scales. If F represents Fahrenheit temperature and C, Celsius temperature, the conversion formula for C in terms of F is

$$C = \frac{5}{9}(F - 32)$$

7. a. Average normal body temperature for humans is given to be 98.6°F. A body temperature 100°F or above usually indicates a fever. A patient in a hospital had a temperature of 39°C. Did she have a fever?

b. Solve the equation $C = \dfrac{5}{9}(F - 32)$ for F.

c. What is the patient's temperature of 39°C in terms of Fahrenheit degrees?

8. Solve each of the following formulas for the indicated variable.

a. $A = lw$, for w

b. $p = c + m$, for m

c. $P = 2l + 2w$, for l

d. $R = 165 - 0.75a$, for a

e. $V = \pi r^2 h$, for h

SUMMARY: ACTIVITY 2.7

To solve a **formula** for a variable, isolate that variable on one side of the equation with all other terms on the "opposite" side.

EXERCISES: ACTIVITY 2.7

1. The profit that a business makes is the difference between its revenue (the money it takes in) and its costs.

a. Write a formula that describes the relationship between the profit, p; revenue, r; and costs, c.

b. It costs a publishing company $85,400 to produce a textbook. The revenue from the sale of the textbook is $315,000. Determine the profit.

   c. The sales from another textbook amount to $877,000, and the company earns a profit of $465,000. Use your formula from part a to determine the cost of producing the book.

2. The distance traveled is the product of the rate (speed) at which you travel and the amount of time you travel at that rate.

   a. Write a formula that describes the relationship between the distance, d; rate, r; and time, t.

   b. A gray whale can swim 20 hours a day at an average speed of approximately 3.5 miles per hour. How far can the whale swim in a day?

   c. A Boeing 747 flies 1950 miles at an average speed of 575 miles per hour. Use the formula from part a to determine the flying time. (Round to nearest tenth of an hour.)

   d. Solve the formula in part a for the variable t. Then use this new formula to rework part c.

3. The speed, s, of an ant (in centimeters per second) is related to the temperature, t (in degrees Celsius), by the formula

$$s = 0.167t - 0.67.$$

   a. If an ant is moving 4 centimeters per second, what is the temperature? (Round to nearest degree.)

b. Solve the equation $s = 0.167t - 0.67$ for t.

c. Use the new formula from part b to answer part a.

4. Air temperature drops about 1°C for each 100-meter rise above ground level. This temperature drop continues to approximately 12 kilometers.

a. If the ground-level temperature is represented by t, write a formula for the temperature, T, at an elevation of h meters above ground level.

b. If the temperature at ground level is 54°C, determine the temperature at a height of 1500 meters.

c. At what height would the temperature drop to 30°C?

d. Solve the formula in part a for h in terms of T and t.

e. Use the formula in part d to determine the height when the ground-level temperature of 54°C has dropped to 30°C.

f. How does the result in part e compare with the result in part c?

5. The pressure, p, of water (in pounds per square foot) at a depth of d feet below the surface is given by the formula

$$p = 15 + \frac{15}{33}d.$$

a. On November 14, 1993, Francisco Ferreras reached a record depth for breath-held diving. During the dive, he experienced a pressure of 201 pounds per square foot. What was his record depth?

b. Solve the equation $p = 15 + \dfrac{15}{33} d$ for d.

c. Use the formula from part b to determine the record depth, and compare your answer with the one you obtained in part a.

Solve each of the following formulas for the specified variable.

6. $E = IR$, for I

7. $C = 2\pi r$, for r

8. $P = 2a + b$, for b

9. $P = 2l + 2w$, for w

10. $R = 143 - 0.65a$, for a

11. $A = P + Prt$, for r

12. $y = mx + b$, for m

13. $m = g - vt^2$, for g

Activity 2.8

Four out of Five Dentists Prefer Crest

Objectives

1. Recognize that equivalent fractions lead to proportions.

2. Use proportions to solve problems involving ratios and rates.

Manufacturers of retail products often conduct surveys to see how well their products are selling compared with competing products. In some cases, they use the results in advertising campaigns.

One company, Procter & Gamble, ran a TV ad in the 1970s claiming that "four out of five dentists surveyed preferred Crest toothpaste over other leading brands." The ad became a classic, and the phrase *four out of five prefer* has become a popular cliché in advertising, in appeals, and in one-line quips.

1. Suppose Procter & Gamble asked 250 dentists what brand toothpaste they preferred and 200 dentists responded that they preferred Crest.

   a. What is the ratio of the number of dentists who preferred Crest to the number who were asked the question? Write your answer in words and as a fraction.

   b. Reduce the ratio in part a to lowest terms, and write the result in words.

Note that part b of Problem 1 shows that the ratio $\frac{200}{250}$ is equivalent to the ratio $\frac{4}{5}$, or $\frac{200}{250} = \frac{4}{5}$. This is an example of a **proportion**.

> ### Definition
>
> The mathematical statement that two ratios are equal is called a **proportion**. In fraction form, a proportion is written $\frac{a}{b} = \frac{c}{d}$.

If one of the four numbers a, b, c, or d in a proportion is unknown and the other three are known, the equation can be solved algebraically for the unknown number.

2. Suppose that Procter & Gamble interviewed 600 dentists and reported that $\frac{4}{5}$ of the 600 dentists preferred Crest. However, the report did not say how many dentists in this survey preferred Crest.

   a. Let x represent the number of dentists who preferred Crest. Write a ratio of the number of dentists who preferred Crest to the number of dentists who were in the survey.

   b. According to the report, the ratio in part a must be equivalent to $\frac{4}{5}$. Write a corresponding proportion.

> To solve a proportion, multiply both sides of the equation by both denominators. This will clear the equation of fractions, leaving an equation you have already learned to solve. Example 1 demonstrates the procedure.

Example 1   *Solve the proportion $\frac{2}{3} = \frac{x}{6000}$.*

SOLUTION

$$6000 \cdot \cancel{3} \cdot \frac{2}{\cancel{3}} = \frac{x}{\cancel{6000}} \cdot \cancel{6000} \cdot 3$$ Multiply both sides of the equation by 6000 and 5

$$12{,}000 = 3x$$ Solve for x

$$400 = x$$

The first step in this solution to a proportion is sometimes called **cross multiplication**. It refers to the fact that $\frac{a}{b} = \frac{c}{d}$ is equivalent to $ad = bc$, so can be used as a shortcut.

3. Use cross multiplication to solve the proportion in Problem 2b.

4. The best season for the New York Mets baseball team was 1986. They won $\frac{2}{3}$ of the games they played and won the World Series, beating the Boston Red Sox. If the Mets played 162 games, how many did they win?

Example 2   *The Los Angeles Dodgers were once the Brooklyn Dodgers team that made it to the World Series seven times from 1941 to 1956, each time opposing the New York Yankees. The Brooklyn Dodgers won only one World Series Championship against the Yankees, and that was in 1955.*

During the 1955 regular season, the Dodgers won about 16 out of every 25 games they played. They won a total of 98 games. How many games did they play during the regular season?

SOLUTION

Let x represent the total number of games the Dodgers played in the 1955 regular season. To write a proportion, the units of the numerators in the proportion must be the same. Similarly, the units of the denominators must be the same. In this case, the proportion is

$$\frac{16 \text{ games won}}{25 \text{ total games played}} = \frac{98 \text{ games won}}{x \text{ total games played}} \quad \text{or} \quad \frac{16}{25} = \frac{98}{x}$$

$$\frac{16}{25} = \frac{98}{x}$$

$$16x = 98 \cdot 25$$

$$x = \frac{98 \cdot 25}{16} \approx 153$$

The Dodgers played 153 games in the 1955 season.

5. During the economic recession, a company was forced to lay off 12 workers, representing $\frac{3}{8}$ of its employees. How many workers had been employed before these layoffs?

Additional Applications

6. As a volunteer for a charity, you were given a job stuffing envelopes for an appeal for donations. After stuffing 240 envelopes, you were informed that you are two-thirds done. How many total envelopes are you expected to stuff?

7. In an effort to increase the education level of their police officers, many municipalities are requiring new recruits to have at least a two-year college degree. A recent survey indicated that 2 out of 5 Houston police officers hold a four-year college degree. There were approximately 3400 officers when the survey was conducted. Determine the number of four-year college degree holders.

8. To estimate the size of the grizzly bear population in a national park, rangers tagged and released 12 grizzly bears into the park. A few months later, 2 out of 21 grizzly bears sighted had tags. Assume this ratio is true for the entire population of grizzly bears in the park. Solve a proportion to estimate the number of grizzly bears in the park.

9. Solve each proportion for x.

a. $\dfrac{7}{10} = \dfrac{x}{24}$

b. $\dfrac{9}{x} = \dfrac{6}{50}$

c. $\dfrac{2}{3} = \dfrac{x}{48}$

d. $\dfrac{5}{8} = \dfrac{120}{x}$

e. $\dfrac{3}{20} = \dfrac{x}{3500}$

SUMMARY: ACTIVITY 2.8

1. The mathematical statement that two ratios are equal is called a **proportion**. In fraction form, a proportion is written as $\dfrac{a}{b} = \dfrac{c}{d}$.

2. The statement $\dfrac{a}{b} = \dfrac{c}{d}$ can be rewritten as $a \cdot d = b \cdot c$. This second statement can be obtained through the process of cross multiplication and the proportion can then be solved using familiar algebraic methods.

EXERCISES: ACTIVITY 2.8

1. A company that offers management solutions to companies that sell online announced in a recent consumer survey that nearly 9 out of 10 customers reported problems with transactions online. The survey sampled 2010 adults in the United States, 18 years and older, who had conducted an online transaction in the past year. Estimate the number of adults who reported problems with transactions online.

2. Solve each proportion for x.

a. $\dfrac{2}{9} = \dfrac{x}{108}$

b. $\dfrac{8}{7} = \dfrac{120}{x}$

c. $\dfrac{x}{20} = \dfrac{70}{100}$

3. The adult hematology/oncology unit at Ruby Memorial Hospital in West Virginia has a bed capacity of 20 with a minimum nurse-to-patient ratio of 1 to 5. How many nurses must be on staff in this unit?

4. The cardiology, cardiothoracic surgery, and vascular surgery unit has a bed capacity of 56 with 13 floor and 43 step-down beds. The minimum nurse-to-patient ratio is 1 to 5 for the floor and 1 to 3 for the step-down. How many nurses must be on staff in this unit?

5. When a house is sold through a real estate broker, the commission is usually 6% of the sale price, paid for by the seller. If the broker earned a commission of $21,000, what was the selling price? How much did the seller receive from the sale of her home?

6. In 2013, the Texas legislature passed a bill that reduced the number of required end-of-course examinations for high school students. The hope is that while the testing load will be reduced, graduation rates will increase and the achievement levels of the students will not be compromised. In your cousin's county last year, 8 out of 10 high school seniors actually graduated.

Exercise numbers appearing in color are answered in the Selected Answers appendix.

If 5400 students in your cousin's county graduated last year, determine the total number of high school seniors in that county last year.

7. Powdered skim milk sells for $6.99 a box in the supermarket. The amount in the box is enough to make 8 quarts of skim milk. What is the price for 12 quarts of skim milk prepared this way?

8. A person who weighs 120 pounds on Earth would weigh 42.5 pounds on Mars. What would be a person's weight on Mars if she weighs 150 pounds on Earth?

9. You want to make up a saline (salt) solution in chemistry lab that has 12 grams of salt per 100 milliliters of water. How many grams of salt will you use if you need 15 milliliters of solution for your experiment?

10. A conservationist can estimate the total number of a certain type of fish in a lake by using a technique called "capture-mark-recapture." Suppose a conservationist marks and releases 100 rainbow trout into a lake. A week later she nets 50 trout in the lake and finds 3 marked fish. An estimate of the number of rainbow trout in the lake can be determined by assuming that the proportion of marked fish in the sample taken is the same as the proportion of marked fish in the total population of rainbow trout in the lake. Use this information to estimate the population of rainbow trout in the lake.

11. The school taxes on a house assessed at $210,000 are $2340. At the same tax rate, what are the taxes (to the nearest dollar) on a house assessed at $275,000?

12. The designers of sport shoes assume that the force exerted on the soles of shoes in a jump shot is proportional to the weight of the person jumping. A 140-pound athlete exerts a force of 1960 pounds on his shoe soles when he returns to the court floor after a jump. Determine the force that a 270-pound professional basketball player exerts on the soles of his shoes when he returns to the court floor after shooting a jump shot.

13. A hybrid car can travel 45 miles on one gallon of gas. Determine the amount of gas needed for a 500-mile trip.

14. The ancient Greeks thought that the most pleasing shape for a rectangle was one for which the ratio of length to width was 8 to 5. This ratio is called the Golden Ratio. If the length of a rectangular painting is 20 inches, determine the width of the painting for the length and width of the painting to have the Golden Ratio.

15. Your college softball team won 80% of the games it played this year. If your team won 20 games, how many games did it play?

Cluster 2   What Have I Learned?

1. Describe how solving the equation $4x - 5 = 11$ for x is similar to solving the equation $4x - 5 = y$ for x.

2. In the formula $d = rt$, assume that the rate, r, is 60 miles per hour. The formula then becomes the equation $d = 60t$.

   a. Which variable is the input variable?

   b. Which is the output variable?

   c. Which variable from the original formula is now a constant?

   d. Discuss the similarities and differences in the equations $d = 60t$ and $y = 60x$.

3. Describe in words the sequence of operations that must be performed to solve for x in each of the following.

   a. $10 = x - 16$

   b. $-8 = \dfrac{1}{2}x$

   c. $-2x + 4 = -6$

   d. $ax - b = c$

4. The area, A, of a triangle is given by the formula $A = \dfrac{1}{2}bh$, where b represents the base and h represents the height. The formula can be rewritten as $b = \dfrac{2A}{h}$ or $h = \dfrac{2A}{b}$. Which formula would you use to determine the base, b, of a triangle, given its area, A, and height, h? Explain.

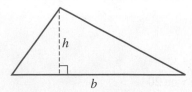

5. Explain how you would solve the proportion equation $\dfrac{5}{x} = \dfrac{35}{168}$.

Cluster 2   How Can I Practice?

1. a. Write a symbolic rule that represents the relationship in which the output variable y is 35 less than the input variable x.

b. What is the output corresponding to an input value of 52?

c. What is the input corresponding to an output value of 123?

2. a. Write a symbolic rule that represents the relationship in which the output variable t is 10 more than 2.5 times the input variable r.

b. What is the output corresponding to an input of 8?

c. What is the input corresponding to an output of -65?

3. Use an algebraic approach to solve each of the following equations for x. In each case, check your result in the original equation.

a. $x + 5 = 2$ **b.** $2x = -20$

c. $x - 3.5 = 12$ **d.** $-x = 9$

e. $13 = x + 15$ **f.** $4x - 7 = 9$

g. $10 = -2x + 3$ **h.** $\dfrac{3}{5}x - 6 = 1$

4. Your car needs a few new parts to pass inspection. The labor cost is $68 an hour, charged by the half hour, and the parts cost a total of $148. Whether you can afford these repairs depends on how long it will take the mechanic to install the parts.

a. Write a statement in words that will enable you to determine a total cost for repairs.

b. Write the symbolic form of the verbal rule in part a, letting x be the input variable and y be the output variable. What does x represent? (Include its units.) What does y represent? (Include its units.)

c. Use the symbolic rule in part b to create a table of values.

HOURS, x	TOTAL COST ($), y

d. How much will it cost if the mechanic works 4 hours?

e. You have $350 available in your budget for car repair. Determine whether you have enough money if the mechanic says that it will take him $3\frac{1}{2}$ hours to install the parts.

f. You decide that you can spend an additional $100. How long can you afford to have the mechanic work?

g. Solve the symbolic rule from part b for the input variable x. Why would it ever be to your advantage to do this?

5. In 1966, the U.S. Surgeon General's health warnings began appearing on cigarette packages. The following data seems to demonstrate that public awareness of the health hazards of smoking has had some effect on consumption of cigarettes.

	1997	1999	2001	2003	2005	2007	2009	2011
% of Total Population 18 And Older Who Smoke	24.7	23.5	22.8	21.6	20.9	19.8	20.6	19.0

Source: U.S. National Center for Health Statistics

The percent P of the total population 18 and older who smoke can be modeled by the formula

$$P = -0.38t + 24.3,$$

where t is the number of years since 1997.

a. Use the formula to predict in which year 18% of the total population 18 and older will smoke.

b. Use the formula to predict the percent of the total population 18 and older that will smoke in 2020.

6. Medical researchers have determined that for exercise to be beneficial, a person's desirable heart rate, R, in beats per minute, can be approximated by the formulas

$$R = 143 - 0.65a \text{ for women}$$
$$R = 165 - 0.75a \text{ for men,}$$

where a (years) represents the person's age.

a. If the desirable heart rate for a woman is 130 beats per minute, how old is she?

b. If the desirable heart rate for a man is 135 beats per minute, how old is he?

7. The basal energy rate is the daily amount of energy (measured in calories) needed by the body at rest to maintain body temperature and the basic life processes of respiration, cell metabolism, circulation, and glandular activity. As you may suspect, the basal energy rate differs for individuals, depending on their gender, age, height, and weight. The formula for the basal energy rate for men is

$$B = 655.096 + 9.563W + 1.85H - 4.676A,$$

where B is the basal energy rate (in calories), W is the weight (in kilograms), H is the height (in centimeters), and A is the age (in years).

a. A male patient is 70 years old, weighs 55 kilograms, and is 172 centimeters tall. A total daily calorie intake of 1000 calories is prescribed for him. Determine whether he is being properly fed.

b. A man is 178 centimeters tall and weighs 84 kilograms. If his basal energy rate is 1500 calories, how old is the man?

8. Solve each of the following equations for the given variable.

a. $d = rt$, for r b. $P = a + b + c$, for b c. $A = P + Prt$, for r

d. $y = 4x - 5$, for x
          e. $w = \dfrac{4}{7}h + 3$, for h

9. Solve each proportion for the unknown quantity.

a. $\dfrac{2}{3} = \dfrac{x}{48}$ b. $\dfrac{5}{8} = \dfrac{120}{x}$

10. According to the local Public Utility Agency, 7 out of every 25 homes in the region are heated by electricity. At this rate, predict how many homes in a community of 12,000 are heated by electricity.

11. A normal 10-cubic centimeter specimen of human blood contains 1.2 grams of hemoglobin. How much hemoglobin would 16 cubic centimeters of the same blood contain?

12. The ratio of the weight of an object on Mars to the weight of an object on Earth is 0.4 to 1. How much would a 170-pound astronaut weigh on Mars?

13. According to industry reports, the total global sales of smartphones (in millions of units) can be modeled by the following graph, where t represents the number of years since 2008 and P represents global smartphone sales measured in millions of units.

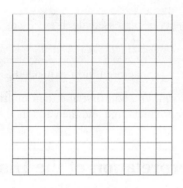

a. What is the input in this situation? What is the output?

b. Describe the general trend in the sale of smartphones.

c. Approximate the amount of global sales of smartphones in 2012.

d. Estimate during which year the global sales of smartphones reached 400 million units.

Cluster 3   More Problem Solving Using Algebra

Activity 2.9

Do It Two Ways

Objectives

1. Apply the distributive property.

2. Use areas of rectangles to interpret the distributive property geometrically.

3. Identify the greatest common factor in an expression.

4. Factor out the greatest common factor in an expression.

5. Recognize like terms.

6. Simplify an expression by combining like terms.

Distributive Property Revisited

1. You earn $10 per hour at your job and are paid every other week. You work 25 hours the first week and 15 hours the second week. Use two different approaches to compute your gross salary for the pay period. Explain in a sentence each of the approaches you used.

Problem 1 demonstrates the **distributive property** of multiplication over addition. In the problem, the distributive property asserts that adding the hours first and then multiplying the sum by $10 produces the same gross salary as does multiplying separately each week's hours by $10 and then adding the weekly salaries.

> The **distributive property** is expressed algebraically as
>
> $$a \cdot (b + c) = a \cdot b + a \cdot c.$$
>
> $\underbrace{\hspace{2cm}}_{\text{factored form}}$ $\underbrace{\hspace{3cm}}_{\text{expanded form}}$
>
> Note that in factored form, you add first and then multiply. In the expanded form, you calculate the individual products first and then add the results.

Geometric Interpretation of the Distributive Property

The distributive property can also be interpreted geometrically. Consider the following diagram:

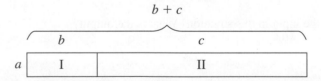

2. **a.** Write an expression for the area of rectangle I in the diagram.

   b. Write an expression for the area of rectangle II in the diagram.

   c. Write an expression for the area of the rectangle having width a and total top length $b + c$.

d. Explain in terms of the areas in the geometric diagram why $a(b + c)$ equals $ab + ac$.

The distributive property is frequently used to transform one algebraic expression into an equivalent expression.

> **Example 1** $2(x + 5)$ *can be transformed into* $2 \cdot x + 2 \cdot 5$*, which is usually written as* $2x + 10$**.**
>
> The *factored form*, $2(x + 5)$, indicates that you start with x, add 5, and then multiply by 2. The *expanded form*, $2x + 10$, indicates that you start with x, multiply by 2, and then add 10.

A few definitions (some you will recall from previous activities) will be helpful for understanding Example 1 and all that follows.

- A **term** is either a number or a product of a number and one or more variables. Terms can be easily identified as the parts of an algebraic expression that are added or subtracted.

  For example, the expression $13x + y - 7$ contains the three terms $13x$, y, and -7. Notice that the subtraction of the last term should be understood as addition of its opposite.

- When two or more mathematical expressions are multiplied to form a product, each of the original expressions is called a **factor** of the product.

  For example, the product $13x$ contains the two factors 13 and x.
  The product $5(x + 2)(x + 4)$ contains the three factors 5, $x + 2$, and $x + 4$.

- A numerical factor that multiplies a variable term is called the **coefficient** of the variable.

  For example, in the expression $13x + y - 7$, the coefficient of the first term is 13. The coefficient of the second term is understood to be 1. Because the third term has no variable factor, we simply call it a **constant** term.

The distributive property transforms a product such as $2(x + 5)$ into a sum of two terms $2x + 10$.

Using the Distributive Property to Expand a Factored Expression

There are two ways in which you can visualize the process of writing an expression such as $4(3x - 5)$ in expanded form using the distributive property.

First, you can make a diagram that looks similar to a rectangular area problem. (See below.) Place the factor 4 on the left of the diagram. Place the terms of the expression $3x - 5$ along the top. Multiply each term along the top by 4, and then add the resulting products. Note that the terms do not actually represent lengths of the sides of the rectangle. The diagram is an organizational tool to help you apply the distributive property properly.

	$3x$	-5
4	$12x$	-20

Therefore, $4(3x - 5) = 12x - 20$.

Second, you can draw arrows to emphasize that each term within the parentheses is multiplied by the factor 4:

$$4(3x - 5) = 4(3x) - 4(5) = 12x - 20$$

3. Use the distributive property to multiply and simplify each of the following expressions, writing in expanded form.

   a. $5(x + 6)$

   b. $-10x(y + 11)$

   c. Note: A negative sign preceding parentheses indicates multiplication by -1. Use this fact to expand $-(2x - 7)$.

   d. $3x(4 - 2x)$

Extension of the Distributive Property

The distributive property can be extended to sums of more than two terms within the parentheses. For example, it can be used to multiply $5(3x - 2y + 6)$.

To help you apply the distributive property, you can use the diagram approach

	$3x$	$-2y$	6
5	$15x$	$-10y$	30

or you can use the arrows approach,

$$5(3x - 2y + 6) = 5 \cdot 3x + 5 \cdot (-2y) + 5 \cdot 6 = 15x - 10y + 30.$$

4. Describe in words the procedure used to multiply $5(3x - 2y + 6)$.

5. Use the distributive property to write each of the expressions in expanded form.

   a. $-3(2x^2 - 4x - 5)$

   b. $2x(3a + 4b - x)$

Using the Distributive Property to Factor an Expanded Expression

Consider the expression $5 \cdot x + 5 \cdot 3$, whose two terms both contain the common factor 5. The distributive property allows you to rewrite this expression in factored form by

   i. dividing each term by the common factor to remove it from the term

   ii. placing the common factor outside parentheses that contain the sum of the remaining factors

Therefore, $5 \cdot x + 5 \cdot 3$ can be written equivalently as $5 \cdot (x + 3)$.

The challenge in transforming an expanded expression into factored form is identifying the common factor.

6. Identify the common factor in each of the following expressions, and then rewrite the expression in factored form.

   a. $2a + 6$

   b. $5x + 3x - 7x$

   c. $3x + 12$

   d. $2xy - 5x$

Greatest Common Factor

A common factor is called a **greatest common factor** if there are no additional factors other than common to the terms in the expression. In the expression $12x + 30$, the numbers 2, 3, and 6 are common factors, but 6 is the greatest common factor. When an expression is written in factored form and the remaining terms in parentheses have no factors in common, the expression is said to be in **completely factored form**. Therefore, $12x + 30$ is written in completely factored form as $6(2x + 5)$.

7. a. What is the greatest common factor of $8x + 20$? Rewrite the expression in completely factored form.

   b. What are common factors in the sum $10x + 6x$? What is the greatest common factor? Rewrite the expression as an equivalent product.

Procedure

Factoring a Sum of Terms Containing a Common Factor

1. Identify the common factor.

2. Divide each term of the sum by the common factor. (Factor out the common factor.)

3. Place the sum of the remaining factors inside the parentheses, and place the common factor outside the parentheses.

8. Factor each of the following completely by factoring out the greatest common factor.

   a. $6x + 18y - 24$

   b. $5x^2 - 3x^2$

   c. $9x - 12x + 15x$

Like Terms

Consider the expression $2x + 5x + 3x$, in which each term contains the identical variable factor x. Because of this common variable factor, the three terms are called like terms. They differ only by their numerical coefficients.

Definition

Like terms are terms that contain identical variable factors, including exponents.

Example 2

a. $4x$ and $6x$ are like terms.

b. $4xy$ and $-10xy$ are like terms.

c. x^2 and $-10x^2$ are like terms.

d. $4x$ and $9y$ are not like terms.

e. $-3x^2$ and $-3x$ are not like terms.

Combining Like Terms

The distributive property provides a way to simplify expressions containing like terms by combining the like terms into a single term. You can do this by factoring out the common variable factor, as follows:

$$2x + 5x + 3x = (2 + 5 + 3)x = 10x,$$

where the coefficient 10 in the simplified term is precisely the sum of the coefficients of the original like terms.

Like terms can be combined by adding their coefficients. This is a direct result of the distributive property.

Example 3 *15xy and $-8xy$ are like terms with coefficients 15 and -8, respectively. Thus, $15xy - 8xy = (15 - 8)xy = 7xy$.*

9. Identify the like terms, if any, in each of the following expressions, and combine them.

a. $3x - 5y + 2z - 2x$

b. $13s^2 + 6s - 4s^2$

c. $2x + 5y - 4x + 3y - x$

d. $3x^2 + 2x - 4x^2 - (-4x)$

When there is no coefficient written immediately to the left of a set of parentheses, the number 1 is understood to be the coefficient of the expression in parentheses. For example, $45 - (x - 7)$ can be understood as $45 - 1(x - 7)$. Therefore, you can use the distributive property to multiply each term inside the parentheses by -1 and then combine like terms.

$$45 - 1(x - 7) = 45 - x + 7 = 52 - x$$

10. Your mathematics instructor brings calculators to class each day. There are 30 calculators in the bag she brings. She never knows how many students will be late for class on a given day. Her routine is to remove 15 calculators and then to remove one additional calculator for each late arrival. If x represents the number of late arrivals on any given day, then the expression $30 - (15 + x)$ represents the number of calculators left in the bag after the late arrivals remove theirs.

    a. Simplify the expression for your instructor so that she can more easily keep track of her calculators.

    b. Suppose there are four late arrivals. Evaluate both the original expression and the simplified expression. Compare your two results.

11. Use the distributive property to simplify and combine like terms.

    a. $20 - (10 - x)$ **b.** $4x - (-2x + 3)$

    c. $2x - 5y - 3(5x - 6y)$ **d.** $2(x - 3) - 4(x + 7)$

SUMMARY: ACTIVITY 2.9

1. A **term** is either a number or a product of a number and one or more variables. Terms can be easily identified as the parts of an algebraic expression that are added or subtracted.

2. When two or more mathematical expressions are multiplied to form a product, each of the original expressions is called a **factor** of the product.

3. A numerical factor that multiplies a variable term is called the **coefficient** of the variable.

4. The **distributive property** is expressed algebraically as $a \cdot (b + c) = a \cdot b + a \cdot c$, where a, b, and c are any real numbers.

5. The **distributive property** can be extended to sums of more than two terms, as follows: $a \cdot (b + c + d) = a \cdot b + a \cdot c + a \cdot d$.

6. The process of rewriting the product $a(b + c)$ as the equivalent sum $ab + ac$ is called **expanding** the expression $a(b + c)$.

7. The process of rewriting the sum $ab + ac$ as the equivalent is product $a(b + c)$ called **factoring** the expression $ab + ac$.

8. A factor common to each term of an algebraic expression is called a **common factor**.

9. The process of dividing each term by a common factor and placing it outside the parentheses containing the sum of the remaining terms is called **factoring out a common factor**.

10. A common factor of an expression involving a sum of terms is called a **greatest common factor** if the terms remaining inside the parentheses have no factor in common other than 1.

11. An algebraic expression is said to be in **completely factored form** when it is written in factored form and none of its factors can be factored any further.

12. Procedure for factoring a sum of terms containing a common factor:

    a. Identify the common factor.

    b. Divide each term of the sum by the common factor. (Factor out the common factor.)

    c. Place the sum of the remaining factors inside the parentheses, and place the common factor outside the parentheses.

13. Like terms are terms that contain identical variable factors, including exponents. They differ only in their numerical coefficients.

14. To combine like terms of an algebraic expression, add or subtract their coefficients.

EXERCISES: ACTIVITY 2.9

1. Use the distributive property to expand the algebraic expression $10(x - 8)$. Then evaluate the factored form and the expanded form for these values of x: 5 and -3. What do you discover about these two algebraic expressions? Explain.

Use the distributive property to expand each of the algebraic expressions in Exercises 2–13.

2. $6(4x - 5)$

3. $-7(t + 5.4)$

4. $2.5(4 - 2x)$

5. $3(2x^2 + 5x - 1)$

6. $-(3p - 17)$

7. $-(-2x - 3y)$

8. $-3(4x^2 - 3x + 7)$

9. $-(4x + 10y - z)$

10. $\dfrac{5}{6}\left(\dfrac{3}{4}x + \dfrac{2}{3}\right)$

11. $-\dfrac{1}{2}\left(\dfrac{6}{7}x - \dfrac{2}{5}\right)$

12. $3x(5x - 4)$

13. $4a(2a + 3b - 6)$

14. Consider the algebraic expression $8x^2 - 10x + 9y - z + 7$.

    a. How many terms are in this expression?   **b.** What is the coefficient of the first term?

    c. What is the coefficient of the fourth term?   **d.** What is the coefficient of the second term?

e. What is the constant term?

f. What are the factors in the first term?

15. In chemistry, the ideal gas law is given by the equation $PV = n(T + 273)$, where P is the pressure, V, is the volume, T, is the temperature, and n, is the number of moles of gas. Write the right-hand side of the equation in expanded form (without parentheses).

16. In business, an initial deposit of P dollars, invested at a simple interest rate, r (in decimal form), will grow after t years, to amount A given by the formula

$$A = P(1 + rt).$$

Write the right side of the equation in expanded form (without parentheses).

17. **a.** The width, w, of a rectangle is increased by 5 units. Write an expression that represents the new width.

b. If l represents the length of the rectangle, write a product that represents the area of the new rectangle.

c. Use the distributive property to write the expression in part b in expanded form (without parentheses).

18. A rectangle has width w and length l. If the width is increased by 4 units and the length is decreased by 2 units, write a formula in expanded form that represents the perimeter of the new rectangle.

19. The manager of a local discount store reduces the regular retail price of a certain cell phone brand by $5.

a. Use y to represent the regular retail price, and write an expression that represents the discounted price of the cell phone.

b. If 12 of these phones are sold at the reduced price, write an expression in expanded form (without parentheses) that represents the store's total receipts for the 12 phones.

20. Identify the greatest common factor and rewrite the expression in completely factored form.

a. $3x + 15$

b. $5w - 10$

c. $3xy - 7xy + xy$

d. $6x + 20xy - 10x$

e. $4 - 12x$

f. $2x^2 + 3x^2y$

g. $4srt^2 - 3srt^2 + 10st$

h. $10abc + 15abd + 35ab$

21. a. How many terms are in the expression $2x^2 + 3x - x - 3$ as written?

   b. How many terms are in the simplified expression $2x^2 + 2x - 3$?

22. Are $3x^2$ and $3x$ like terms? Explain.

23. Simplify the following expressions by combining like terms.

   a. $5a + 2ab - 3b + 6ab$

   b. $3x^2 - 6x + 7$

   c. $100r - 13s^2 + 4r - 18s^3$

   d. $3a + 7b - 5a - 10b$

   e. $2x^3 - 2y^2 + 4x^2 + 9y^2$

   f. $7ab - 3ab + ab - 10ab$

   g. $xy^2 + 3x^2y - 2xy^2$

   h. $9x - 7x + 3x^2 + 5x$

   i. $2x - 2x^2 + 7 - 12$

   j. $3mn^3 - 2m^2n + m^2n - 7mn^3 + 3$

   k. $5r^2s - 6rs^2 + 2rs + 4rs^2 - 3r^2s + 6rs - 7$

24. Simplify the expression $2x - (4x - 8)$.

25. Use the distributive property, and then combine like terms to simplify the expressions.

   a. $30 - (x + 6)$

   b. $18 - (x - 8)$

   c. $2x - 3(25 - x)$

   d. $27 - 6(4x + 3y)$

   e. $12.5 - (3.5 - x)$

   f. $3x + 2(x + 4)$

   g. $x + 3(2x - 5)$

   h. $4(x + 2) + 5(x - 1)$

   i. $7(x - 3) - 2(x - 8)$

   j. $11(0.5x + 1) - (0.5x + 6)$

   k. $2x^2 - 3x(x + 3)$

   l. $y(2x - 2) - 3y(x + 4)$

26. You will be entering a craft fair with your latest metal wire lawn ornament. It is metal around the exterior and hollow in the middle, in the shape of a bird, as illustrated. However, you can produce different sizes, and all will be in proportion, depending on the length of the legs, x.

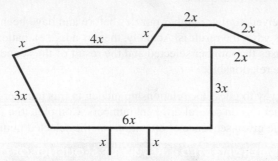

a. Write an algebraic expression that represents the amount of wire you need to create the bird. Be sure to include its legs.

b. Simplify the expression by combining like terms.

c. The amount of wire you need for each lawn ornament is determined from the expression in part b plus an extra 2 inches for the eye. If you have a spool containing 400 inches of metal wire, write an expression showing how much wire you will have left after completing one ornament.

d. Exactly how much wire is needed for an ornament whose legs measure 3 inches? How much wire will be left on the spool?

Activity 2.10

Decoding

Objectives

1. Recognize an algebraic expression as a code of instruction.

2. Simplify algebraic expressions.

You receive an e-mail from your cousin that contains the following puzzle.

> Select a number. Multiply the number by 4. Add 12 to the product. Divide the sum by 2. Subtract 6 from the quotient.

You have received similar number puzzles before and have been curious as to how a person always knows what the result is. Actually, there is a hidden pattern in the sequence of operations that causes the number selected and the result of the sequence of operations to always have the same relationship.

1. a. One way to guess the relationship hidden in this number trick is to use the code of instructions on several different numbers. Complete the following table by performing the given sequence of operations on the selected numbers.

NUMBER SELECTED	RESULT OF SEQUENCE OF OPERATIONS
2	
5	
10	
15	

b. A hidden pattern in the sequence of operations causes the original number and the result of the sequence of operations always to have the same relationship. What relationship do you observe between the number and the result?

> Based on the specific examples in part a, you might conclude that if you follow the given sequence of operations on *any* number, the result will always be twice the original number. Arriving at a general conclusion from specific examples is a type of reasoning called **inductive reasoning** or **induction**. However, this type of reasoning can often be misleading and cannot be used to prove that the general conclusion is true for all numbers.

2. a. To show how and why this trick works, you generalize the situation by choosing a variable to represent the number selected. Use x as your variable, and translate the first instruction into an algebraic expression. Then simplify it by removing parentheses (if necessary) and combining like terms.

b. Now use the result from part a and translate the second instruction into an algebraic expression, and simplify where possible. Continue until you complete all the steps.

c. Use your result from part b to interpret in words the relationship between the number selected and the result. Also explain how to obtain the original number.

> The type of reasoning used in Problem 2 is called **deductive reasoning** or **deduction**. Deductive reasoning is used to prove conjectures true or false. The deductive reasoning process is the tool used to assess the truth of every proposition in all branches of mathematics and logic.

3. Rather than simplifying after each step, you can wait until after you have written a single algebraic expression using all the steps. As in Problem 2, the resulting expression will represent the algebraic code for the number trick.

    a. Write all the steps for this number trick as a single algebraic expression without simplifying at each step.

    b. Simplify the expression you obtained in the final step of part a. How does this simplified expression compare with the result you obtained in Problem 2b?

4. Here is another trick you can try on a friend. The instructions for your friend are as follows:

Think of a number.

Double it.

Subtract 5.

Multiply by 3.

Add 9.

Divide by 6.

Add 1.

Tell me your result.

    a. Try this trick with a friend several times. Record the result of your friend's calculations and the numbers he or she originally selected.

        i. Start with 5:

        ii. Start with 8:

        iii. Start with 10:

    b. Explain how you know your friend's numbers.

    c. Show how your trick works no matter what number your friend chooses. Use an algebraic approach as you did in either Problem 2 or Problem 3.

5. Simplify the algebraic expressions. Simplify inside the brackets first by applying the distributive property and combining like terms.

    a. $[3 - 2(x - 3)] \div 2 + 9$

b. $[3 - 2(x - 1)] - [-4(2x - 3) + 5] + 4$

c. $\dfrac{2(x - 6) + 8}{2} + 9$

SUMMARY: ACTIVITY 2.10

Procedure for simplifying algebraic expressions:

1. Simplify the expression from the innermost parentheses outward using the order of operations convention.

2. Use the distributive property when it applies.

3. Combine like terms.

EXERCISES: ACTIVITY 2.10

1. You want to boast to a friend about the stock that you own without telling him how much money you originally invested in the stock. Let your original stock value be represented by x dollars. You watch the market once a month for 4 months and record the following.

Month	1	2	3	4
Stock Value	Increased $50	Doubled	Decreased $100	Tripled

a. Use x to represent the value of your original investment, and write an algebraic expression to represent the value of your stock after the first month.

b. Use the result from part a to determine the value at the end of the second month, simplifying when possible. Continue until you determine an expression that represents the value of your stock at the end of the fourth month.

c. Do you have good news to tell your friend? Explain to him what happened to the value of your stock over 4 months.

Exercise numbers appearing in color are answered in the Selected Answers appendix.

d. Instead of simplifying expressions after each step, write a single algebraic expression that represents the 4-month period.

e. Simplify the expression in part d. How does the simplified algebraic expression compare with the result in part b?

2. a. Write a single algebraic expression for the following sequence of operations. Begin with a number represented by n.

Multiply by -2.

Add 4.

Divide by 2.

Subtract 5.

Multiply by 3.

Add 6.

b. Simplify the algebraic expression $3(-n + 2 - 5) + 6$.

3. Show how you would simplify the expression $2\{3 - [4(x - 7) - 3] + 2x\}$.

4. Simplify the following algebraic expressions.

a. $5x + 2(4x + 9)$　　　　　　　　**b.** $2(x - y) + 3(2x + 3) - 3y + 4$

c. $-(x - 4y) + 3(-3x + 2y) - 7x$　　　**d.** $2 + 3[3x + 2(x - 3) - 2(x + 1) - 4x]$

e. $6[3 + 2(x - 5)] - [2 - (x + 1)]$

f. $\dfrac{5(x - 2) - 2x + 1}{3}$

5. a. Complete the following table by performing the following sequence of operations on the input value to produce a corresponding output value: Multiply the input by 3, add 8 to the product, subtract the input, divide by 2, and subtract 4 from the quotient.

INPUT VALUE	OUTPUT VALUE
2	
4	
10	
15	

b. There is a direct connection between each input number and corresponding output obtained by performing the sequence of operations in part a. Describe it.

c. Confirm your observation in part c algebraically by performing the sequence of operations given in part a. Use the variable x to represent the input. Simplify the resulting algebraic expression.

d. Does your result in part c confirm the pattern you observed in part a?

6. In your part-time job, your hours vary each week. Last month you worked 22 hours the first week, 25 hours the second week, 14 hours the third week, and 19 hours the fourth week. Your gross pay for those four weeks was $960.

a. Let p represent your hourly pay rate. Write the equation for your gross pay over these four weeks.

b. Solve the equation in part a to determine your hourly pay rate.

7. You are a sales associate in a large retail electronics store. Over the first three months of the year, you sold 12, 15, and 21 ProPix digital cameras. The total in sales was $18,960.

    a. Let p represent the price of one ProPix digital camera. Write the equation for your total in sales over these three months.

    b. Solve the equation in part a to find the retail price of the ProPix digital camera.

Activity 2.11

Comparing Energy Costs

Objectives

1. Develop mathematical models to solve problems.

2. Write and solve equations of the form $ax + b = cx + d$, where $a \neq 0$ and $c \neq 0$.

3. Use the distributive property to solve equations involving grouping symbols.

4. Solve formulas for a specified variable.

An architect is hired to design a 3000-square-foot home. She obtains installation and annual operating costs for two types of heating and cooling systems: solar and electric.

TYPE OF HEATING AND COOLING SYSTEM	INSTALLATION COST	OPERATING COST PER YEAR
Solar	$40,000	$100
Electric	$9000	$1200

You will use this information to compare the cost for each system over a period of years. The following questions will guide you in making this comparison.

1. a. If you select the solar system, you will be eligible to receive 30% off the total installation cost from the federal government in the form of a Federal Solar Tax Credit. Determine the amount of credit.

b. What is the new installation cost of the solar system after the credit is applied?

2. a. Determine the total cost of the solar heating system after 5 years of use.

b. Write a verbal rule that describes how to calculate the total cost of the solar heating system in terms of the number of years of use.

c. Let x represent the number of years of use and c represent the total cost of the solar heating system. Translate the statement in part b into a symbolic rule.

d. Use the symbolic rule from part c to complete the following table.

Number of Years in Use, x	5	10	15	20
Total Cost, c				

3. a. Determine the total cost of the electric heating system after 5 years of use.

b. Write a verbal rule that describes how to calculate the total cost of the electric heating system in terms of the number of years of use.

c. Let x represent the number of years of use and c represent the total cost of the electric heating system. Translate the statement in part b into a symbolic rule.

d. Use the symbolic rule from part c to complete the following table:

Number of Years in Use, x	5	10	15	20
Total Cost, c				

The installation cost of solar heating is more than that of the electric system, but the operating cost per year of the solar system is much lower. Therefore, it is reasonable to think that the total cost for the electric system will eventually catch up and surpass the total cost of the solar system.

4. Compare the table values for total heating costs in Problems 2d and 3d. Estimate in what year the total cost for electric heating will catch up and surpass the total cost for solar heating. Explain.

5. The year in which the total costs of the two heating systems are equal can be determined algebraically. Write an equation you can solve to determine when the total heating costs are the same by setting the expressions in the symbolic rules in Problems 2c and 3c equal to each other.

The equation in Problem 5 is challenging to solve because the variable appears on both sides of the equation. By subtracting one of the variable terms from both sides of the equation, you will be able to combine like terms and proceed to solve the equation as usual.

Example 1 illustrates this procedure in solving a similar equation.

Example 1   *Solve for x:* $2x + 14 = 8x + 2.$

SOLUTION

$$2x + 14 = 8x + 2$$ Subtract $8x$ from both sides, and combine like terms.
$$\underline{-8x \qquad\quad -8x}$$
$$-6x + 14 = 2$$
$$\underline{\qquad\quad -14 = -14}$$ Subtract 14 from each side, and combine like terms.
$$-6x \qquad = -12$$
$$\frac{-6x}{-6} = \frac{-12}{-6}$$ Divide each side by -6, the coefficient of x.
$$x = 2$$

Check: $2(2) + 14 = 8(2) + 2$
$$4 + 14 = 16 + 2$$
$$18 = 18$$

6. In Example 1, the variable terms are combined on the left side of the equation. To solve the equation $2x + 14 = 8x + 2$, combine the variable terms on the right side. Does it matter on which side you choose to place the variable terms?

7. a. Solve the equation in Problem 5 for x.

   b. Interpret what your answer in part a represents in the context of the two heating systems.

8. Your architect tells you that instead of an annual operating expense of $100 you may actually realize a net surplus of $500 per year by using the solar system.

   a. Assuming this net surplus, write a new symbolic rule that describes the total cost of the solar heating system in terms of the number of years of use.

   b. Write and solve a new equation to determine when the total heating costs are the same.

Purchasing a Car

You are interested in purchasing a new car and have narrowed the choice to a Honda Accord LX (4 cylinder) and a Passat GLS (4 cylinder). Being concerned about the value of the car depreciating over time, you search the Internet and obtain the following information.

MODEL OF CAR	MARKET SUGGESTED RETAIL PRICE (MSRP) ($)	ANNUAL DEPRECIATION ($)
Accord LX	20,925	1730
Passat GLS	24,995	2420

9. a. Complete the following table.

YEARS THE CAR IS OWNED	VALUE OF ACCORD LX ($)	VALUE OF PASSAT GLS ($)
1		
2		
3		

   b. Will the value of the Passat GLS ever be lower than the value of the Accord LX? Explain.

   c. Let v represent the value of the car after x years of ownership. Write a symbolic rule to determine v in terms of x for the Accord LX.

d. Write a symbolic rule to determine v in terms of x for the Passat GLS.

e. Write an equation to determine when the value of the Accord LX will equal the value of the Passat GLS.

f. Solve the equation in part e.

NBA Basketball Court

You and your friend are avid professional basketball fans and discover that your mathematics instructor shares your enthusiasm for basketball. During a mathematics class, your instructor tells you that the perimeter of an NBA basketball court is 288 feet and the length is 44 feet more than its width. He challenges you to use your algebra skills to determine the dimensions of the court. To solve the problem, you and your friend use the following plan.

10. a. Let w represent the width of the court. Write an expression for the length in terms of the width, w.

b. Use the formula for the perimeter of a rectangle, $P = 2l + 2w$, and the information given to obtain an equation containing just the variable w.

c. To solve the equation you obtained in part b, apply the distributive property and then combine like terms in the expression involving w.

d. What are the dimensions of an NBA basketball court?

11. Solve each of the following formulas for the specified variable.

a. $P = 2l + 2w$ for w

b. $V(P + a) = k$ for P

c. $w = 110 + \dfrac{11}{2}(h - 60)$ for h

12. Solve each of the following equations for x. Remember to check your result in the original equation.

a. $2x + 9 = 5x - 12$ **b.** $21 - x = -3 - 5x$

c. $2(x - 3) = -8$ **d.** $2(x - 4) + 6 = 4x - 7$

SUMMARY: ACTIVITY 2.11

General strategy for solving equations for an unknown quantity, such as x:

1. If necessary, apply the distributive property to remove parentheses.

2. Combine like terms that appear on the same side of the equation.

3. Isolate x so that only a single term in x remains on one side of the equation and all other terms, not containing x, are moved to the other side. This is generally accomplished by adding (or subtracting) appropriate terms to both sides of the equation.

4. Solve for x by dividing each side of the equation by its coefficient.

5. Check your result by replacing x with this value in the original equation. A correct solution will produce a true statement.

EXERCISES: ACTIVITY 2.11

1. You need to rent a van, so you contact two local rental companies and acquire the following information for the one-day cost of renting a van.

   Company 1: $60 per day, plus $0.75 per mile

   Company 2: $30 per day, plus $1.00 per mile

   Let x represent the number of miles driven in one day and C represent the total daily rental cost ($).

   a. Write an equation that represents the total cost of renting the van for one day from company 1.

   b. Write an equation that represents the total cost of renting the van for one day from company 2.

   c. Write an equation to determine for what mileage the one-day rental cost would be the same.

   d. Solve the equation you obtained in part c.

   e. For which mileage does company 2 have the lower price?

2. You are considering installing a basic security system in your new house. You gather the following information from the Internet about similar security systems from two local home security dealers.

   Lease: $99 to install and $35 per month monitoring fee

   Purchase: $450 to purchase equipment and install and $15 per month for monitoring

   You see that the initial cost of purchasing the system is much higher than leasing it, but the monitoring fee is lower. You want to determine which is the better system for your needs.

   Let x represent the number of months you have the security system and C represent the cumulative cost ($).

   a. Write an equation that represents the total cost of leasing the system in terms of the number of months you have the system.

   b. Write an equation that represents the total cost of purchasing the system in terms of the number of months you have the system.

   c. Write an equation to determine the number of months for which the total cost of the systems will be equal.

Exercise numbers appearing in color are answered in the Selected Answers appendix.

d. Solve the equation in part c.

e. If you plan to live in the house and use the system for 5 years, which system will be less expensive?

3. You get three summer jobs to help you save for college expenses. In your job as a cashier, you work 20 hours per week and earn $9.50 per hour. Your second and third jobs are at a local hospital. There, you earn $9.00 per hour as a payroll clerk and $7.00 per hour as an aide. You always work 10 hours less per week as an aide than you do as a payroll clerk. Your total weekly salary depends on the number of hours you work at each job.

a. Determine the input and output variables for this situation.

b. Explain how you calculate the total amount earned each week.

c. If x represents the number of hours you work as a payroll clerk, represent the number of hours you work as an aide in terms of x.

d. Write an equation that describes the total amount you earn each week. Use x to represent the input variable and y to represent the output variable. Simplify the expression as much as possible.

e. If you work 12 hours as a payroll clerk, how much will you make in one week?

f. What are the practical replacement values for x? Would 8 hours at your payroll job be a realistic replacement value? What about 50 hours?

g. When you don't work as an aide, what is your total weekly salary?

h. If you plan to earn a total of $505 in one week from all of the jobs, how many hours must you work at each job? Is the total number of hours worked realistic? Explain.

i. Solve the equation in part d for x in terms of y. When would it be useful to have the equation in this form?

4. A florist sells roses for $1.50 each and carnations for $0.85 each. Suppose you purchase a bouquet of 1 dozen flowers consisting of roses and carnations.

a. Let x represent the number of roses purchased. Write an expression in terms of x that represents the number of carnations purchased.

b. Write an expression that represents the cost of purchasing x roses.

c. Write an expression that represents the cost of purchasing the carnations.

d. What does the sum of the expressions in parts b and c represent?

e. Suppose you are willing to spend $14.75. Write an equation that can be used to determine the number of roses that can be included in a bouquet of 1 dozen flowers consisting of roses and carnations.

f. Solve the equation in part e to determine the number of roses and the number of carnations in the bouquet.

5. The viewing window of a certain calculator is in the shape of a rectangle.

a. Let w represent the width of the viewing window in centimeters. If the window is 5 centimeters longer than it is wide, write an expression in terms of w for the length of the viewing window.

b. Write a symbolic rule that represents the perimeter, P, of the viewing window in terms of w.

c. If the perimeter of the viewing window is 26 centimeters, determine the dimensions of the window.

Solve the equations in Exercises 6–19.

6. $5x - 4 = 3x - 6$

7. $3x - 14 = 6x + 4$

8. $0.5x + 9 = 4.5x + 17$

9. $4x - 10 = -2x + 8$

10. $0.3x - 5.5 = 0.2x + 2.6$

11. $4 - 0.025x = 0.1 - 0.05x$

12. $5t + 3 = 2(t + 6)$

13. $3(w + 2) = w - 14$

14. $21 + 3(x - 4) = 4(x + 5)$

15. $2(x + 3) = 5(2x + 1) + 4x$

16. $500 = 0.75x - (750 + 0.25x)$

17. $1.5x + 3(22 - x) = 70$

18. You are asked to grade some of the questions on a skills test. Here are five results you are asked to check. If an example is incorrect, find the error and show the correct solution.

a. $34 = 17 - (x - 5)$

$34 = 17 - x - 5$

$34 = 12 - x$

$22 = -x$

$x = -22$

b. $-47 = -6(x - 2) + 25$

$-47 = -6x + 12 + 25$

$-47 = -6x + 37$

$-6x = 84$

$x = 14$

c. $-93 = -(x - 5) - 13x$

$-93 = -x + 5 - 13x$

$-93 = -14x + 5$

$-14x = -98$

$x = 7$

d. $3(x + 1) + 9 = 22$

$3x + 4 + 9 = 22$

$3x + 13 = 22$

$3x = 9$

$x = 3$

e. $83 = -(x + 19) - 41$

$83 = -x - 19 - 41$

$83 = -x - 50$

$133 = -x$

$x = -133$

19. Explain how to solve for y in terms of x in the equation $2x + 4y = 7$.

20. Solve each formula for the specified variable.

a. $y = mx + b$, for x

b. $A = \dfrac{B + C}{2}$, for B

c. $A = 2\pi r^2 + 2\pi rh$, for h

d. $F = \dfrac{9}{5}C + 32$, for C

e. $3x - 2y = 5$, for y

f. $12 = -x + \dfrac{y}{3}$, for y

g. $A = P + Prt$, for P
(*Hint:* Factor first.)

h. $z = \dfrac{x - m}{s}$, for x

Project 2.12

Summer Job Opportunities

Objective

1. Use problem-solving skills to make decisions based on solutions of mathematical models.

It can be very difficult keeping up with college expenses, so it is important for you to find a summer job that pays well. Luckily, the classified section of your newspaper lists numerous summer job opportunities in sales, road construction, and food service. The advertisements for all these positions welcome applications from high school students. All positions involve the same period during the summer.

Sales

A new electronics store opened recently. Several sales associate positions pay an hourly rate of $7.50 plus a 5% commission based on total weekly sales. You would be guaranteed at least 30 hours of work per week, but not more than 40 hours.

Construction

Your state highway department hires college students every summer to help with road construction projects. The hourly rate is $11.75 with the possibility of up to 10 hours per week in overtime, for which you are paid time and a half. Of course, the work is dependent on good weather; so the number of hours you work per week can vary.

Restaurants

Local restaurants experience an increase in business during the summer. There are several positions for waitstaff. The hourly rate is $4.90, and the weekly tip total ranges from $200 to $750. You are told that you can expect a weekly average of approximately $450 in tips. You will be scheduled to work five dinner shifts of 6.5 hours each, for a total of 32.5 hours per week. However, on slow nights, you may be sent home early, perhaps after working only 5 hours. Thus, your total weekly hours may be fewer than 32.5.

All of the jobs would provide an interesting summer experience. Your personal preferences might favor one position over another. Keep in mind all of your college expenses.

1. At the electronics store, sales associates average $7000 in sales each week.

   a. Based on the expected weekly average of $7000 in sales, calculate your gross weekly paycheck (before any taxes or other deductions) if you worked a full 40-hour week in sales.

   b. Use the average weekly sales figure of $7000 to write an equation for your weekly earnings, s, where x represents the total number of hours you work.

   c. What will be your gross paycheck for the week if you worked 30 hours and still managed to sell $7000 in merchandise?

   d. You are told that you will typically work 35 hours per week if your total electronic sales do average $7000. Calculate your typical gross paycheck for a week.

   e. You calculate that to pay college expenses for the upcoming academic year, you need to gross at least $550 a week. How many hours must you work in sales each week? Assume that you would sell $7000 in merchandise.

2. In the construction job, you would average a 40-hour workweek.

   a. Calculate your gross paycheck for a typical 40-hour workweek.

   b. Write an equation for your weekly salary, s, for a week with no overtime. Let x represents the total number of hours worked.

   c. If the weather is ideal for a week, you can expect to work 10 hours of overtime (over and above the regular 40-hour workweek). Determine your total gross pay for a week with 10 hours of overtime.

   d. The equation in part b can be used to determine the weekly salary, s, when x, the total number of hours worked, is less than or equal to 40 (no overtime). Write an equation to determine your weekly salary, s, if x is greater than 40 hours.

   e. Suppose it turns out to be a gorgeous summer, and your supervisor tells you that you can work as many hours as you want. If you can gross $800 a week, you will be able to afford to buy a computer. How many hours must you work each week to achieve your goal?

3. The restaurant job involves working a maximum of five dinner shifts of 6.5 hours each.

   a. Calculate your gross paycheck for an exceptionally busy week of five 6.5-hour dinner shifts and $750 in tips.

   b. Calculate your gross paycheck for an exceptionally slow week of five 5-hour dinner shifts and only $200 in tips.

   c. Calculate what your gross paycheck would be for a typical week of five 6.5-hour dinner shifts and $450 in tips.

   d. Use $450 as your typical weekly total for tips, and write an equation for your gross weekly salary, s, where x represents the number of hours.

e. Calculate what your gross paycheck would be for a 27-hour week and $450 in tips.

f. During the holiday week of July 4, you would be asked to work an extra dinner shift. You are told to expect $220 in tips for that night alone. Assuming a typical workweek for the rest of the week, would working that extra dinner shift enable you to gross at least $850? Explain.

4. You would like to make an informed decision in choosing one of the three positions. Based on all the information you have about the three jobs, fill in the following table.

	LOWEST WEEKLY GROSS PAYCHECK	TYPICAL WEEKLY GROSS PAYCHECK	HIGHEST WEEKLY GROSS PAYCHECK
Sales Associate			
Construction Worker			
Waitstaff			

5. Money may be the biggest factor in making your decision. But it is summer, and it would be nice to enjoy what you are doing. Discuss the advantages and disadvantages of each position. What would your personal choice be? Why?

6. You decide that you would prefer an indoor job. Use the algebraic rules you developed for the sales job in Problem 1b and for the restaurant position in Problem 3d to calculate the number of hours you would have to work in each job to receive the same weekly salary.

Cluster 3   What Have I Learned?

1. a. Explain the difference between the two expressions $-x^2$ and $(-x)^2$. Use an example to illustrate your explanation.

   b. What role does the negative sign to the left of the parentheses play in simplifying the expression $-(x - y)$? Simplify this expression.

2. a. Are $2x$ and $2x^2$ like terms? Why or why not?

   b. A student simplified the expression $6x^2 - 2x + 5x$ and obtained $9x^2$. Is he correct? Explain.

3. Can the distributive property be used to simplify the expression $3(2xy)$? Explain.

4. a. Is the expression $2x(5y + 15x)$ completely factored? Explain.

   b. One classmate factored the expression $12x^2 - 18x + 12$ as $2(6x^2 - 9x + 6)$, and another factored it as $6(2x^2 - 3x + 2)$. Which is correct?

5. For extra credit on exams, your mathematics instructor permits you to locate and circle your errors and then correct them. To get additional points, you are to show all correctly worked steps alongside the incorrect ones. On your last math test, the following problem was completed incorrectly. Show the work necessary to obtain the extra points.

$$2x - 5 = x + 7$$
$$3x = 12$$
$$x = 4$$
$$2(x - 3) = 5x + 3x - 7(x + 1)$$
$$2x - 5 = 8x - 7x + 7$$

Exercise numbers appearing in color are answered in the Selected Answers appendix.

Cluster 3   How Can I Practice?

1. Complete the following table. Then determine numerically and algebraically which of the following expressions are equivalent.

   a. $13 + 2(5x - 3)$ **b.** $10x + 10$ **c.** $10x + 7$

x	13 + 2(5x − 3)	10x + 10	10x + 7
1			
5			
10			

2. Use the distributive property to expand each of the following algebraic expressions.

   a. $6(x - 7)$ **b.** $3x(x + 5)$ **c.** $-(x - 1)$

   d. $-2.4(x + 1.1)$ **e.** $4x(a - 6b - 1)$ **f.** $-2x(3x + 2y - 4)$

3. Write each expression in completely factored form.

   a. $5x - 30$ **b.** $6xy - 8xz$ **c.** $-6y - 36$

   d. $2xa - 4xy + 10xz$ **e.** $2x^2 - 6x$

4. Consider the expression $5x^3 + 4x^2 - x - 3$.

   a. How many terms are there?

   b. What is the coefficient of the first term?

   c. What is the coefficient of the third term?

   d. If there is a constant term, what is its value?

   e. What are the factors of the second term?

5. a. Let n represent the input. The output is described by the following verbal phrase.

   Six times the square of the input, decreased by twice the input and then increased by eleven

   Translate the above phrase into an algebraic expression.

   b. How many terms are in this expression?

   c. List the terms in the expression from part a.

6. Combine like terms to simplify the following expressions.

   a. $5x^3 + 5x^2 - x^3 - 3$

   b. $xy^2 - x^2y + x^2y^2 + xy^2 + x^2y$

   c. $3ab - 7ab + 2ab - ab$

7. For each of the following algebraic expressions, list the specific operations indicated, in the order in which they are to be performed.

   a. $10 + 3(x - 5)$
       i.
       ii.
       iii.

   b. $(x + 5)^2 - 15$
       i.
       ii.
       iii.

   c. $(2x - 4)^3 + 12$
       i.
       ii.
       iii.
       iv.

8. Simplify the following algebraic expressions.

   a. $4 - (x - 2)$

   b. $4x - 3(4x - 7) + 4$

   c. $x(x - 3) + 2x(x + 3)$

   d. $2[3 - 2(a - b) + 3a] - 2b$

   e. $3 - [2x + 5(x + 3) - 2] + 3x$

   f. $\dfrac{7(x - 2) - (2x + 1)}{5}$

9. You own 25 shares of a certain stock. The share price at the beginning of the week is x dollars. By the end of the week, the share's price doubles and then drops $3. Write a symbolic rule that represents the total value, V, of your stock at the end of the week.

10. A volatile stock began the last week of the year worth x dollars per share. The following table shows the changes during that week. If you own 30 shares, write a symbolic rule that represents the total value, V, of your stock at the end of the week.

Day	1	2	3	4	5
Change in Value/Share	Doubled	Lost 10	Tripled	Gained 12	Lost half its value

11. Physical exercise is most beneficial for burning fat when it increases a person's heart rate to a target level. The symbolic rule

$$T = 0.6(220 - a)$$

describes how to calculate an individual's target heart rate, T, measured in beats per minute, in terms of age, a, in years.

   a. Use the distributive property to rewrite the right-hand side of this symbolic rule in expanded form.

   b. Use both the factored and expanded rules to determine the target heart rate for an 18-year-old person during aerobics.

12. The manager of a clothing store decides to reduce the price of a leather jacket by $25.

   a. Use x to represent the regular cost of the jacket, and write an expression that represents the discounted price of the leather jacket.

   b. If eight jackets are sold at the reduced price, write an expression in factored form that represents the total receipts for the jackets.

   c. Write the expression in part b as an equivalent expression without parentheses (expanded form).

13. You planned a trip with your best friend from college. You had only 4 days for your trip and planned to travel x hours each day.

The first day, you stopped for sightseeing and lost 2 hours of travel time. The second day, you gained 1 hour because you did not stop for lunch. On the third day, you traveled well into the night and doubled your planned travel time. On the fourth day, you traveled only a fourth of the time you had planned because your friend was sick. You averaged 45 miles per hour for the first 2 days and 48 miles per hour for the last 2 days.

   a. How many hours, in terms of x, did you travel the first 2 days?

b. How many hours, in terms of x, did you travel the last 2 days?

c. Express the total distance, D, traveled over the 4 days as a symbolic rule in terms of x.
Simplify the rule.
Recall that distance = average rate · time.

d. Write a symbolic rule that expresses the total distance, y, you would have traveled had you
traveled exactly x hours each day at the average speeds indicated above. Simplify the rule.

e. If you had originally planned to travel 7 hours each day, how many miles did you actually
travel?

f. How many miles would you have gone had you traveled exactly 7 hours each day?

14. You read about a full-time summer position in sales at the Furniture Barn. The job pays $280 per
week plus 20% commission on sales over $1000.

a. Explain how you would calculate the total amount earned each week.

b. Let x represent the dollar amount of sales for the week. Write a symbolic rule that expresses
your earnings, E, for the week in terms of x.

c. Write an equation to determine how much furniture you must sell to have a gross salary of
$600 for the week.

d. Solve the equation in part c.

e. Is the total amount of sales reasonable?

15. As a prospective employee in an electronics store, you are offered a choice of salary. The following table shows your options.

Option 1	$200 per week	Plus 30% of all sales
Option 2	$350 per week	Plus 15% of all sales

a. Write a symbolic rule to represent the total salary, S, for option 1 if the total sales are x dollars per week.

b. Write a symbolic rule to represent the total salary, S, for option 2 if the total sales are x dollars per week.

c. Write an equation that you could use to determine how much you would have to sell in a week to earn the same salary under both plans.

d. Solve the equation in part c. Interpret your result.

e. What is the common salary for the amount of sales found in part d?

16. A triathlon includes swimming, long-distance running, and cycling.

a. Let x represent the number of miles the competitors swim. If the long-distance run is 10 miles longer than the distance swum, write an expression that represents the distance the competitors run in the event.

b. The distance the athletes cycle is 55 miles longer than they run. Use the result in part a to write an expression in terms of x that represents the cycling distance of the race.

c. Write an expression that represents the total distance of all three phases of the triathlon. Simplify the expression.

d. If the total distance of the triathlon is 120 miles, write and solve an equation to determine x. Interpret the result.

e. What are the lengths of the running and cycling portions of the race?

The bracketed numbers following each concept indicate the activity in which the concept is discussed.

CONCEPT/SKILL	DESCRIPTION	EXAMPLE
Variable [2.1]	A quantity or quality that may change, or vary, in value from one particular instance to another, usually represented by a letter. When a variable describes an actual quantity, its values must include the unit of measurement of that quantity.	The number of miles you drive in a week is a variable. Its value may (and usually does) change from one week to the next. x and y are commonly used to represent variables.
Input [2.1]	The value that is given first in an input/output relationship.	Your weekly earnings depend on the number of hours you work. There are two variables in this relationship: hours worked is the input;
Output [2.1]	The second number in an input/output relationship. It is the number that corresponds to or is matched with the input.	Total earnings is the output in the weekly earnings situation.
Replacement values for the input [2.1]	The replacement values for the input is the collection of all numbers for which a corresponding output value can be determined.	The amount of lawn you can mow depends on whether you can mow for 2, 3, or 4 hours. The set of input values is $\{2, 3, 4\}$.
Input/output relationship [2.1]	Relationship between two variables that can be represented numerically by a table of values (ordered pairs) or graphically as plotted points in a rectangular coordinate system.	<table><tr><th>NUMBER OF HOURS WORKED</th><th>EARNINGS</th></tr><tr><td>15</td><td>$90</td></tr><tr><td>25</td><td>$150</td></tr></table>
Algebraic expression [2.1]	An algebraic expression is a shorthand code for a sequence of arithmetic operations to be performed on a variable.	The algebraic expression $2x + 3$ indicates that you start with a value of x, multiply by 2, and then add 3.
Evaluate an algebraic expression [2.1]	To evaluate an algebraic expression, replace the variable(s) with its (their) assigned value(s) and perform the indicated arithmetic operation(s).	Evaluate $3x^2 - 2x + 4$ when $x = 2$. $3(2)^2 - 2(2) + 4$ $= 3 \cdot 4 - 4 + 4$ $= 12 - 4 + 4$ $= 12$
Constant [2.1]	A constant is a quantity that does not change in value within the context of a problem.	A number such as 2 or a symbol such as k that is understood to have a constant value in a problem.
Numerical coefficient [2.1]	A numerical coefficient is a number that multiplies a variable or an expression.	The number 8 in $8x$.
Terms [2.1]	Terms are the parts of an expression that are separated by plus or minus signs.	The expression $3x + 6y - 8z$ contains three terms.

CONCEPT/SKILL	DESCRIPTION	EXAMPLE
Horizontal axis [2.2]	In graphing an input/output relationship, the input is referenced on the horizontal axis.	Output
Vertical axis [2.2]	In graphing an input/output relationship, the output is referenced on the vertical axis.	Input
Rectangular coordinate system [2.2]	Allows every point in the plane to be identified by an ordered pair of numbers, the coordinates of the point, determined by the distance of the point from two perpendicular number lines (called coordinate axes) that intersect at their respective 0 values, the origin.	y — (x, y) 0 x
Scaling [2.2]	Setting the same distance between each pair of adjacent tick marks on an axis.	$-4\ -3\ -2\ -1\ \ 0\ \ 1\ \ 2\ \ 3\ \ 4$ Scale is 1
Point in the plane [2.2]	Points are identified by an ordered pair of numbers (x, y) in which x represents the horizontal distance from the origin and y represents the vertical distance from the origin.	$(2, 30)$ are the coordinates of a point in the first quadrant located 2 units to the right and 30 units above the origin.
Verbal rule [2.3]	A statement that describes in words the arithmetic relationship between the input and output variables.	Your payment (in dollars) for mowing a lawn is 8 times the number of hours worked.
Equation [2.3]	An equation is a statement that two algebraic expressions are equal.	$2x + 3 = 5x - 9$
Symbolic rule [2.3]	A symbolic rule is a mathematical statement that defines an output variable as an algebraic expression in terms of the input variable. The symbolic rule is essentially a recipe that describes how to determine the output value corresponding to a given input value.	The symbolic rule $y = 3x - 5$ indicates that the output, y, is obtained by multiplying x by 3 and then subtracting 5.
Equivalent expressions [2.4]	Equivalent expressions are two algebraic expressions that always produce the same output for identical inputs.	$4(x + 6)$ and $4x + 24$ are equivalent expressions. For any value of x, adding 6 and then multiplying by 4 *always* gives the same result as multiplying by 4 and then adding 24.

CONCEPT/SKILL	DESCRIPTION	EXAMPLE
Solution of an equation [2.5]	The solution of an equation is a replacement value for the variable that makes both sides of the equation equal in value.	3 is a solution of the equation $4x - 5 = 7$.
Solve equations of the form $ax + b = c, a \neq 0$, algebraically [2.6]	The expression $ax + b$ indicates the arithmetic sequence: start with x, multiply by a and then add b. To solve the related equation $ax + b = c$ you must reverse this sequence, replacing each operation by its inverse. That is, subtract b from both sides and then divide by a to obtain the value of x.	To solve $3x - 6 = 15$ for x, add 6 and then divide by 3 as follows: $$3x - 6 = 15$$ $$3x - 6 + 6 = 15 + 6$$ $$3x = 21$$ $$\frac{3x}{3} = \frac{21}{3}$$ $$x = 7$$
Solve a formula for a given variable [2.7]	To solve a formula, isolate the term containing the variable of interest on one side of the equation by performing the appropriate sequence of operations. None of the terms on the opposite side should contain that variable. Then divide both sides of the equation by the coefficient of the variable.	$2(a + b) = 8 - 5a$, for a $$2a + 2b = 8 - 5a$$ $$2a + 5a = 8 - 2b$$ $$7a = 8 - 2b$$ $$a = \frac{8 - 2b}{7}$$
Proportion [2.8]	An equation stating that two ratios are equivalent.	$\frac{a}{b} = \frac{c}{d}$
Cross multiplication [2.8]	Cross multiplication is the result of multiplying both sides of a proportion by the 2 denominators. It is equivalent to multiplying the numerator of each fraction by the denominator of the other and then setting these products equal to one another.	$\frac{15}{x} = \frac{3}{20}$ becomes $15 \cdot 20 = 3x$, so $x = 100$.
Factors [2.9]	Factors are numbers, variables, and/or expressions that are multiplied together to form a product.	$2x(x - 1)$ Here, 2, x, and $x - 1$ are all factors.
Distributive property [2.9]	$a(b + c) = ab + ac$	$4(x + 6) = 4x + 4 \cdot 6$ $= 4x + 24$
Geometric interpretation of the distributive property [2.9]	The area of the large rectangle is equal to the sum of the areas of the two smaller rectangles.	$a(b + c) = ab + ac$

CONCEPT/SKILL	DESCRIPTION	EXAMPLE
Factored form of an algebraic expression [2.9]	An algebraic expression is in factored form when it is written as a product of factors.	The expression $4(x + 6)$ is in factored form. The two factors are 4 and $x + 6$.
Expanded form of an algebraic expression [2.9]	An algebraic expression is in expanded form when it is written as a sum of distinct terms.	The expression $4x + 24$ is in expanded form. The two distinct terms are $4x$ and 24.
Extension of the distributive property [2.9]	The distributive property extended to sums or differences of more than two terms is $$a(b + c + d) = a \cdot b + a \cdot c + a \cdot d.$$	$$4(3x + 5y - 6)$$ $$= 4 \cdot 3x + 4 \cdot 5y - 4 \cdot 6$$ $$= 12x + 20y - 24$$
Factoring [2.9]	Factoring is the process of writing a sum of distinct terms equivalently as a product of factors.	$10xy + 15xz = 5x(2y + 3z)$
Common factor [2.9]	The common factor is the factor contained in every term of an algebraic expression.	In the expression $12abc + 3abd - 21ab$, 3, a, and b are common factors.
Factoring out a common factor [2.9]	This is the process of dividing each term by a common factor and placing this factor outside parentheses containing the sum of remaining terms.	$$12abc + 3abd - 21ab$$ $$= 3b(4ac + ad - 7a)$$
Greatest common factor [2.9]	The greatest common factor is a common factor such that there are no additional factors (other than 1) common to the terms in the expression.	The greatest common factor in the expression $12abc + 6abd - 21ab$ is $3ab$.
Completely factored form [2.9]	An algebraic expression is in completely factored form when none of its factors can be factored any further.	$4x(2y + 10z)$ is not in completely factored form because the expression in parentheses has a common factor of 2. $8x(y + 5z)$ is the completely factored form of this expression.
Procedure for factoring an algebraic expression whose terms contain a common factor [2.9]	The procedure for factoring an algebraic expression whose terms contain a common factor: 1. Identify the common factor. 2. Divide each term by the common factor. 3. Place the sum of the remaining factors inside the parentheses, and place the common factor outside the parentheses.	Given $20x + 35y - 45$: The common factor of the terms in this expression is 5. Dividing each term by 5 yields remaining terms $4x$, $7y$, -9. The factored form is therefore $5(4x + 7y - 9)$.
Like terms [2.9]	Like terms are terms that contain identical variable factors, including exponents. They differ only in their numerical coefficients.	$5y^2$ and $5y$ are not like terms. $7xy$ and $-3xy$ are like terms.

CONCEPT/SKILL	DESCRIPTION	EXAMPLE
Combining like terms [2.9]	To combine like terms into a single term, add or subtract the coefficients of like terms.	$12cd^2 - 5c^2d + 10 +$ $3cd^2 - 6c^2d$ $= 15cd^2 - 11c^2d + 10$
Procedure for simplifying algebraic expressions [2.10]	Steps for simplifying algebraic expressions: 1. Simplify the expression from the innermost grouping outward using the order of operations convention. 2. Use the distributive property when applicable. 3. Combine like terms.	$[5 - 2(x - 7)] + 5(3x - 10)$ $= [5 - 2x + 14] + 5(3x - 10)$ $= -2x + 19 + 5(3x - 10)$ $= -2x + 19 + 15x - 50$ $= 13x - 31$
Algebraic goal of solving an equation [2.11]	The goal of solving an equation is to isolate the variable on one side of the equation.	$3x - 5 = x - 2$ $2x = 3$ $x = \dfrac{3}{2}$
General strategy for solving equations algebraically [2.11]	The goal in solving an equation algebraically is to perform the appropriate operations in the correct sequence so that the unknown quantity is isolated on one side and its value is given on the other side. In a formula, the unknown quantity should be isolated on one side; the other side should contain only the remaining variables and numerical quantities.	$3(2x - 5) + 2x = 9$ $6x - 15 + 2x = 9$ $8x - 15 = 9$ $8x = 24$ $x = 3$ Check: $3(2 \cdot 3 - 5) + 2 \cdot 3$ $= 3(6 - 5) + 6$ $= 3 \cdot 1 + 6$ $= 9$

1. Let x represent any number. Translate each of the following phrases into an algebraic expression.

   a. 5 more than x

   b. x less than 18

   c. Double x

   d. Divide 4 by x

   e. 17 more than the product of 3 and x

   f. 12 times the sum of 8 and x

   g. 11 times the difference of 14 and x

   h. 49 less than the quotient of x and 7

2. Determine the values requested in the following.

   a. If $y = 2x$ and $y = 18$, what is the value of x?

   b. If $y = -8x$ and $x = 3$, what is the value of y?

   c. If $y = 15$ and $y = 6x$, what is the value of x?

   d. If $y = -8$ and $y = 4 + x$, what is the value of x?

   e. If $x = 19$ and $y = x - 21$, what is the value of y?

   f. If $y = -71$ and $y = x - 87$, what is the value of x?

   g. If $y - 36$ and $y - \dfrac{x}{4}$, what is the value of x?

   h. If $x = 5$ and $y = \dfrac{x}{6}$, what is the value of y?

   i. If $y = \dfrac{x}{3}$ and $y = 24$, what is the value of x?

3. Phil Mickelson had scores of 63, 68, and 72 for three rounds of a golf tournament.

   a. Let x represent his score on the fourth round. Write a symbolic rule that expresses his average score after four rounds of golf.

   b. To be competitive in the tournament, Mickelson must maintain an average of about 66. Use the symbolic rule from part a to determine what he must score on the fourth round to achieve a 66 average for the tournament.

4. Determine the values requested in the following.

   a. If $y = 2x + 8$ and $y = 18$, what is the value of x?

   b. If $x = 14$ and $y = 6x - 42$, determine the value of y.

   c. If $y = -33$ and $y = -5x - 3$, determine the value of x.

   d. If $y = -38$ and $y = 24 + 8x$, what is the value of x?

e. If $x = 72$ and $y = -54 - \dfrac{x}{6}$, what is the value of y?

f. If $y = 66$ and $y = \dfrac{2}{3}x - 27$, what is the value of x?

g. If $y = 39$ and $y = -\dfrac{x}{4} + 15$, what is the value of x?

h. If $y = 32.56x + 27$ and $x = 0$, what is the value of y?

5. You must drive to Syracuse to take care of some legal matters. The cost of renting a car for a day is $25, plus 15 cents per mile.

a. Identify the input variable.

b. Identify the output variable.

c. Use x to represent the input and y to represent the output. Write a symbolic rule that describes the daily rental cost in terms of the number of miles driven.

d. Complete the following table.

Input, x (mi)	100	200	300	400	500
Output, y ($)					

e. Plot the points from the table in part d. Then draw the line through all five points. (Make sure you label your axes and use appropriate scaling.)

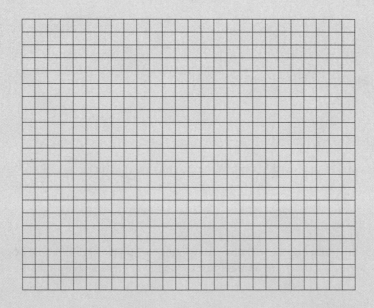

f. The distance from Buffalo to Syracuse is 153 miles. Estimate from the graph in part e how much it will cost to travel from Buffalo to Syracuse and back.

g. Use the symbolic rule in part c to determine the exact cost of this trip.

h. You have budgeted $90 for car rental for your day trip. Use your graph to estimate the greatest number of miles you can travel in the day and not exceed your allotted budget. Estimate from your graph.

i. Use the symbolic rule in part c to determine the exact number of miles you can travel in the day and not exceed $90.

j. What are realistic replacement values for the input if you rent the car for only 1 day?

6. As a real estate salesperson, you earn a small salary, plus a percentage of the selling price of each house you sell. If your salary is $100 a week, plus 3.5% of the selling price of each house sold, what must your total annual home sales be for you to gross $30,000 in 1 year? Assume that you work 50 weeks per year.

7. Use the formula $I = Prt$ to evaluate I, given the following information.

a. $P = \$2000, r = 5\%, t = 1$

b. $P = \$3000, r = 6\%, t = 2$

8. Use the formula $P = 2(w + l)$ to evaluate P for the following information.

a. $w = 2.8$ and $l = 3.4$

b. $w = 7\dfrac{1}{3}$ and $l = 8\dfrac{1}{4}$

9. Complete the following table and use your results to determine numerically which, if any, of the following expressions are equivalent.

a. $(4x - 3)^2$ **b.** $4x^2 - 3$ **c.** $(4x)^2 - 3$

x	$(4x - 3)^2$	$4x^2 - 3$	$(4x)^2 - 3$
−1			
0			
3			

10. Simplify the following expressions.

   a. $3(x + 1)$

   b. $-6(2x^2 - 2x + 3)$

   c. $x(4x - 7)$

   d. $6 - (2x + 14)$

   e. $4 + 2(6x - 5) - 19$

   f. $5x - 4x(2x - 3)$

11. Factor the following expressions completely.

   a. $4x - 12$

   b. $18xz + 60x - xy$

   c. $-12x - 20$

12. Simplify the following expressions.

   a. $4x^2 - 3x - 2 + 2x^2 - 3x + 5$

   b. $(3x^2 - 7x + 8) - (2x^2 - 4x + 1)$

13. Use the distributive property to expand each of the algebraic expressions.

   a. $5(2x - 7y)$

   b. $\dfrac{1}{3}(6a + 3b - 9)$

   c. $-10(x - 2w + z)$

   d. $4\left(\dfrac{1}{2}c - \dfrac{1}{4}\right)$

14. Solve the given equations, and check your results.

   a. $\dfrac{5}{28} = \dfrac{x}{49}$

   b. $\dfrac{18}{x} = \dfrac{90}{400}$

   c. $4(x + 5) - x = 80$

   d. $-5(x - 3) + 2x = 6$

   e. $38 = 57 - (x + 32)$

   f. $-13 + 4(3x + 5) = 7$

   g. $5x + 3(2x - 8) = 2(x + 6)$

   h. $-4x - 2(5x - 7) + 2 = -3(3x + 5) - 4$

   i. $-32 + 6(3x + 4) = -(-5x + 38) + 3x$

15. Your sister has just found the perfect dress for a wedding. The price of the dress is reduced by 30%. She is told that there will be a huge sale next week, so she waits to purchase it. When she goes to buy the dress, she finds that it has been reduced again by 30% of the already reduced price.

    a. If the original price of the dress is $400, what is the price of the dress after the first reduction?

    b. What is the price of the dress after the second reduction?

    c. Let x represent the original price (input). Determine an expression that represents the price of the dress after the first reduction. Simplify this expression.

    d. Use the result from part c to write an expression that represents the price of the dress after the second reduction. Simplify this expression.

    e. Use the result from part d to determine the price of the dress after the two reductions if the original price is $400. How does this compare with your answer to part b?

    f. She sees another dress that is marked down to $147 after the same two reductions. She wants to know the original price of the dress. Write and solve the equation to determine the original price.

16. The proceeds from your college talent show to benefit a local charity totaled $1550. Because the seats were all taken, you know that 500 people attended. The cost per ticket was $2.50 for students and $4.00 for adults. Unfortunately, you misplaced the ticket stubs that would indicate how many students and how many adults attended. You need this information for accounting purposes and future planning.

    a. Let n represent the number of students who attended. Write an expression in terms of n to represent the number of adults who attended.

    b. Write an expression in terms of n to represent the proceeds from the student tickets.

    c. Write an expression in terms of n to represent the proceeds from the adult tickets.

    d. Write an equation that indicates that the total proceeds from the student and adult tickets totaled $1550.

    e. How many student tickets and how many adult tickets were sold?

. You have an opportunity to be the manager of a day camp for the summer. You know that your fixed costs for operating the camp are $1200 per week, even if there are no campers. Each camper who attends costs the management $25 per week. The camp charges each camper $60 per week.

Let x represent the number of campers.

a. Write a symbolic rule in terms of x that represents the total cost, C, of running the camp per week.

b. Write a symbolic rule in terms of x that represents the total income (revenue), R, from the campers per week.

c. Write a symbolic rule in terms of x that represents the total profit, P, from the campers per week.

d. How many campers must attend for the camp to break even with revenue and costs?

e. The camp would like to make a profit of $620. How many campers must enroll to make that profit?

f. How much money would the camp lose if only 20 campers attended?

18. Solve each of the following equations for the specified variable.

a. $I = Prt$, for P

b. $f = v + at$, for t

c. $2x - 3y = 7$, for y

19. You contact the local print shop to produce a commemorative booklet for your college theater group. It is the group's twenty-fifth anniversary, and in the booklet, you want a short history along with a description of all the theater productions for the past 25 years. It costs $750 to typeset the booklet and $0.25 for each copy produced.

a. Write a symbolic rule that gives the total cost, C, in terms of the number, x, of booklets produced.

b. Use the symbolic rule from part a to determine the total cost of producing 500 booklets.

c. How many booklets can be produced for $1000?

d. Suppose the booklets are sold for 75 cents each. Write a symbolic rule for the total revenue, R, from the sale of x booklets.

e. How many booklets must be sold to break even? That is, for what value of x is the total cost of production equal to the total amount of revenue?

f. How many booklets must be sold to make a $500 profit?

20. You live 7.5 miles from work, where you have free parking. Some days, you must drive to work. On other days, you can take the bus. It costs you $0.30 per mile to drive the car and $2.50 round trip to take the bus. Assume that there are 22 working days in a month.

a. Let x represent the number of days that you take the bus. Express the number of days that you drive in terms of x.

b. Express the cost of taking the bus in terms of x.

c. Express the cost of driving in terms of x.

d. Express the total cost of transportation in terms of x.

e. How many days can you drive if you budget $70 a month for transportation?

f. How much should you budget for the month if you would like to take the bus only half of the time?

Chapter

3

Function Sense and Linear Functions

Chapter 3 continues the study of relationships between input and output variables. The focus here is on functions, which are special relationships between input and output variables. You will learn how functions are represented verbally, numerically, graphically, and symbolically. You will also study a special type of function—the linear function.

Cluster 1 — Function Sense

Activity 3.1

Gold, Silver, and Bronze

Objectives

1. Use the input-output concept to define a function.

2. Identify a functional relationship between two variables.

3. Identify the independent and dependent variables.

4. Use function notation to express the input-output correspondence of a function.

5. Identify the domain and range of a function.

6. Expand the rectangular coordinate system to include all four quadrants in the plane.

7. Represent functions in tabular and graphical forms.

As you observed in Chapter 2, a key step in the problem-solving process is to look for the relationships and connections between the variable quantities in a given situation. Problems encountered in the world around us, including the environment, medicine, economics, architecture, and the Internet, are often very complicated and contain several variable quantities. In many of these situations, the variables will have a special relationship called a **function**. In this text, you will deal primarily with functional situations that contain two variables.

Functions

> **Definition**
>
> An input-output relationship in which every input value is assigned one and only one output value is called a function. The **output variable** is a **function of the input variable**. For all functions, the input variable is called the **independent** variable and the output variable is called the **dependent** variable.

This correspondence can be described in words, displayed in table form, shown through a graph, or defined by a symbolic rule.

Tabular (Numerical) Representation of a Function

Table 1 describes the number of Summer and Winter Olympic medals (gold, silver, and bronze) won over the 7 Olympics from Summer 2000 Sydney through Summer 2012 London as a **function** of athlete's age. The input-output pairing given in the table is said to define the Olympic medal function **numerically**. Note that the table data stops at age 40, although there were a small number of athletes above that age who won a medal at the olympics.

Table 1.

AGE	NUMBER OF MEDALS		AGE	NUMBER OF MEDALS
15	23		28	701
16	80		29	572
17	104		30	486
18	155		31	377
19	252		32	345
20	382		33	331
21	520		34	193
22	720		35	133
23	757		36	116
24	807		37	56
25	763		38	55
26	765		39	43
27	724		40	34

1. Identify the input and output variables in this functional relationship.

2. a. Determine the output corresponding to an input of 22. Write a sentence that describes this correspondence using the context of the olympic medal situation.

b. How many medals were won by 32-year-olds? Identify the input and output values.

c. For each value of the input (age), how many different corresponding outputs (number of medals) are there?

3. Explain how the input-output pairs in Table 1 fit the definition of a function.

Function Notation

You can use a compact symbolic notation to indicate a function's input-output correspondence. In this example, you can assign A as the input (independent) variable, age, and M as the output (dependent) variable, number of medals won.

For the input 20 and corresponding output 382, the function notation is written as

$$M(20) = 382.$$

This is read as "M of 20 equals 382." The parentheses here *do not* indicate multiplication. The notation simply indicates that in this particular function, the output value corresponding to an input of 20 is 382. The number *inside* the parentheses is always an *input* value.

The symbol "$M(20)$" is called the "function value at 20" and is understood to represent the corresponding *output* and has the value 382.

4. Use function notation to represent the input output correspondence for age 28.

5. Use Table 1 to determine each of the following. Interpret their meaning in context of this situation.

    a. $M(16)$ **b.** $M(38)$

In applications, the name of a function is often associated with the output (dependent) variable. Using function notation, the output of the medal function is written as $M(A)$, where M is also used as the name of the function.

Domain and Range of a Function

> ### Definition
>
> The collection of all replacement values of the input (independent) variable is called the **domain** of a function. The **range** of a function is the collection of the corresponding output values.

Because the medal function M is defined by a table of values, its domain consists of the collection of numbers in the input column of the table, the consecutive integers from 15 through 40. The range consists of the set of corresponding output values, the collection of the specific numbers in the second column of the table.

6. Determine whether the following numbers are in the domain of function M?

    a. 6   **b.** 56   **c.** 32

7. Determine whether the following numbers are in the range of function M?

    a. 10   **b.** 193   **c.** 70

Representing Functions Graphically

As you observed in Chapter 2, a graph is a visual display of data that is often helpful in revealing trends or other information about the two variable quantities that are not apparent in a table.

8. Following is a plot of the data in Table 1 on an appropriately scaled and labeled rectangular coordinate system. Remember that the input values are placed along the horizontal axis and the output values are along the vertical axis. Each input-output data pair in Table 1 corresponds to a plotted point on the graph.

Note: The slash marks (//) indicate that the horizontal axis has been cut out between 0 and 15.

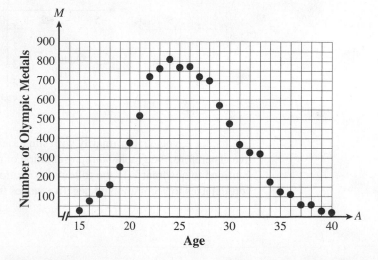

a. Without referring to Table 1, use the graph to estimate the total number of medals won by 35-year-olds.

b. Write the estimate in part b in ordered pair notation. Write the estimate using function notation.

c. Use the graph as a guide to describe how the number of medals won changes with age.

d. From which representation, table or graph, do you find it easier to determine at which age athletes won the most medals?

e. From which representation, table or graph, do you find it easier to determine how many medals were won by a 26 year old?

f. From which representation, table or graph, do you find it easier to determine at which age(s) the number of medals won significantly decreased?

Rectangular Coordinate System Revisited

In the rectangular (Cartesian) coordinate system, the horizontal axis (commonly called the **x-axis**) and the vertical axis (commonly called the **y-axis**) are number lines that intersect at their respective zero values at a point called the **origin**. The two perpendicular coordinate axes divide the plane into four **quadrants**. The quadrants are labeled counterclockwise, using Roman numerals, with quadrant I being the upper-right quadrant.

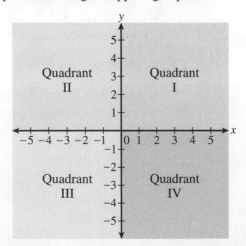

In Chapter 2, you dealt primarily with situations in which the input and output were nonnegative. Therefore, the graphical representation of the relationship between the two variables was confined to quadrant I. In the following situation, there is a need to extend the horizontal and vertical axes to accommodate negative values for the input and output.

Earth's Temperature Represented Graphically

Living on Earth's surface, you experience a relatively narrow range of temperatures. You may know what −20°F (or −28.9°C) feels like on a bitterly cold winter day. Or you may have sweated through 100°F (or 37.8°C) during summer heat waves. If you were to travel below Earth's surface and above Earth's atmosphere, you would discover a wider range of temperatures. The following graph represents the temperature, *T*, in degrees Celsius, as a function of altitude, *A*, in kilometers, measured from the surface of the Earth.

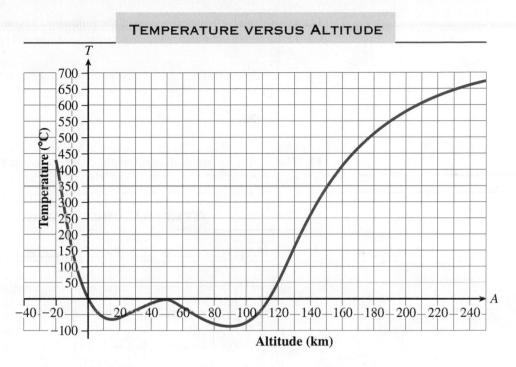

TEMPERATURE VERSUS ALTITUDE

9. a. What is the domain of this function as indicated by its graph? What are the units of this quantity?

b. What is the practical significance when this quantity has positive values?

c. What is the practical significance when this quantity has negative values?

d. How many kilometers are represented between the tick marks on the horizontal axis?

10. a. What is the range of this function as indicated by its graph? What are the units of this quantity?

b. What is the practical significance of the positive values of this quantity?

c. What is the practical significance of the negative values of this quantity?

d. How many degrees Celsius are represented between the tick marks on the vertical axis?

11. Consult the graph to determine at which elevations or depths the temperature is above 100°C.

12. Use the graph to estimate each of the following. In your own words, interpret this value in the context of the Earth's temperature situation using correct units of measure.

a. $T(120)$

b. $T(190)$

c. $T(-20)$

13. The following graph displays eight points selected from the temperature graph on page 285.

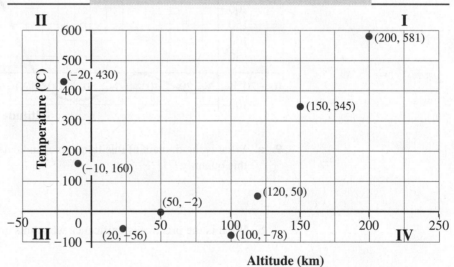

Altitude (km)

a. What is the practical meaning of the point with coordinates (150, 345)? Write the ordered pair using function notation.

b. What is the practical meaning of the point with coordinates (100, −78)?

c. What is the practical meaning of the point with coordinates (−20, 430)?

14. a. In which quadrant are the points (120, 50), (150, 345), and (200, 581) located?

   b. In which quadrant are the points $(-10, 160)$ and $(-20, 430)$ located?

   c. In which quadrant are the points $(20, -56)$ and $(100, -78)$ located?

   d. Are there any points located in quadrant III? What is the significance of your answer?

The activities of this chapter will examine the connections between the various representations of functions—numerical (table), graphical (plotted points) and algebraic (symbolic rule). In addition, you will explore the properties of different types of functions, including linear (Chapter 3) and nonlinear functions (Chapter 4).

SUMMARY: ACTIVITY 3.1

1. A **function** is a correspondence relating an input variable and output variable that assigns one and only one output value to each input value. In such a case, you state that the output is a function of the input.

2. For all functions, the input variable is called the **independent** variable and the output variable is called the **dependent** variable.

3. A function can be defined **numerically** as a list of **ordered pairs**, often displayed in a two-column (vertical) table or a two-row (horizontal) table. A function can also be defined as a set of ordered pairs.

4. When a function is defined **graphically**, the input (independent) variable is referenced on the horizontal axis and the output (dependent) variable is referenced on the vertical axis.

5. A function can be named by a symbol or letter, with the independent variable in parentheses. For example, if F is the name of the function and x is the independent variable, then in function notation $F(x)$, read as "F of x", represents the dependent variable. $F(10)$ represents the output value when the input is 10."

6. Associated with any function is a specific set of input values. The collection of all possible replacement values for the input variable is called the **domain** of the function.

7. The range of a function is the collection of all corresponding output values.

8. Input-output data pairs can be represented graphically as plotted points on a grid called a **rectangular coordinate system**.

9. The rectangular coordinate system consists of two perpendicular number lines (called **coordinate axes**) that intersect at their respective zero values. The point of intersection is called the **origin**.

10. Each point in the plane is identified by its horizontal and vertical directed distances from the axes. The distances are listed as an **ordered pair** of numbers (x, y) in which the horizontal coordinate, x, is written first and the vertical coordinate, y, second. The origin has coordinates $(0, 0)$.

11. The two perpendicular coordinate axes divide the plane into four **quadrants**. The quadrants are labeled counterclockwise using roman numerals, with quadrant I being the upper-right quadrant.

EXERCISES: ACTIVITY 3.1

1. The weights and heights of six mathematics students are given in the following table.

WEIGHT (lb.)	HEIGHT (cm)
165	172
123	157
212	183
175	178
165	163
147	167

a. In the statement "Height is a function of weight," which is the independent variable and which is the dependent variable?

b. Is height a function of weight for the six students? Explain using the definition of function.

c. In the statement "Weight is a function of height," which is the independent variable and which is the dependent variable?

d. Is weight a function of height for the six students? Explain using the definition of function.

e. In general, is weight a function of height? Explain.

2. Each of the following tables defines a relationship between an input and an output. Which of the relationships represent functions? Explain your answers.

a.

INPUT	−8	−3	0	6	9	15	24	38	100
OUTPUT	24	4	9	72	−14	−16	53	24	7

b.

INPUT	−8	−5	0	6	9	15	24	24	100
OUTPUT	24	4	9	72	14	−16	53	29	7

3. Identify the input and output variables in each of the following. Then determine whether the statement is true. Give a reason for your answer.

a. The letter grade in a course is a function of the numerical grade.

b. The numerical grade is a function of the letter grade.

4. a. Suppose that all you know about a function f is that $f(2) = -4$. Use this information to complete the following.
   i. If the input value for f is 2, then the corresponding output value is ____.
   ii. One point on the graph of f is _____.

b. The statement in part a indicates that $f(x) = -4$ for $x = 2$. Is it possible for f to have an output of −4 for another value of x, such as $x = 5$? Explain.

5. Suppose the number of hours using the Internet is the input variable, and the monthly cost for the Internet service is $39.95, regardless of the number of hours of usage. This input-output relationship is represented by the following table. Does this relationship represent a function?

NUMBER OF HOURS	MONTHLY COST
10	$39.95
50	$39.95
75	$39.95
100	$39.95

6. The following table presents the average recommended weights for given heights for 25- to 29-year-old medium-framed women. Consider height to be the input variable and weight to be the output variable. As ordered pairs height and weight take on the form (h, w). Designate the horizontal (input) axis as the h-axis, and the vertical (output) axis as the w-axis. Because all values of the data are positive, the points will lie in quadrant I only.

h, Height (in.)	58	60	62	64	66	68	70	72
w, Weight (lb)	115	119	125	131	137	143	149	155

a. Plot the ordered pairs in the height–weight table on the following grid. Note the consistent spacing between tick marks on each axis. The distance between tick marks on the horizontal axis represents 2 inches. On the vertical axis the distance between tick marks represents 5 pounds.

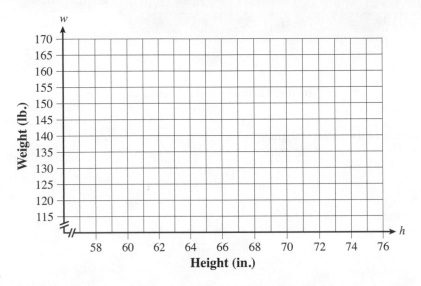

Recall that the slash marks (//) near the origin indicate that the interval from 0 to 56 on the horizontal axis and the interval from 0 to 115 on the vertical axis are not shown. That is, only the part of the graph containing the plotted points is shown.

b. Is the weight, w, a function of the height, h? Explain.

c. Identify the independent and dependent variables.

d. Determine $w(64)$. Interpret the result in the context of this situation.

e. Use the table to determine the value of h for which $w(h) = 149$. Interpret the result in the context of this situation.

7. Consider ordered pairs of the form (x, y). Determine the sign (positive or negative) of the x- and y-coordinates of a point in each quadrant. For example, any point located in quadrant I has a positive x-coordinate and a positive y-coordinate.

QUADRANT	SIGN (+ OR −) OF x-COORDINATE	SIGN (+ OR −) OF y-COORDINATE
I		
II		
III		
IV		

8. In the following coordinate system, the horizontal axis is labeled the *x*-axis and the vertical axis is labeled the *y*-axis. Determine the coordinates (*x*, *y*) of points *A* to *N*. For example, the coordinates of *A* are (30, 300).

A   (30, 300)   *B* _____   *C* _____

D _____   *E* _____   *F* _____

G _____   *H* _____   *I* _____

J _____   *K* _____   *L* _____

M _____   *N* _____

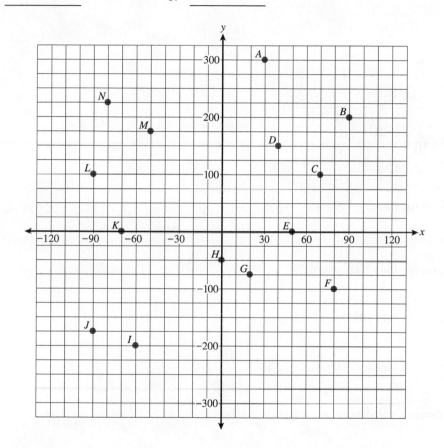

9. Which points (*A* to *N*) from Problem 8 are located

  a. in the first quadrant?

  b. in the second quadrant?

  c. in the third quadrant?

  d. in the fourth quadrant?

Points that lie on either the horizontal or vertical axis are not in any of the four quadrants. They lie on a boundary between quadrants.

10. a. Which points (*A* to *N*) from Problem 8 are on the horizontal axis? What are their coordinates?

  b. What is the *y*-value of any point located on the *x*-axis?

c. Which points are on the vertical axis? What are their coordinates?

d. What is the *x*-value of any point located on the *y*-axis?

11. You work for the National Weather Service and are asked to study the average daily temperatures in Fairbanks, Alaska. You calculate the mean of the average daily temperatures for each month. You decide to place the information on a graph in which the date is the input and the temperature is the output. You also decide that January 2000 will correspond to the month zero. Determine the quadrant in which you would plot the points that correspond to the following data.

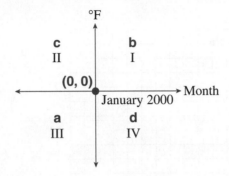

a. The average daily temperature for January 1990 was −13°F.

b. The average daily temperature for June 2005 was 62°F.

c. The average daily temperature for July 1995 was 63°F.

d. The average daily temperature for January 2009 was −13°F.

12. Use the graph on the right of function *f* to determine each of the following.

a. $f(2)$

b. $f(3.5)$

c. $f(-2)$

d. $f(-5)$

e. For what value of *x* is $f(x) = 1$?

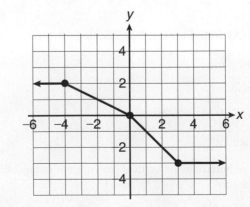

Activity 3.2

How Fast Did You Lose?

Objective

1. Determine the average rate of change of an output variable with respect to the input variable.

You are a member of a health and fitness club. The club's registered dietitian and your personal trainer helped you develop a special 8-week diet and exercise program. The data in the following table represents your weight, w, as a function of time, t, over an 8-week period.

Time (wk.)	0	1	2	3	4	5	6	7	8
Weight (lb.)	140	136	133	131	130	127	127	130	126

1. a. Plot the data points using ordered pairs of the form (t, w). For example, $(3, 131)$ is a data point that represents your weight at the end of the third week.

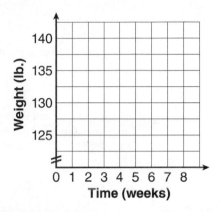

Recall that the resulting graph of points in part a is called a **scatterplot**.

b. What is the domain of this function?

c. What is the range of this function?

2. a. What is your weight at the beginning of the program?

b. What is your weight at the end of the first week?

3. To see how the health program is working for you, you analyze your weekly weight changes during the 8-week period.

a. During which week(s) does your weight increase?

b. During which week(s) does your weight decrease?

c. During which week(s) does your weight remain unchanged?

Average Rate of Change

4. Your weight decreases during each of the first 5 weeks of the program.

    a. Determine the actual change in your weight over the first 5 weeks of the program by subtracting your initial weight from your weight at the end of the first 5 weeks.

    b. What is the sign (positive or negative) of your answer? What is the significance of this sign?

    c. Determine the change in the input value over the first 5 weeks, that is, from $t = 0$ to $t = 5$.

    d. Write the quotient of the change in weight from part a divided by the change in time in part c. Interpret the meaning of this quotient.

The change in weight describes how much weight you have gained or lost, but it does not tell how quickly you shed those pounds. That is, a loss of 3 pounds in 1 week is more impressive than a loss of 3 pounds over a month's time. Dividing the change in weight by the change in time gives an average rate at which you have lost weight over this period of time. The units for this rate are output units per input unit—in this case, pounds per week.

The rate in Problem 4d,

$$\frac{-13 \text{ lb.}}{5 \text{ wk.}} = -2.6 \text{ lb./week,}$$

is called the **average rate of change** of weight with respect to time over the first 5 weeks of the program. It can be interpreted as an average loss of 2.6 pounds each week for the first 5 weeks of the program.

$$\textbf{average rate of change} = \frac{\text{change in output}}{\text{change in input}},$$

where

    i. change in output is calculated by subtracting the first (initial) output value from the second (final) output value

    ii. change in input is calculated by subtracting the first (initial) input value from the second (final) input value

Delta Notation

The rate of change in output with respect to a corresponding change in input is so important that special symbolic notation has been developed to denote it.

The uppercase Greek letter delta, Δ (Greek "D" for difference), is used with a variable's name to represent a change in the value of the variable from a starting point to an ending point.

For example, in Problem 4 the notation Δw represents the change in value of the output variable weight, w. It is calculated by subtracting the initial value of w, denoted by w_1, from the final value of w, denoted by w_2. Symbolically, this change is represented by

$$\Delta w = w_2 - w_1.$$

In the case of Problem 4a, the change in weight over the first 5 weeks can be calculated as

$$\Delta w = w_2 - w_1 = 127 - 140 = -13 \text{ lb.}$$

In the same way, the change in value of the input variable time, t, is written as Δt and is calculated by subtracting the initial value of t, denoted by t_1, from the final value of t, denoted by t_2. Symbolically, this change is represented by

$$\Delta t = t_2 - t_1.$$

In Problem 4c, the change in the number of weeks can be written using Δ notation as follows:

$$\Delta t = t_2 - t_1 = 5 - 0 = 5 \text{ weeks}$$

The average rate of change of weight over the first 5 weeks of the program can now be symbolically written as follows:

$$\frac{\Delta w}{\Delta t} = \frac{-13}{5} = -2.6 \text{ lb./week}$$

5. Use Δ notation and determine the average rate of change of weight over the last 4 weeks of the program.

Graphical Interpretation of the Average Rate of Change

6. a. On the graph in Problem 1, connect the points (0, 140) and (5, 127) with a line segment. Does the line segment rise, fall, or remain horizontal as you follow it from left to right?

b. Recall that the average rate of change over the first 5 weeks was -2.6 pounds per week. What does the average rate of change tell you about the line segment drawn in part a?

7. a. Determine the average rate of change of your weight over the time period from $t = 5$ to $t = 7$ weeks. Include the appropriate sign and units.

b. Interpret the rate in part a with respect to your diet.

c. On the graph in Problem 1, connect the points $(5, 127)$ and $(7, 130)$ with a line segment. Does the line segment rise, fall, or remain horizontal as you follow it from left to right?

d. How is the average rate of change of weight over the given 2-week period related to the line segment you drew in part c?

8. a. At what average rate is your weight changing during the sixth week of your diet, that is, from $t = 5$ to $t = 6$?

b. Interpret the rate in part a with respect to your diet.

c. Connect the points $(5, 127)$ and $(6, 127)$ on the graph with a line segment. Does the line segment rise, fall, or remain horizontal as you follow it from left to right?

d. How is the average rate of change in part a related to the line segment drawn in part c?

9. a. What is the average rate of change of your weight over the period from $t = 4$ to $t = 7$ weeks?

b. Explain how the rate in part a reflects the progress of your diet over those 3 weeks.

The **average rate of change over an interval** only reflects the change between the left endpoint value and the right endpoint value. It does not reflect any changes that have occurred in between. Therefore a rate of change of zero does not indicate that the quantity has remained constant. It only indicates that the value at the "end" of the interval is the same as the value at the "beginning."

10. At the beginning of the diet and exercise program and once a week thereafter, you are tested on the treadmill. The test measures how many minutes it takes you to walk, jog, or run 3 miles on the treadmill. The following data gives your time, t, over an 8-week period.

End of Week, w	0	1	2	3	4	5	6	7	8
Time, t (min.)	45	42	40	39	38	38	37	39	36

Note that $w = 0$ corresponds to the first time on the treadmill, $w = 1$ is the end of the first week, $w = 2$ is the end of the second week, etc.

a. Plot the data points using ordered pairs of the form (w, t).

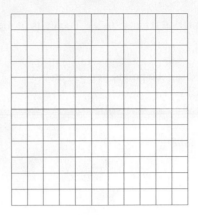

b. Determine the average rate of change of t with respect to w during the sixth and seventh weeks (from the point where $w = 5$ to the point where $w = 7$).

c. What is the significance of the positive sign of the average rate of change in this situation?

d. Connect the data points $(5, 38)$ and $(7, 39)$ on your graph from part a, using a line segment. Is the output increasing, decreasing, or remaining constant on the interval?

e. At what average rate did your time change during the fifth week (from $w = 4$ to $w = 5$)?

f. Interpret your answer from part e.

g. Connect the data points $(4, 38)$ and $(5, 38)$ on the graph from a using a line segment. Is the output increasing, decreasing, or remaining constant over this interval?

SUMMARY: ACTIVITY 3.2

1. Let y_1 represent the corresponding output value for the input value x_1 and y_2 represent the corresponding output value for the input value x_2. As the independent variable x changes in value from x_1 (initial value) to x_2 (final value),

   a. the change in input is represented by $\Delta x = x_2 - x_1$.

   b. the change in output is represented by $\Delta y = y_2 - y_1$.

2. The quotient $\dfrac{\Delta y}{\Delta x} = \dfrac{y_2 - y_1}{x_2 - x_1}$ is called the **average rate of change** of y (output) with respect

to x (input) over the x-interval from x_1 to x_2. The units of measurement of the quantity $\dfrac{\Delta y}{\Delta x}$

are *output units* per *input unit*.

3. The line segment connecting the points (x_1, y_1) and (x_2, y_2)

  a. rises from left to right if $\dfrac{\Delta y}{\Delta x}$ is positive.

  b. falls from left to right if $\dfrac{\Delta y}{\Delta x}$ is negative.

  c. remains constant (horizontal) if $\dfrac{\Delta y}{\Delta x} = 0$

EXERCISES: ACTIVITY 3.2

The following table of data from the U.S. Census Bureau gives the median age of an American man at the time of his first marriage.

Year	1910	1920	1930	1940	1950	1960	1970	1980	1990	2000	2010
Median Age	25.1	24.6	24.3	24.3	22.8	22.8	23.2	24.7	26.1	26.8	27.5

Use this data to answer Exercises 1–6.

1. a. Determine the average rate of change in median age per year from 1950 to 2010.

  b. Describe what the average rate of change in part a represents in this situation.

2. Determine the average rate of change in median age per year from 1930 to 1960.

3. What is the average rate of change over the 100-year period described in the table?

4. During what ten-year period did the average age increase the most?

5. a. What does it mean in this situation if the average rate of change is negative?

b. Determine at least one ten-year period when the average rate of change is negative.

c. What trend would you observe in the graph if the average rate of change were negative? That is, would the graph go up, go down, or remain constant?

6. a. Is the average rate of change zero over any ten-year period? If so, when?

b. What does a rate of change of zero mean in this situation?

c. What trend would you observe in the graph during this period? That is, would the graph go up, go down, or be horizontal?

7. Between 1960 and 2010, the size and shape of automobiles in the United States changed almost annually. The amount of fuel consumed by these vehicles also changed. The following table describes the average fuel consumption, measured in gallons of gas, of a passenger car as a function of year.

Year, t	1960	1970	1980	1990	1995	2000	2005	2010
Gallons Consumed per Passenger Car (average), g	668	760	576	520	530	547	567	453

a. Determine the average rate of change, in gallons of fuel used per passenger car, from 1960 to 1970.

b. Determine the average rate of change, in gallons of gas per year, from 1960 to 1990.

c. Determine the average rate of change, in gallons of gas per year, from 1995 to 2005.

d. Determine the average rate of change, in gallons of gas per year, between 1960 and 2010.

e. What does the result in part d mean in this situation?

8. The National Weather Service recorded the following temperatures one February day in Chicago.

Time of Day	10 A.M.	12 NOON	2 P.M.	4 P.M.	6 P.M.	8 P.M.	10 P.M.
Temperature (°F)	30	35	36	36	34	30	28

a. Determine the average rate of change (including units and sign) of temperature with respect to time over the entire 12-hour period given in the table.

b. Over which period(s) of time is the average rate of change zero? What, if anything, can you conclude about the actual temperature fluctuation within this period?

c. What is the average rate of change of temperature with respect to time over the evening hours from 6 P.M. to 10 P.M.? Interpret this value (including units and sign) in a complete sentence.

d. Write a brief paragraph describing the temperature and its fluctuations during the 12-hour period in the table.

In a mathematics setting, functions are usually described by symbolic rules. In this activity, you will investigate three functions defined symbolically and use the symbolic rules to create tables of values and graphs (visual representations) of these functions. These functions will not represent any specific contextual situation.

Typically, in such a case where functions are described by symbolic rules without a specific real-world context, their input variable is denoted by x and the function name is denoted by the letter f (for *function*). The output is not necessarily assigned a variable name; it can simply be denoted by the symbol $f(x)$, read "f of x." However, the output variable frequently is denoted by y, and its dependence on the independent variable x is written $y = f(x)$. Note that y and $f(x)$ are different names for the same quantity—they both represent the **output** corresponding to input x—and are used interchangeably.

If a problem contains more than one function, the second function is often called g; the third, h. The outputs in these functions are denoted $g(x)$ and $h(x)$, respectively.

In this section, you will explore three functions f, g, and h defined by the following symbolic rules:

$$f(x) = x^2 \qquad g(x) = 4x \qquad h(x) = \frac{20x}{x + 1}$$

Evaluating Functions

The instruction "evaluate the function f when $x = 5$" is asking you to determine the output corresponding to an input value 5. This instruction can also be stated equivalently as evaluate $f(5)$. The output is obtained by replacing x with 5 and performing the sequence of operations

$$f(5) = (5)^2 = 25.$$

The output, denoted by $f(5)$, has the value 25, written $f(5) = 25$ and read "f of 5 equals 25." This function correspondence for input value 5 and output value 25 can also be expressed using the ordered pair notation $(5, 25)$.

1. For the function defined by $f(x) = x^2$, determine the following output values.

   a. $f(0)$ **b.** $f(-5)$ **c.** $f(10)$ **d.** $f(-1)$

2. For the function defined by $g(x) = 4x$, determine the following output values.

   a. $g(0)$ **b.** $g(-1)$ **c.** $g(4)$ **d.** $g(-2)$

3. For the function defined by $h(x) = \dfrac{20x}{x + 1}$, determine the output value in each ordered pair.

   a. $(0, \quad)$ **b.** $(3, \quad)$ **c.** $(9, \quad)$ **d.** $(-5, \quad)$

The Connections among the Symbolic (Algebraic), Numerical (Tabular), and Graphical (Visual) Representations of Functions

In this section, the three functions f, g, and h will be restricted to the domain $0 \le x \le 4$. This notation means that the replacement values for x can be any real number *between* 0 and 4, including 0 and 4. The collection of numbers $0 \le x \le 4$ is called an **interval**.

4. Which of the following numbers are contained in the domain $0 \le x \le 4$?

   a. -2 **b.** 0 **c.** 3 **d.** 4.01 **e.** 0.95 **f.** $\dfrac{8}{3}$

Activity 3.3

Symbolically Defined Functions and their Graphs

Objectives

1. Define functions by symbolic rules.

2. Recognize that an output value is determined by evaluating the symbolic rule.

3. Recognize that an input value is determined by solving an associated equation.

4. Understand the connection between the input-output pairs generated by the symbolic rule and the plotted points that form its graph.

5. Understand the connection between the average rates of change of a function and the shape of its graph.

6. Explore and compare the algebraic and graphical methods for evaluating a function.

7. Explore and compare the algebraic and graphical methods for solving an associated equation.

5. Is it possible to list all the numbers in the domain $0 \leq x \leq 4$? Explain why or why not. How many input-output pairs can be generated from each function rule?

6. Complete the following table by evaluating the three functions, f, g, and h, at input values $x = 0$ and $x = 4$.

x	$f(x) = x^2$	$g(x) = 4x$	$h(x) = \dfrac{20x}{x+1}$
0			
4			

7. Determine the average rate of change of each of the following functions over the interval from $x = 0$ to $x = 4$.

  a. function f

  b. function g

  c. function h

  d. Compare the results in parts a–c.

8. a. The table in Problem 6 contains two input-output pairs for each function. Write them here using ordered pair notation. f: g: h:

  b. Plot the ordered pairs for functions f, g, and h on the indicated grids.

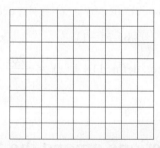

function f

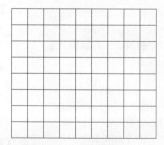

function g

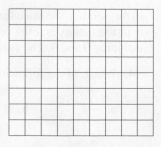

function h

9. In looking at the table (Problem 6), average rates of change (Problem 7), and graphs (Problem 8) of the functions f, g, and h, are you able distinguish one function from another?

10. Expand the table in Problem 6 by evaluating the three functions, f, g, and h, at the additional input value $x = 2$.

x	$f(x) = x^2$	$g(x) = 4x$	$h(x) = \dfrac{20x}{x+1}$
0			
2			
4			

11. a. Determine the average rate of change of f over

    i. the interval from $x = 0$ to $x = 2$.

    ii. the interval from $x = 2$ to $x = 4$.

  b. Determine the average rate of change of g over

    i. the interval from $x = 0$ to $x = 2$.

    ii. the interval from $x = 2$ to $x = 4$.

  c. Determine the average rate of change of h over

    i. the interval from $x = 0$ to $x = 2$.

    ii. the interval from $x = 2$ to $x = 4$.

12. a. Write the new input-output pair at $x = 2$ for each function, using ordered-pair notation. f: g: h:

  b. Plot each of these new ordered pairs on the appropriate grid in Problem 8.

  c. What is the connection between the average rates of change for function f (Problem 11a) and the *steepness* of the graph formed by plotting the additional three points?

  d. What is the connection between the average rates of change for function g (Problem 11b) and the *steepness* of the graph formed by plotting the additional three points?

  e. What is the connection between the average rates of change for function h (Problem 11c) and the *steepness* of the graph formed by plotting the additional three points?

13. a. Are the graphs in Problem 12 "good" overall visual representations of each of these three functions?

b. Suggest a way to improve these visual representations.

c. How many additional input-output pairs are needed to obtain a "good" visual representation of each function?

14. Expand the table below, that was used in Problem 10, by evaluating the three functions, f, g, and h, using increments of 0.5 for x. Plot each new ordered pair on the appropriate grid in Problem 8.

x	$f(x) = x^2$	$g(x) = 4x$	$h(x) = \dfrac{20x}{x+1}$
0	0	0	0
0.5			
1			
1.5			
2.0	4	8	13.33

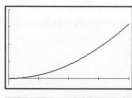

```
Plot1  Plot2  Plot3
\Y₁▤X²
\Y₂▤4X
\Y₃▤(20X)/(X+1)
\Y₄=
\Y₅=
\Y₆=
\Y₇=
\Y₈=
```

```
WINDOW
 Xmin=0
 Xmax=4
 Xscl=1
 Ymin=-2
 Ymax=20
 Yscl=5
 Xres=1
 △X=.01515151515151
 TraceStep=.03030303030303
```

You should now be able to see the shape of each graph emerging. Connect the plotted points in each graph as smoothly as you can to sketch a curve that displays the shape. A graph that contains *all* the input-output pairs of the function is called a **complete graph**.

15. a. In your own words, describe the shape of the complete graph of:

i. the function f.

ii. the function g.

iii. the function h.

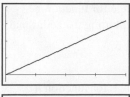

b. What seems to be the connection between the average rates of change and the shape of

i. the graph of f?

ii. the graph of g?

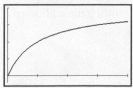

iii. the graph of h?

16. a. Does the point $(1, 4)$ lie on any of your graphs? Which graph(s) contains it? Explain.

b. Which function's graph contains the point $(3, 9)$? Explain.

c. What is the y-coordinate of the point $(1, y)$ that lies on the graph of function h?

d. Does the point $(2, 10)$ lie on any of the three graphs? Explain.

In this section, you have seen that when a function is defined by a symbolic rule,

- the rule generates *all* the input-output pairs of the function

- each input-output pair can be written in ordered-pair notation and viewed as the coordinates of a point in a rectangular coordinate system

- the entire collection of plotted input-output pairs forms the complete graph of the function

- every input-output pair of the function is represented by a specific point on its graph

- the coordinates of every point on the graph represent an input-output pair of the function

- the average rates of change of the function are reflected in the shape of its graph

  - if the average rates of change increase as you move along the graph, the graph "bends" upward

  - if the average rates of change decrease as you move along the graph, the graph "bends" downward

  - if the average rates of change remain the same, the graph is a line

The symbolic (algebraic) and complete graphical (visual) representations of a function contain identical information, but presented in different forms. These forms are often used interchangeably. Certain questions about a function can be answered more easily using the symbolic rule; others, by using the complete graph. The following section of this activity will explore this interchangeability more closely.

Evaluating Functions and Solving Equations Algebraically and Graphically

As you observed throughout Chapter 2, there are two types of calculations that are important in input-output relationships:

- Given an input value, determine the corresponding output value

- Given an output value, determine the corresponding input value(s)

If the input-output relationship is defined symbolically, both of these calculations can be done either by algebraic methods using the symbolic rule OR by visual methods using the corresponding graph of the symbolic rule. See Items 2 and 3 in the **Summary** on page 307 for a review of each method.

Problems 17 and 18 illustrate these methods for a function defined symbolically.

17. Consider the function defined by $F(x) = 2x - 3.5$.

   a. Given input value $x = 4$, determine the corresponding output of the function F algebraically by evaluating its symbolic rule for $x = 4$.

   b. Use the graph below to estimate the corresponding output of the function graphically for the input $x = 4$.

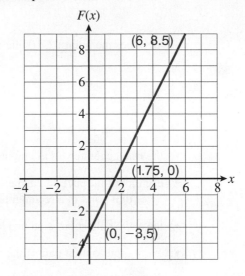

   c. How do the results from parts a and b compare? Are they the same? Should they be the same?

   d. Which method, algebraic or graphical, seems simpler or faster for each function?

   e. Which method produces the more accurate values?

18. a. Now, suppose you are given the output value 10. Using the symbolic rule for the function F in Problem 17, write an equation whose solution provides the corresponding input value for function F.

   b. Use algebraic methods to solve the equation in part a.

   c. Use graphical methods to solve the equation in part a (see graph in Problem 17b).

   d. Which method, algebraic or graphical, seems simpler or faster for solving the equation?

   e. Which method produces the more accurate value?

SUMMARY: ACTIVITY 3.3

1. When a function is defined by a symbolic rule,

- the rule generates *all* the input-output pairs of the function

- each input-output pair can be written in ordered-pair notation and viewed as the coordinates of a point in a rectangular coordinate system

- the entire collection of plotted input-output pairs forms the complete graph of the function

- every input-output pair of the function is represented by a specific point on its graph

- the coordinates of every point on the graph represent an input-output pair of the function

- the average rates of change of the function are reflected in the shape of its graph

  - if the average rates of change increase as you move along the graph, the graph "bends" upward
  - if the average rates of change decrease as you move along the graph, the graph "bends" downward
  - if the average rates of change remain the same, the graph is a line.

2. For a given input value, the corresponding output value can be determined algebraically or graphically.

- Algebraically: Evaluate the symbolic rule at the given input value.

- Graphically: Locate the given input value on the horizontal axis. Move vertically (up or down) until you reach the point on the graph. The y-coordinate of the point represents the corresponding output value.

3. For a given output value, the corresponding input value(s) can be determined algebraically or graphically.

- Algebraically: Construct and solve the appropriate equation.

- Graphically: Locate the given output value on the vertical axis. Move horizontally to the point(s) on the graph; there may be more than one. The x-coordinate of each point represents the input value you are seeking.

EXERCISES: ACTIVITY 3.3

1. Suppose $f(x) = 3x^2 - 4x + 5$.

Evaluate: $f(1)$ $f(-1)$ $f(2)$ $f(-5)$

2. a. Let $f(x) = 2x - 1$. Evaluate $f(4)$.

b. Let $g(n) = 3n + 5$. Evaluate $g(-3)$.

c. Let $h(m) = 2m^2 + 3m - 1$. Evaluate $h(-2)$.

Exercise numbers appearing in color are answered in the Selected Answers appendix.

d. Let $p(x) = 3x^2 - 2x + 4$. Evaluate $p(5)$.

3. a. Let f be a function defined by $f(x) = x + 1$. Determine the value of x for which $f(x) = 3$.

b. Let g be a function defined by $g(x) = 5x - 4$. Determine the value of x for which $g(x) = 11$.

c. Let h be a function defined by $h(t) = \dfrac{t}{4}$. Determine the value of t for which $h(t) = 12$.

4. Without drawing the graph, explain which of the following functions have graphs that contain the origin.

$$f(x) = 2 + x \qquad g(x) = 2x \qquad h(x) = \dfrac{x}{1 + x} \qquad k(x) = \dfrac{1}{1 + x^2}$$

5. Without drawing the graph, explain which of the following functions have graphs that contain the point $(-1, 2)$.

$$f(x) = 3 + x \qquad g(x) = 3 - x \qquad h(x) = 2x^2 \qquad k(x) = \dfrac{x - 5}{x - 2}$$

6. Suppose $g(x) = 5 - \dfrac{1}{2}x$. Determine the y-coordinate of the following points on its graph.

$$(-4, \quad) \qquad (0, \quad) \qquad (3, \quad) \qquad (10, \quad)$$

7. Suppose $f(x) = \dfrac{x}{1 + x^2}$. Which of the following points lie on its graph? Explain.

$$(1, 2) \qquad (2, 5) \qquad \left(\dfrac{1}{2}, \dfrac{2}{5}\right) \qquad (0, 1) \qquad \left(-2, -\dfrac{2}{5}\right)$$

8. Consider the two functions $f(x) = 4x$ and $g(x) = \dfrac{12x}{1 + x}$, each restricted to the domain $0 \le x \le 2$.

a. Complete the following table.

x	$f(x) = 4x$	$g(x) = \dfrac{12x}{1 + x}$
0		
$\frac{1}{2}$		
1		
$\frac{3}{2}$		
2		

b. The input values in the table are equally spaced with spacing $\frac{1}{2}$.

    i. What do you notice about the corresponding spacing of the output values in f?

    ii. What does this tell you about the average rate of change of f over each input interval?

    iii. What "shape" graph do you expect f to have?

    iv. What do you notice about the corresponding spacing of the output values in g?

    v. What does this tell you about the average rate of change of g ?

    vi. What "shape" graph do you expect g to have?

c. i. Write the five input-output pairs for each function using ordered-pair notation.

    ii. Use an appropriate scale to plot the ordered pairs for each function (use a different color or symbol for each function) on the grid below. If necessary, determine several additional ordered pairs to obtain a good visualization of each function. Connect the plotted points in each graph as smoothly as you can to obtain a reasonable sketch of the complete graph. Are the shapes of each graph what you expected?

d. i. Solve the equation $4x = 5$ algebraically.

    ii. Solve the equation $4x = 5$ graphically using the graph in part c.

e. i. Given output value 5, write the equation whose solution provides the corresponding input for function g.

    ii. Use algebraic methods to solve the equation in part a. (Hint: Use cross multiplication)

    iii. Use **graphical** methods to solve the equation in part e.i.

9. The following graphs represent the results of three diet plans. Describe how the graphs are the same and how they are different.

Plan 1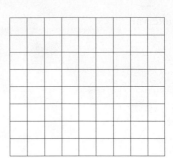

Plan 2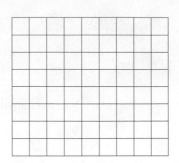

Plan 3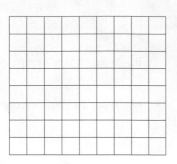

Activity 3.4

Course Grade

Objectives

1. Determine the symbolic rule that defines a function.

2. Identify the practical domain and range of a function.

The semester is drawing to a close, and you are concerned about your grade in your anthropology course. During the semester, you have already taken four exams and scored 82, 75, 85, and 93. Your score on exam 5 will determine your final average for the anthropology course.

1. a. Identify the input and output variables.

b. Four possible exam 5 scores are listed in the following table. Calculate the final average corresponding to each one, and record your answers.

EXAM 5 SCORE, input	FINAL AVERAGE, output
100	
85	
70	
60	

Recall from Activity 3.1 that a function relates the input to the output in a special way. For any specific input value, there is one and only one output value.

c. Is the final average a function of the score on exam 5? Explain.

d. Plot the (input, output) pairs from the preceding table.

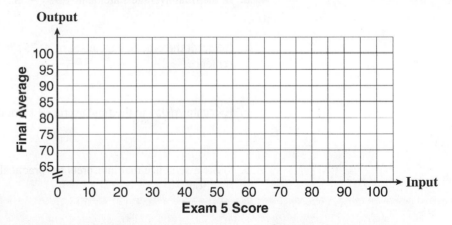

Representing Functions Verbally and Symbolically

There are several ways to represent a function. So far, you have seen that a function may be presented *numerically* by a table and *graphically* on a grid, as in Problem 1. A function can also be given *verbally* by stating how the output value is determined for a given input value.

2. a. Describe in words how to obtain the final average for any given score on the fifth exam.

b. Let *A* represent the final average and *s* represent the score on the fifth exam. Translate the verbal rule in part a into a symbolic rule that expresses *A* in terms of *s*.

3. a. The symbolic rule in Problem 2b represents a fourth way to represent a function. Use the symbolic rule to determine the final average for a score of 75 on exam 5.

    b. Use the symbolic rule to determine the score you will need on exam 5 to earn a final average of 87.

Using function notation, the formula that defines the final average function is written as

$$A(s) = \frac{335 + s}{5}.$$

The symbol $A(100)$ denotes the final average for a score of 100 on the fifth exam. To determine the value of $A(100)$, substitute 100 for s in the expression $\frac{335 + s}{5}$ and evaluate the expression:

$$A(100) = \frac{335 + 100}{5} = \frac{435}{5} = 87$$

You interpret the symbolic statement $A(100) = 87$ as "The final average for a fifth exam score of 100 is 87."

  4. a. In the final average statement $A(75) = 82$, identify the input value and output value.

    b. Interpret the practical meaning of $A(75) = 82$. Write your answer as a complete sentence.

    c. Evaluate $A(95)$ in the final average function.

    d. Write a sentence that interprets the practical meaning of $A(95)$ in the final average function.

Practical Domain and Range

> Recall that the collection of all possible replacement values of the input variable is called the **domain** of the function. The **practical domain** is the collection of replacement values of the input variable that makes practical sense in the context of a particular problem.

  5. a. Determine the practical domain of the final average function. Assume that no fractional part of a point can be given and that the exam has a total of 100 points.

b. What is the domain of the function defined by $A(s) = \dfrac{335 + s}{5}$, where neither A nor s has any contextual significance?

> The collection of all possible values of the output variable is the **range** of the function. The collection of all possible values of the output variable using the practical domain is called the **practical range**.

6. Show or explain how you would determine the practical range of the final average function, and interpret the meaning of this range.

7. You are on your way to take the fifth exam in the anthropology course. The gas gauge on your car indicates that you are almost out of gas. You stop to fill your car with gas.

a. Two input variables determine the cost (output) of a fill-up. What are they? Be specific.

b. Assume that you need 12.6 gallons to fill up your car. Now one of the input variables in part a will become a constant. The value of a constant will not vary throughout the problem. The cost of a fill-up is now dependent on only one variable, the price per gallon. Complete the following table:

Price Per Gallon	3.00	3.50	4.00	4.50	5.00
Cost of Fill-up					

c. Is the cost of a fill-up a function of the price per gallon? Explain.

d. Write a verbal statement that describes how the cost of a fill-up is determined.

e. Let p represent the price of a gallon of gasoline pumped (input) and c represent the cost of the fill-up (output). Translate the verbal statement in part a into a symbolic statement (an equation) that expresses c in terms of p.

f. Using function notation, write the cost if the price is $2.85 per gallon and evaluate. Write the result as an ordered pair.

8. a. Can any real number be substituted for the input variable p in the cost-of-fill-up function? Describe the values of p that make sense, and explain why they do.

   b. Determine the practical domain of the cost-of-fill-up function.

   c. Determine the domain for the general function defined by $c = 12.6p$ with no connection to the context of the situation.

9. a. What is the practical range for the cost function defined by $c(p) = 12.6p$ if the practical domain is 2 to 5?

   b. What is the range of this function if it has no connection to the context of the situation?

SUMMARY: ACTIVITY 3.4

1. Functions can be defined by a **verbal rule** (in words) or **symbolically** (by an equation that indicates the sequence of operations performed on the input to obtain the corresponding output value).

2. In functions that result from contextual situations, the domain consists of input values that make sense within the context. Such a domain is often called a **practical domain**.

3. In functions that result from contextual situations, the **practical range** is the set of output values assigned to each element of the practical domain.

EXERCISES: ACTIVITY 3.4

1. a. You have a part-time job. You work between 0 and 25 hours per week. If you earn $10.50 per hour, write an equation to determine the gross pay, g, for working h hours.

   b. What is the independent variable? What is the dependent variable?

   c. Complete the following table using your graphing calculator.

Number of Hours, h	0	5	10	15	20	25
Gross Pay, g (dollars)						

   d. Using f for the name of the function, the output variable g can be written as $g = f(h)$.

   Rewrite the equation in part a using the function notation $f(h)$ for gross pay.

e. What are the practical domain and the practical range of the function? Explain.

f. Evaluate $f(14)$ and write a sentence describing its meaning.

g. Plot the ordered pairs you determined in part c on an appropriately scaled and labeled set of axes.

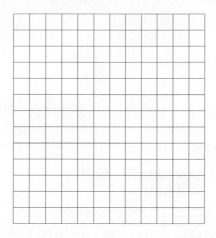

2. Companies often use the method of straight-line depreciation to reduce the value of their assets by a fixed amount each year. The amount of the reduction depends on the useful life of the asset. This is generally determined by the company. Suppose a business selling computer hardware has obtained a fleet of new hybrid cars for its sales force. Each car costs $26,700.

   a. The business estimates that the useful life of each vehicle is 5 years. Using the straight-line method, determine the amount of depreciation per year.

   b. Determine the value of each car after 3 years.

   c. In this situation, the value of each car, v, is a function of the age of the car, a. Identify the independent and dependent variables.

   d. Use function notation to write a function rule for the value $v(a)$ of each car in terms of its age a.

   e. What is the practical domain and practical range of the depreciation function?

   f. Determine $v(4)$. Interpret the result in the context of this situation.

   g. Use the function rule to determine the age of each car if the value is $16,020.

3. To help pay for college, you have accepted a summer position as a sales associate in a family-owned men's and women's specialty clothing store. You are offered a weekly salary of $200 plus 5% commission on all of your sales.

   a. Identify the input and output variables in this situation.

   b. Write a statement that describes how the output in part a is determined for a given input value.

   c. Does the relationship in part b define a function? Explain.

   d. Let $s(t)$ represent the total weekly earning and t represent total sales. Write a function rule for $s(t)$ in terms of t.

   e. What are the practical domain and practical range of the total earnings function?

   f. Evaluate $s(2000)$ and write a sentence explaining the meaning of $s(2000)$ in the context of the situation.

   g. Use the function rule to determine total sales needed to earn a total weekly earning of $400.

4. Your roommate has asked you for help in preparing for the Pharmacy Technician Certification Board (PTCB) exam. One of the responsibilities of a pharmacy technician is to calculate the proper pediatric dose of a medical drug for children. Most technicians use Young's Rule.

   a. In words, Young's Rule states that the pediatric dose of a particular medical drug for a child is the product of the child's age times the adult dosage divided by the sum of the child's age and 12. If A represents the adult dosage and a represents the age of the child in years, write a formula for the child's dosage, denoted by C, in terms of A and a.

   b. The average adult dosage of amoxicillin is 500 milligrams. Substitute this value into the formula in part a. The resulting formula gives the proper dose of amoxicillin for children of age a.

c. Determine the dose for a typical

  i. 3-year-old child

  ii. 4.5-year-old child

d. The pediatric dose of amoxicillin is a function of the age of the child. Identify the independent and dependent variables.

e. Write Young's Rule for amoxicillin using function notation.

f. Determine the value of $C(2)$. Interpret the result in the context of the situation.

5. a. The distance you travel while hiking is a function of how fast you hike and how long you hike at this rate. You usually maintain a speed of 3 miles per hour while hiking. Write a statement that describes how the distance you travel is determined.

b. Identify the independent and dependent variables of this function.

c. Write the statement in part a using function notation. Let t represent the independent variable. Let h represent the function and $h(t)$ represent the dependent variable.

d. Use the equation from part c to determine the distance traveled in 4 hours.

e. Evaluate $h(7)$ and write a sentence describing its meaning. Write the result as an ordered pair.

f. Determine the domain and range of the general function.

g. Determine the practical domain and the practical range of the function.

h. Use technology to generate a table of values beginning at zero with an increment of 0.5.

t						
$h(t)$						

6. Let k be a function defined by $k(w) = 0.4w$. Determine the value of w for which $k(w) = 12$.

Cluster 1   What Have I Learned?

1. What is the mathematical definition of a function? Give a real-life example, and explain how this example satisfies the definition of a function.

2. Explain the meaning of the symbolic statement $H(5) = 100$.

3. The sales tax rate in Ann Arbor, Michigan, is 6%. Consider the function that associates the sales tax (output) paid on an item with the price of the item (input). Represent this function in each of the four formats discussed in this Cluster: verbally, symbolically, numerically, and graphically.

Verbal Definition of the Function:

Symbolic Definition:

Numerical Definition:

Cost of Item (dollars)									
Sales Tax ($)									

Graphical Definition:

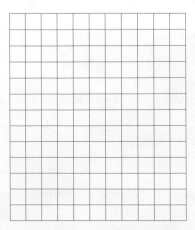

4. A function is defined by the rule $f(x) = 5x - 8$. What does $f(1)$ refer to on the graph of f?

5. The notation $g(t)$ represents the weight (in grams) of a melting ice cube t minutes after being removed from the freezer. Interpret the meaning of $g(10) = 4$.

6. What is true about the sign of the average rate of change between any two points on the graph of an increasing function?

7. You are told that the average rate of change of a particular function is always negative. What can you conclude about the graph of that function and why?

8. Describe a step-by-step procedure for determining the average rate of change between any two points on the graph of a function. Use the points represented by $(85, 350)$ and $(89, 400)$ in your explanation.

9. Give an example of a function from your major field of study or area of interest.

10. Give at least two examples of functions from your daily life. Explain how each fits the description of a function. Be sure to identify the input variable and output variable.

Cluster 1   How Can I Practice?

1. Students at one community college pay $129 per credit hour when taking fewer than 12 credits, provided they are state residents. For 12 or more credit hours, they pay $1550 per semester.

a. Determine the tuition cost for a student taking the given number of credit hours.

NUMBER OF CREDIT HOURS	TUITION COST ($)
3	
6	
10	
12	
16	
18	

b. Is the tuition cost a function of the number of credit hours for the values in your completed table from part a? Explain. Be sure to identify the input and output variables in your explanation.

c. What is the practical domain of the tuition cost function? Assume that there are no half-credit courses. However, there are 1- and 2-credit courses available.

d. Use the table in part a to help graph the tuition cost at this college as a function of the number of credit hours taken.

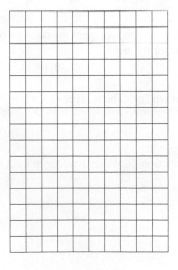

e. Suppose you have saved $700 for tuition. Use the graph to estimate the most credit hours you can take.

f. Let *h* represent the number of credit hours (input) and *C* represent the tuition cost (output). Write a symbolic rule for the cost of part-time tuition in terms of the number of credit hours taken.

g. Use your equation from part f to verify the tuition cost for the credit hours given in the table in part a.

2. The cost of a history club trip includes $78 for transportation rental plus a $2 admission charge for each student participating.

a. The total cost of the trip depends on the number of students participating. Identify the input and output variables.

b. Write a verbal rule to determine the total cost of the trip in terms of the number of students participating.

c. Complete the following table.

NUMBER OF STUDENTS	COST OF TRIP ($)
10	
15	
20	
25	

d. Write the symbolic rule using function notation. Use *n* to represent the number of students and *C* to represent the name of the cost function for the trip.

e. Use function notation to write a symbolic statement for "The total cost is $108 if 15 students go on the trip."

f. Choose another input-output pair for the cost function *C*. Write this pair in both ordered-pair notation and in function notation.

g. What is the practical domain if the club's transportation is a bus?

h. Determine $C(37)$.

i. Describe, in words, what information about the cost of the trip is given by the statement
$C(24) = 126$.

3. You bought a company in 2011 and have tracked the company's profits and losses from its beginning in 2005 to the present. You decide to graph the information where the number of years since
2011 is the x-variable and profit or loss for the year is the y-variable. Note that the year 2011
corresponds to zero on the x-axis. Determine the quadrant or axis on which you would plot the
points that correspond to the following data. If your answer is on an axis, indicate between which
quadrants the point is located.

a. The loss in 2008 was $1500.

b. The profit in 2012 was $6000.

c. The loss in 2014 was $1000.

d. In 2007, there was no profit or loss.

e. The profit in 2005 was $500.

f. The loss in 2011 was $800.

4. The following table shows the total number of points accumulated by each student and the
numerical grade in a course.

STUDENT	TOTAL POINTS	NUMERICAL GRADE
TOM	432	86.4
JEN	394	78.8
KATHY	495	99
MICHAEL	330	66
BRADY	213	42.6

a. Is the numerical grade a function of the total number of points? Explain.

b. Is the total number of points for these five students a function of the numerical grade?
Explain.

c. Using the total points as the input and the numerical grade as the output, write the ordered
pairs that represent each student. Call this function f, and write it as a set of ordered pairs.

d. Plot the ordered pairs you determined in part c on an appropriately scaled and labeled set of axes.

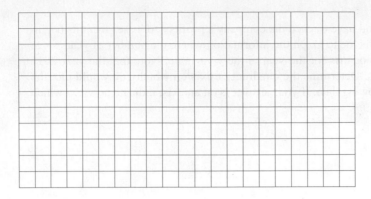

e. What is the value of $f(394)$?

f. What is the practical meaning of $f(394)$?

g. What is the value of $f(213)$?

h. What is the practical meaning of $f(213)$?

i. Determine the numerical value n, given that $f(n) = 66$.

5. In parts a–g, determine which of the given relationships represent functions.

   a. The money you earn at a fixed hourly rate is a function of the number of hours you work.

   b. Your heart rate is a function of your level of activity.

   c. The cost of daycare depends on the number of hours a child stays at the facility.

   d. The number of children in a family is a function of the parents' last name.

   e. $\{(2, 3), (4, 3), (5, -5)\}$

   f. $\{(-3, 4), (-3, 6), (2, 6)\}$

   g.

x	−3	5	7
$f(x)$	0	−5	9

6. Let f be defined by the set of two ordered pairs $\{(2, 3), (0, -5)\}$.

   a. List the set of numbers that constitute the domain of f.

   b. List the set of numbers that constitute the range of f.

7. You decide to lose weight and will cut down on your calories to lose 2 pounds per week. Suppose that your present weight is 180 pounds. Sketch a graph covering 20 weeks showing your projected weight loss. Describe your graph. If you stick to your plan, how much will you weigh in 3 months (13 weeks)?

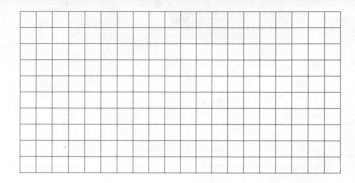

8. A taxicab driver charges a flat rate of $2.50 plus $2.20 per mile. The fare F (in dollars) is a function of the distance driven, x (in miles). The driver wants to display a table for her customers to show approximate fares for different locations within the city.

   a. Write a symbolic rule for F in terms of x.

   b. Use the symbolic rule to complete the following table.

x (mi.)	0.25	0.5	0.75	1.0	1.5	2.0	3.0	5.0	10.0
F(x)									

9. Let $f(x) = -3x + 4$. Determine $f(-5)$.

10. Let $g(x) = (x + 3)(x - 2)$. Determine $g(-4)$.

11. Let $m(x) = 2x^2 + 6x - 7$. Determine $m(3)$.

12. Let $h(s) = (s - 1)^2$. Determine $h(-3)$.

13. Let f be a function defined by $f(x) = x - 6$. Determine the value of x for which $f(x) = 10$.

14. Let h be a function defined by $h(x) = 2x + 1$. Determine the value of x for which $h(x) = 13$.

15. As part of your special diet and exercise program, you record your weight at the beginning of the program and each week thereafter. The following data gives your weight, w, over a 5-week period.

Time, t (week)	0	1	2	3	4	5
Weight, w (lb.)	196	183	180	177	174	171

a. Sketch a graph of the data on appropriately scaled and labeled axes.

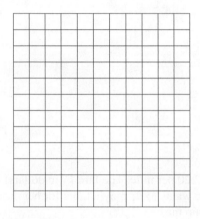

b. Determine the average rate of change of your weight during the first 3 weeks. Be sure to include the units of measurement of this rate.

c. Determine the average rate of change during the 5-week period.

d. On the graph in part a, connect the points $(0, 196)$ and $(3, 177)$ with a line segment. Does the line segment rise, fall, or remain horizontal as you follow the line left to right?

e. What is the practical meaning of the average rate of change in this situation?

f. What can you say about the average rate of change of weight during any of the time intervals in this situation?

16. The total amount of rainfall in a given area can vary widely from year to year. The following table gives information on the total rainfall received over a recent 7-year period for the area.

ANNUAL RAINFALL							
Year, t	1	2	3	4	5	6	7
Rainfall, r (in.)	45.49	41.88	39.63	32.91	37.47	50.08	37.54

a. Plot the data points using ordered pairs of the form (t, r).

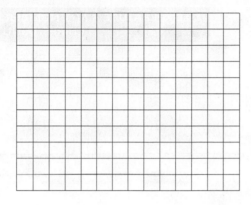

b. Determine the average rate of change of annual rainfall in the area from year 4 to year 7.

c. Determine the average rate of change of annual rainfall in the area from year 1 to year 4.

d. Compare the average rate of change from year 1 to year 4 to the average rate of change from year 4 to year 7.

e. When the average rate of change is negative, what trend do you observe in the graph? What does that mean in this situation?

Cluster 2   Introduction to Linear Functions

Activity 3.5

The Snowy Tree Cricket

Objectives

1. Identify linear functions by a constant average rate of change of the output variable with respect to the input variable.

2. Interpret slope as an average rate of change.

3. Determine the slope of a nonvertical line drawn through two points.

4. Identify increasing and decreasing linear functions using slope.

5. Determine horizontal and vertical intercepts of a linear function from its graph.

6. Interpret the meaning of horizontal and vertical intercepts in a contextual situation.

One of the more familiar late-evening sounds during the summer is the rhythmic chirping of a male cricket. Of particular interest is the snowy tree cricket, sometimes called the temperature cricket. It is very sensitive to temperature, speeding up or slowing down its chirping as the temperature rises or falls. The following data shows the number of chirps per minute of the snowy tree cricket as a function of temperature.

t, TEMPERATURE (°F)	$N(t)$, NUMBER OF CHIRPS/MINUTE
55	60
60	80
65	100
70	120
75	140
80	160

1. Crickets are usually silent when the temperature falls below 55°F. What is a possible practical domain for the snowy tree cricket function?

2. **a.** Determine the average rate of change of the number of chirps per minute with respect to temperature as the temperature increases from 55°F to 60°F.

   b. What are the units of measure of this rate of change?

3. **a.** How does the average rate of change determined in Problem 2 compare with the average rate of change as the temperature increases from 65°F to 80°F?

   b. Determine the average rate of change of number of chirps per minute with respect to temperature for the temperature intervals given in the following table. The results from Problems 2 and 3 are already recorded. Add several more of your own choice. List all your results in the table.

TEMPERATURE INCREASES	AVERAGE RATE OF CHANGE (number of chirps/minute per degree F)
From 55° to 60°F	4
From 65° to 80°F	4
From 55° to 75°F	
From 60° to 80°F	

c. What can you conclude about the average rate of increase in the number of chirps per minute for any particular increase in temperature?

4. Because the average rate of change in the number of chirps per minute with respect to temperature is constant, what type of graph do you expect?

5. Plot the data pairs (temperature, chirps per minute) from the table preceding Problem 1. Does the pattern of points confirm what you expected? Explain.

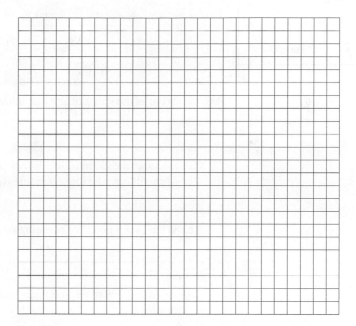

Linear Functions

If the average rate of change in output with respect to input remains constant (stays the same) for *any* two points in a data set, then all the points will lie on a single line. That is, the output is a **linear** function of the input. Conversely, if all the points of a data set lie on a line when graphed, the average rate of change of output with respect to input will be constant for any two data points.

Slope of a Line

The average rate of change for a linear function determines the steepness of the line and is called the *slope* of the line.

Definition

The **slope** of a line is a measure of its steepness. It is the average rate of change between any two points on the line. Symbolically, the letter m is used to denote slope:

$$\text{slope} = m = \frac{\text{change in output}}{\text{change in input}} = \frac{\Delta y}{\Delta x} = \frac{y_2 - y_1}{x_2 - x_1},$$

where (x_1, y_1) and (x_2, y_2) are any two points on the line and $x_1 \neq x_2$.

Example 1 *Determine the slope of the line containing the points $(1, -2)$ and $(3, 8)$.*

SOLUTION

Let $x_1 = 1, y_1 = -2, x_2 = 3$, and $y_2 = 8$; so

$$m = \frac{\Delta y}{\Delta x} = \frac{y_2 - y_1}{x_2 - x_1} = \frac{8 - (-2)}{3 - 1} = \frac{10}{2} = \frac{5}{1} = 5.$$

6. a. What is the slope of the line in the snowy tree cricket situation?

b. Because the slope of this line is positive, what can you conclude about the direction of the line as the input variable (temperature) increases in value?

c. What is the practical meaning of slope in this situation?

d. For any 7° increase in temperature, what is the expected increase in chirps per minute?

Geometric Meaning of Slope

On a graph, slope can be understood as the ratio of two distances. For example, in the snowy tree cricket situation, the slope of the line between the points $(55, 60)$ and $(56, 64)$ as well as between $(56, 64)$ and $(57, 68)$ is $\frac{4}{1}$, as shown on the following graph. The change in the input, 1, represents a horizontal distance (the run) in going from one point to another point on the same line. The change in the output, 4, represents a vertical distance (the rise) between the same points. The graph on the right illustrates that a horizontal distance (run) of 2 and a vertical distance (rise) of 8 from $(55, 60)$ will also locate the point $(57, 68)$ on the line.

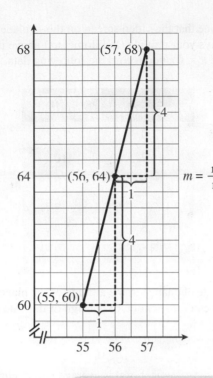

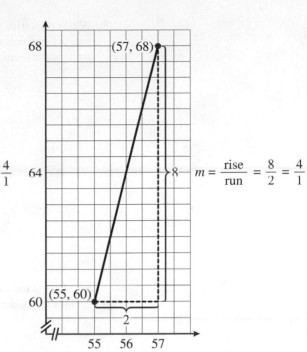

From its geometric meaning, slope can be determined using the vertical and horizontal distances between any two points on a line by the formula

$$m = \text{slope} = \frac{\text{rise}}{\text{run}} = \frac{\text{distance up } (+) \text{ or down } (-)}{\text{distance right } (+) \text{ or left } (-)}.$$

Because all slopes can be written in fraction form, including fractions having denominator 1, you can extend the geometric meaning of slope by forming equivalent fractions.

Note that a positive slope of $\frac{5}{3}$ can be interpreted as $\frac{+5}{+3} = \frac{\text{up } 5}{\text{right } 3}$ or $\frac{-5}{-3} = \frac{\text{down } 5}{\text{left } 3}$ or as any other equivalent fraction, such as $\frac{-10}{-6}$.

The slope can be used to locate additional points on a graph.

7. Consider the line containing the point $(-6, 4)$ and having slope $\frac{3}{4}$. Plot the point, and then determine two additional points on the line. Draw a line through these points.

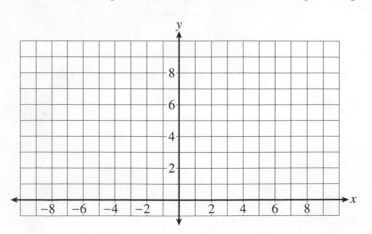

8. While on a trip, you notice that the video screen on the airplane, in addition to showing movies and news, displays your altitude (in kilometers) above the ground. As the plane starts its descent (at time $t = 0$), you record the following data.

TIME, t (min)	ALTITUDE, $A(t)$ (km)
0	12
2	10
4	8
6	6
8	4
10	2

a. What is the average rate of change in the altitude of the plane from 2 to 6 minutes into the descent? Pay careful attention to the sign of this rate of change.

b. What are the units of measurement of this average rate of change?

c. Determine the average rate of change over several other input intervals.

d. What is the significance of the sign of these average rates of change?

e. Based on your calculation in parts a and c, do you think that the data lie on a single straight line? Explain.

f. What is the slope of the line? What is the practical meaning of slope in this situation?

9. a. What is the domain of the descent function?

b. Graph the data points from the table in Problem 8.

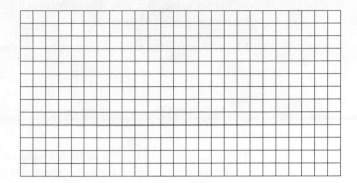

c. Is the descent function an increasing or decreasing function?

Horizontal and Vertical Intercepts

10. a. From the time the plane (in Problems 8 and 9) begins its descent, how many minutes does it take to reach the ground?

b. Use a straightedge to connect the data points on your graph in Problem 9. Extend the line so that it crosses both axes.

> ### Definition
>
> A **horizontal intercept** of a graph is a point at which the graph crosses (or touches) the horizontal (input) axis.

A horizontal intercept is clearly identified by noting that it is the point at which the output value is zero. The ordered pair notation for all horizontal intercepts always has the form $(a, 0)$, where a is the input value. Because it is understood that its output value is 0, the horizontal intercept is occasionally referred to simply by a.

11. a. Identify the horizontal intercept of the line in Problem 9 from the graph.

b. How is the horizontal intercept related to the answer you obtained in Problem 10a? That is, what is the practical meaning of the horizontal intercept?

> ### Definition
>
> A **vertical intercept** of a graph is a point at which the graph crosses (or touches) the vertical (output) axis.

A vertical intercept is clearly identified by noting that it is the point at which the input value is zero. The ordered-pair notation for all vertical intercepts has the form $(0, b)$, where b is the output value. Because it is understood that its input value is 0, the vertical intercept is occasionally referred to simply by b.

12. a. Identify the vertical intercept of the descent function from its graph in Problem 9.

b. What is the practical meaning of this intercept?

If x represents the independent variable and y represents the dependent variable, then the horizontal intercept is called the **x-intercept** and the vertical intercept is called the **y-intercept**.

13. For each of the following lines, determine
   i. the slope.
   ii. the horizontal (x) intercept.
   iii. the vertical (y) intercept.

a.

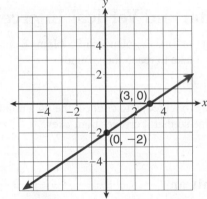

b.

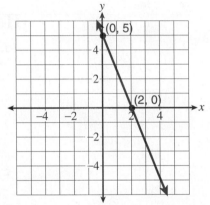

SUMMARY: ACTIVITY 3.5

1. A **linear function** is one whose average rate of change of output with respect to input from any one data point to any other data point is always the same (constant) value.

2. The **graph** of a linear function is a line whose slope is the constant rate of change of the function.

3. The **slope of a line segment** joining two points (x_1, y_1) and (x_2, y_2) is denoted by m and can be calculated using the formula $m = \dfrac{\Delta y}{\Delta x} = \dfrac{y_2 - y_1}{x_2 - x_1}$, where $x_1 \neq x_2$. Geometrically, Δy represents a vertical distance (rise) and Δx represents a horizontal distance (run).

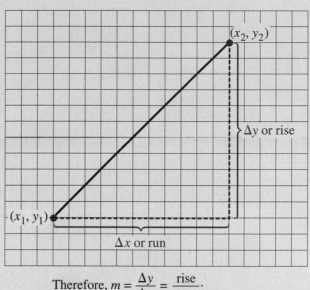

Therefore, $m = \dfrac{\Delta y}{\Delta x} = \dfrac{\text{rise}}{\text{run}}$.

4. The graph of every linear function with **positive slope** is a line rising to the right. The output values increase as the input values increase. The function is then said to be an **increasing function**.

5. The graph of every linear function with **negative slope** is a line falling to the right. The output values decrease as the input values increase. The function is said to be a **decreasing function**.

6. The **horizontal intercept** is the point at which the graph crosses or touches the horizontal (input) axis. Its ordered-pair notation is $(a, 0)$; that is, the output value is equal to zero. If the input is denoted by x, then the horizontal intercept is referred to as the x-intercept.

7. The **vertical intercept** is the point at which the graph crosses or touches the vertical (output) axis. Its ordered-pair notation is $(0, b)$; that is, the input value is equal to zero. If the output is denoted by y, then the intercept is referred to as the y-intercept.

EXERCISES: ACTIVITY 3.5

1. Consider the following data regarding the growth of the U.S. national debt from 1950 to 2008.

Number of Years since 1950	0	10	20	30	40	50	58
National Debt (billions of dollars)	257	291	381	909	3113	5662	10,700

a. Compare the average rate of increase in the national debt from 1950 to 1960 with that from 1980 to 1990. Is the average rate of change constant? Explain using the data.

b. Plot the data points. If the points are connected to form a smooth curve, is the graph a straight line? Are the input and output variables in this problem related linearly?

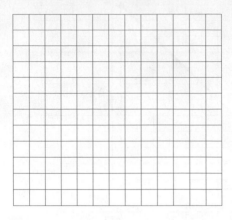

2. Calculate the average rate of change between consecutive data points to determine whether the output in each table is a linear function of the input.

a.

Input	−5	0	5	8
Output	−45	−5	35	59

b.

Input	2	7	12	17
Output	0	10	16	18

3. For each of the following, determine two additional points on the line. Then sketch a graph of the line.

a. A line contains the point $(-5, 10)$ and has slope $\frac{2}{3}$.

b. A line contains the point $(3, -4)$ and has slope 5.

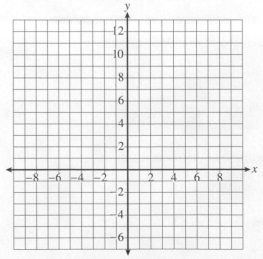

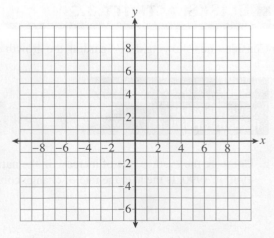

4. The concept of slope arises in many practical applications. When designing and building roads, engineers and surveyors need to be concerned about the grade of the road. The grade, usually expressed as a percent, is one way to describe the steepness of the finished surface of the road.

For example, a 5% grade means that the road has a slope (rise over run) of $0.05 = \dfrac{5}{100}$.

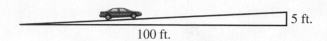

100 ft.

5 ft.

a. If a road has a 5% upward grade over a 1000-foot run, how much higher will you be at the end of that run than at the beginning?

b. What is the grade of a road that rises 26 feet over a distance of 500 feet?

5. The American National Standards Institute (ANSI) requires that the slope for a wheelchair ramp not exceed $\dfrac{1}{12}$.

a. Does a ramp that is 160 inches long and 10 inches high meet the requirements of ANSI? Explain.

b. A ramp for a wheelchair must be 20 inches high. Determine the minimum horizontal length of the ramp so that it meets the ANSI requirement.

6. In a science lab, you collect the following sets of data. Which of the four data sets are linear functions? If linear, determine the slope.

a.

Time (sec.)	0	10	20	30	40
Temperature (°C)	12	17	22	27	32

b.

Time (sec.)	0	10	20	30	40
Temperature (°C)	41	23	5	−10	−20

c.

Time (sec.)	3	5	8	10	15
Temperature (°C)	12	16	24	28	36

d.

Time (sec.)	3	9	12	18	21
Temperature (°C)	25	23	22	20	19

7. a. A special diet and exercise program has been developed for you by a registered dietitian and your personal trainer. You weigh 181 pounds and would like to lose 2 pounds every week. Complete the following table of values for your desired weight each week.

N, Number of Weeks	0	1	2	3	4
W(N), Desired Weight (lb.)					

b. Plot the data points.

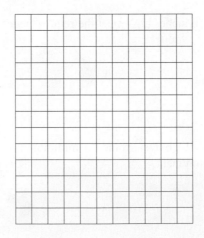

c. Explain why your desired weight is a linear function of time. What is the slope of the line containing the five data points?

d. What is the practical meaning of slope in this situation?

e. How long will it take you to reach your ideal weight of 168 pounds?

8. To receive the full physical benefit from exercising, your heart rate must be maintained at a certain level for at least 12 minutes. The proper exercise heart rate for a healthy person, called the target heart rate, is determined by the person's age. The relationship between these two quantities is illustrated by the data in the following table.

A, Age (yr.)	20	30	40	50	60
B(A), Target Heart Rate (beats/min.)	140	133	126	119	112

a. Does the data in the table indicate that the target heart rate is a linear function of age? Explain.

b. What is the slope of the line for this data? What are the units?

c. What are suitable replacement values (practical domain) for age, A?

d. Plot the data points on coordinate axes where both axes start with zero, with 10 units between grid lines. Connect the points with a line.

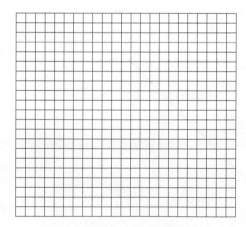

e. Extend the line to locate the horizontal and vertical intercepts. Do these intercepts have a practical meaning in the problem? Explain.

9. a. Determine the slope of each of the following lines.

i.

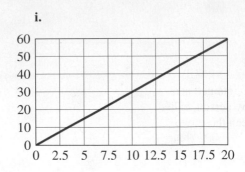

ii.

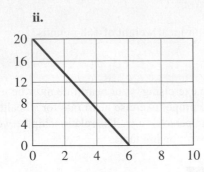

iii.

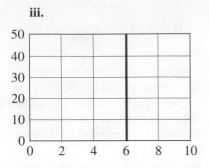

b. Determine the horizontal and vertical intercept of each of the lines in part a.

INTERCEPT	GRAPH i	GRAPH ii	GRAPH iii
Horizontal			
Vertical			

10. Each question refers to the graph that accompanies it. The graphed line in each grid represents the total distance a car travels as a function of time (in hours).

a. How fast is the car traveling? Explain how you obtained your result.

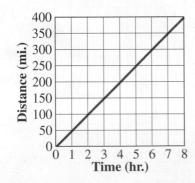

b. How can you determine visually from the following graph which car is going faster? Verify your answer by calculating the speed of each car.

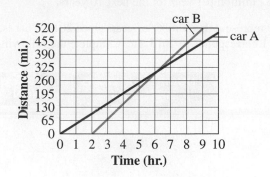

c. Describe in words the movement of the car that is represented by the following graph.

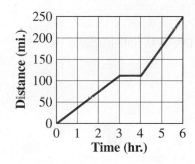

11. a. Determine the slope of each of the following lines.

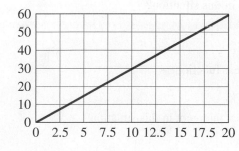 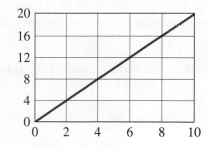 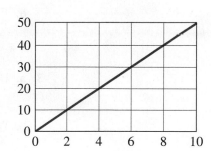

b. At first glance, the three graphs in part a may appear to represent the same line. Do they?

Activity 3.6

Software Sales

Objectives

1. Identify whether a situation can be represented by a linear function.

2. Write an equation of a line in slope-intercept form.

3. Use the y-intercept and the slope to graph a linear function.

4. Determine horizontal intercepts of linear functions using an algebraic approach.

5. Use intercepts to graph a linear function.

6. Identify parallel lines.

You have been hired by a company that sells computer software products. In 2014, the company's total (annual) sales were \$16 million. Its marketing department projects that sales will increase by \$2 million per year for the next 10 years.

1. a. Let t represent the number of years since 2014. That is, $t = 0$ corresponds to 2014, $t = 1$ corresponds to 2015, etc. Complete the following table.

t, NUMBER OF YEARS SINCE 2014	s, TOTAL SALES IN MILLIONS OF DOLLARS
0	
1	
2	
5	

b. Write a symbolic rule that would express the total sales, s, in terms of the number of years, t, since 2014.

c. Does the symbolic rule define a function?

d. What is the practical domain of this total sales function?

e. Is the total sales function linear? Explain.

2. a. Determine the slope of the total annual sales function. What are the units of measurement of the slope?

b. What is the practical meaning of the slope in this situation?

3. a. Determine the vertical intercept of this linear function.

b. What is the practical meaning of the vertical intercept in this situation?

c. Determine $s(6)$. Interpret the meaning of the result.

d. Use the symbolic rule $s(t) = 2t + 16$ for the total sales function to approximate the year in which total sales will reach \$32 million. What ordered pair on the graph conveys the same information?

e. Use the given symbolic rule $s(t) = 2t + 16$ to determine the total sales in 2019. What ordered pair conveys the same information?

Slope-Intercept Form of an Equation of a Line

The total sales function defined by $s(t) = 2t + 16$ has a symbolic form that is representative of *all* linear functions. That is, the symbolic form of a linear function consists of the sum of two terms:

- a *variable term* (the input variable multiplied by its coefficient)

- and a *constant term* (a fixed number)

4. a. Identify the variable term in the symbolic rule $s(t) = 2t + 16$. What is its coefficient?

b. Identify the constant term in this symbolic rule.

c. What characteristic of the linear function graph does the coefficient of the input variable t represent?

d. What characteristic of the linear function graph does the constant term represent?

5. a. Consider a line defined by the equation $y = 2x + 7$. Use the equation to complete the following table.

x	−2	−1	0	1	2
y					

b. Use the slope formula to determine the slope of the line. How does the slope compare to the coefficient of x in the equation?

c. Determine the vertical (y-) intercept. How does it compare to the constant term in the equation of the line?

The answers to Problems 4c and d and Problem 5 generalize to all linear functions.

When the horizontal and vertical axes are labeled, respectively, as the x- and y-axes, the coordinates of every point (x, y) on a nonvertical line will satisfy the equation $y = mx + b$, where m is the slope of the line and $(0, b)$ is its y-intercept. This equation is often called the **slope-intercept** form of the equation of a line.

Example 1 *Given the linear function rule $y = 3x + 1$, identify the slope and y-intercept.*

SOLUTION

The slope of the line is 3, the coefficient of x; the y-intercept is $(0, 1)$.

Example 2 *Identify the slope and y-intercept of each of the following.*

LINEAR FUNCTION RULE	m, SLOPE	$(0, b)$, y-INTERCEPT
$y = 5x + 3$	5	$(0, 3)$
$y = -2x + 7$	-2	$(0, 7)$
$y = \frac{1}{2}x - 4$	$\frac{1}{2}$	$(0, -4)$
$y = 3x + 0$, or simply $y = 3x$	3	$(0, 0)$
$y = 10 + 6x$	6	$(0, 10)$
$y = 85 - 7x$	-7	$(0, 85)$

6. Identify the slope and vertical intercept of each of the following linear functions.

a. $y = -3x + 8$

b. $y = -5x$

c. $y = 16 + 4x$

d. $p = -12 + 2.5n$

7. For each of the following, determine the equation of the line having the given slope and y-intercept.

    a. The slope is 3, and the y-intercept is (0, 4).

    b. The slope is −1 and the y-intercept is (0, 0).

    c. The slope is $\frac{2}{3}$, and the y-intercept is (0, 6).

    d. The slope is 0 and the y-intercept is (0, −5).

Graphing Linear Functions Using the Vertical Intercept and Slope

Example 3   *Use the vertical intercept and slope to plot the total sales function defined by $s = 2t + 16$.*

SOLUTION

Locate and mark the s-intercept, (0, 16).

Write the slope $m = 2$ as a fraction $\frac{2}{1}$, which indicates a move $\frac{up\ 2}{right\ 1}$ from the starting point (0, 16) to the point (1, 18). Mark this point.

Finally, use a straightedge to draw the line through (0, 16) and (1, 18).

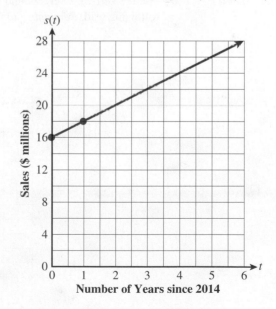

8. a. Start at the s-intercept (0, 16), and explain how to use the slope, $m = 2$, to calculate the coordinates of the point (1, 18) without actually moving on the graph.

b. Interpret the practical meaning of the ordered pair (1, 18) in terms of the total sales situation.

9. Use the slope once again to reach a third point on the line. Interpret the practical meaning of this new ordered pair in terms of the total sales situation.

10. a. Use the slope to determine the change in total sales over any 6-year period.

b. Use the result of part a to determine the coordinates of the point corresponding to the year 2020.

11. Sales of a start-up software company were $6 million in 2015, and the company anticipates sales to increase by $3 million per year for the next several years.

a. Write a symbolic rule to express the total sales, *s*, in terms of the number of years, *t*, since 2015.

b. Use the vertical intercept and slope to draw the graph of the symbolic rule on the following grid. *Remember to first label the axes with appropriate scales.*

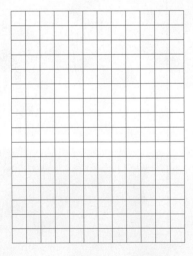

c. Use the slope to locate two more points on the line. Interpret the meanings of the two points you locate.

Graphing Linear Functions Using Intercepts

Another way to graph a linear function is to plot its vertical and horizontal intercepts and then use a straightedge to draw the line containing these two points.

12. You purchase a laptop computer so that you can use the software products you acquired through your job. The initial cost of the computer is $1350. You expect that the computer will depreciate (lose value) at the rate of $450 per year.

    a. Write a symbolic rule that will determine the value, $v(t)$, of the computer in terms of the number of years, t, that you own it.

    b. Write the ordered pairs that represent the vertical and horizontal intercepts.

       i. Because the vertical intercept is a point on the vertical axis, what is the value of its horizontal coordinate? Use this information to determine the value of the vertical intercept and record your result in the following table.

      ii. Because the horizontal intercept is a point on the horizontal axis, what is the value of its vertical coordinate? Use this information to determine the value of the horizontal intercept and record your result in the table.

INTERCEPTS	t, NUMBER OF YEARS	v, VALUE OF COMPUTER ($)
Vertical		
Horizontal		

    c. On the following grid, plot the intercept points you determined in part b. Then use a straightedge to draw a straight line through the intercepts.

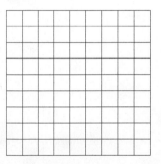

    d. What is the practical meaning of the vertical intercept in this situation?

    e. What is the practical interpretation of the horizontal intercept in this situation?

    f. What portion of the line in part c can be used to represent the computer value situation? (Hint: What is the practical domain of this function?)

13. Determine the vertical and horizontal intercepts for each of the following. Then sketch a graph of the line using the intercepts. Use technology, if available, to check your results.

    a. $y = -3x + 6$ **b.** $f(x) = \dfrac{1}{2}x - 8$

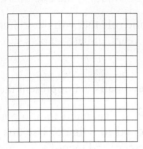 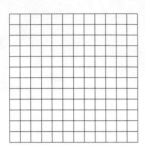

14. a. Solve the equation $3x + 4y = 12$ for y. What do you notice about the form of the resulting equation? What can you conclude about its graph?

    b. Determine the intercepts and slope of the line in part a.

Parallel Lines

15. a. Graph the following linear functions on the same coordinate axes.

          $y = 2x - 3$ $y = 2x + 2$ $y = 2x + 5$

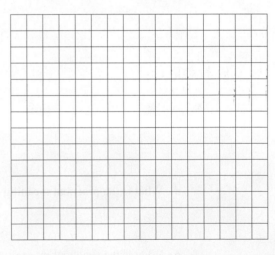

    b. In what ways are the lines similar? In what ways are the lines different?

> **Definition**
>
> Two lines are **parallel** if they have the same slope, but different y-intercepts.

16. Write the equation of a line that is parallel to the line having equation $y = -3x + 4$.

SUMMARY: ACTIVITY 3.6

1. The symbolic rule for a linear function, also called the **slope-intercept form** of the line, is given by $y = mx + b$, where m is the slope and $(0, b)$ is the y- (vertical) intercept of the line.

2. To **plot** a **linear function** using slope-intercept form,

   - plot the vertical intercept on the vertical axis.

   - write the slope in fractional form as $\dfrac{\text{change in output}}{\text{change in input}}$

   - start at the vertical intercept. Move up or down as many units as the numerator indicates, and then move to the right or left as many units as the denominator indicates. It is often helpful to move multiples of numerator and denominator units so that the second point is a larger distance from the vertical intercept. Mark the point you have reached.

   - use a straightedge to draw a line between the two points.

3. Given an equation of a line, determine its **y-intercept** by setting $x = 0$ and calculating the corresponding y-value.

4. Given an equation of a line, determine its **x-intercept** by setting $y = 0$ and calculating the corresponding x-value.

5. To plot a linear function using its intercepts,

   - determine the horizontal and vertical intercepts, and then use a straightedge to draw the line containing the two points.

6. Two lines are parallel if they never intersect, no matter how far you extend the lines in either direction. Parallel lines have equal slopes, but different y-itercepts.

EXERCISES: ACTIVITY 3.6

1. Consider the line having equation $y = -3x + 1.5$.

   a. Complete the following table.

x	-2	-1	0	1	2
y					

   b. Use the slope formula to determine the slope of the line. How does the slope compare to the coefficient of the input variable x in the equation of the line?

   c. How does the y-intercept you determined in part a compare to the constant term in the equation of the line?

Exercise numbers appearing in color are answered in the Selected Answers appendix.

2. Identify the slope and vertical intercept of each of the following.

LINEAR FUNCTION RULE	m, SLOPE	(0, b), VERTICAL INTERCEPT
$y = 3x - 2$		
$y = -2x + 5$		
$y = \dfrac{1}{2}x + 3$		
$y = -2x$		
$y = 6$		
$y = 9 + 3.5x$		
$w = -25 + 8x$		
$v = 48 - 32t$		
$z = -15 - 6u$		

3. Housing prices in your neighborhood have been increasing steadily since you purchased your home in 2009. The relationship between the market value, V, of your home and the length of time, x, you have owned your home is modeled by the symbolic rule

$$V(x) = 2500x + 125{,}000,$$

where $V(x)$ is measured in dollars and x in years.

a. Explain how you know that the graph of this symbolic rule is a line. What is the slope of this line? What is the practical meaning of slope in this situation?

b. Determine the vertical intercept. What is the practical meaning of this intercept in the context of this problem?

c. Determine and interpret the value $V(8)$.

4. The value of a car decreases (depreciates) immediately after it is purchased. The value of a car you recently purchased can be modeled by the symbolic rule

$$V(x) = -1350x + 18{,}500,$$

where $V(x)$ is the market value (in dollars) and x is the length of time you own your car (in years).

a. Explain how you know that the graph of this relationship is a line. Determine the slope of this line. What is the practical meaning of slope in this situation?

b. Determine the vertical intercept. What is the practical meaning of this intercept?

c. Determine and interpret the value $V(3)$.

5. a. Given the linear function rule $y = -x + 4$, identify the slope and y-intercept.

b. Determine which of the following points lie on the line: $(-1, 5), (3, 1)$, and $(5, 9)$.

6. a. Determine an equation of the line whose slope is 2 and whose y-intercept is $(0, -3)$.

b. Determine an equation of the line whose slope is -3 and whose y-intercept is $(0, 0)$.

c. Determine an equation of the line whose slope is $\dfrac{3}{4}$ if it contains the point $(0, 1)$.

In Exercises 7–12, determine the slope, y-intercept, and x-intercept of each line. Then sketch each graph, labeling and verifying the coordinates of each intercept. Use technology to check your results.

7. $y = 3x - 4$ **8.** $y = -5x + 2$

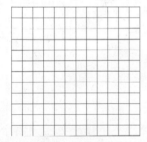 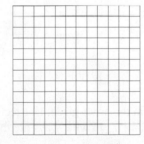

9. $y = 8$ **10.** $y = \dfrac{x}{2} + 5$

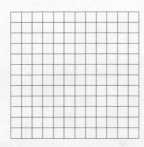 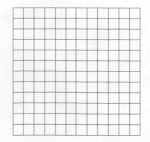

11. $2x - y = 3$

12. $3x + 2y = 1$

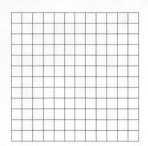

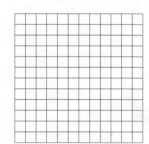

13. Graph the following linear functions in the order given. Use technology to verify your answers. In what ways are the lines similar? In what ways are they different?

a. $y = x - 4$

b. $g(x) = x$

c. $y = x + 2$

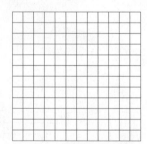

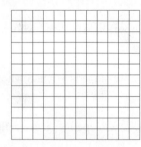

14. Graph the following linear functions in the order given. Use technology to verify your answers. In what ways are the lines similar? In what ways are they different?

a. $y = -4x + 2$

b. $h(x) = -2x + 2$

c. $y = 2$

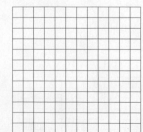

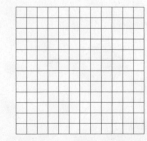

d. $g(x) = 2x + 2$ **e.** $y = 4x + 2$

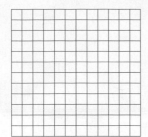 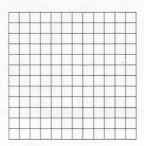

15. What is the equation of the linear function with slope 12 and y-intercept $(0, 3)$?

16. a. You start with \$20 in your savings account and add \$10 every week. At what rate does the amount in your account, excluding interest, change from week to week?

b. Write an equation that models your savings, $s(t)$, as a function of time, t (in weeks).

17. a. What is the slope of the line that goes through the points $(0, 5)$ and $(2, 11)$?

b. What is the equation (symbolic rule) of the line through these two points?

18. a. What is the slope of the line that goes through the points $(0, -43.5)$ and $(-1, 13.5)$?

b. What is the equation of the line through these two points?

19. Determine the horizontal and vertical intercepts of each of the following. Use the intercepts to sketch a graph of the function.

a. $y = -3x + 12$

b. $y = \dfrac{1}{2}x + 6$

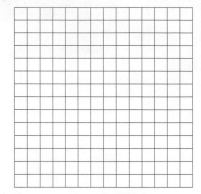
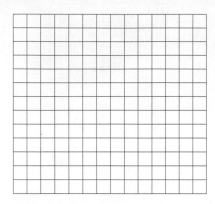

20. After applying the brakes a car travelling at 60 miles per hour continues moving for 3 seconds before stopping 132 feet from where the brakes were applied. The information about its speed at different times is summarized in the table.

Time from Application of Brakes, t, (s)	0	3
Speed, v, (MPH)	60	0

a. Assume that the speed of the car, v, is a linear function of the time, t, since the brakes were applied. Determine the slope of the line containing the points (0, 60) and (3, 0). What is the practical meaning of slope in this situation?

b. Determine the equation of the line in part a.

c. Determine the speed of the car half a second before it finally stops.

Activity 3.7

Predicting Population

Objectives

1. Use the slope-intercept form of linear equations to solve problems.

2. Determine the relative error in a measurement or prediction using a linear model.

The U.S. Census Bureau keeps historical records on populations in the United States from the year 1790 onward. The bureau's records show that from 1940 to 1950, the yearly changes in the national population were quite close to being constant. Therefore, for the objectives of this activity, you may assume that the average rate of change of population with respect to time is a constant value in this decade. In other words, the relationship between time and population may be considered linear from 1940 to 1950.

1. a. According to the U.S. Census Bureau, the population of the United States was approximately 132 million in 1940 and 151 million in 1950. Write the data as ordered pairs of the form $(t, P(t))$, where t is the number of years since 1940 and $P(t)$ is the corresponding population, in millions.

b. Plot the two data points, and draw a straight line through them. Label the horizontal axis from 0 to 25 and the vertical axis from 130 to 180, compressing the axis between 0 and 130.

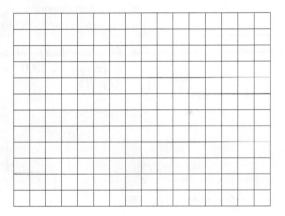

2. a. What is the average rate of change of population from $t = 0$ (1940) to $t = 10$ (1950)?

b. What is the slope of the line connecting the two points in part a? What is the practical meaning of the slope in this situation?

3. What is the vertical intercept of this line? What is the practical meaning of the vertical intercept in this situation?

4. a. Use the slope and vertical intercept from Problems 2 and 3 to write an equation for the line.

b. Assume that the average rate of change you determined in Problem 2 stays the same through 1960. Use the equation in part a to predict the U.S. population in 1960. Also estimate the population in 1960 from the graph.

The **error** in a prediction is the difference between the observed value (actual value that was measured) and the predicted value. The **relative error** of a prediction is the ratio of the error to the observed value. That is,

$$\text{relative error} = \frac{\text{observed value} - \text{predicted value}}{\text{observed value}} = \frac{\text{error}}{\text{observed value}}.$$

5. a. The relative error is usually reported as a percent. The actual U.S. population in 1960 was approximately 179 million. What is the *relative error* (expressed as a percent) in your prediction?

   b. What do you think was the cause of your prediction error?

6. You want to develop a population model based on more recent data. The U.S. population was approximately 281 million in 2000 and 309 million in 2010.

   a. Plot these data points using ordered pairs of the form $(t, P(t))$, where t is the number of years since 2000 (now $t = 0$ corresponds to 2000). Draw a line through the points.

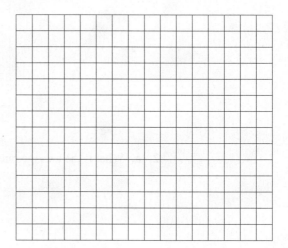

   b. Determine the slope of the line in part a. What is the practical meaning of the slope in this situation? How does this slope compare with the slope in Problem 2?

   c. In which decade, 1940–1950 or 2000–2010, did the U.S. population increase more rapidly? Explain your answer in terms of slope.

   d. Determine the P-intercept of the line in part a.

   e. Write the equation of the line in part a.

7. a. Use the linear model $P(t) = 2.8t + 281$ developed in Problem 6 to predict the population in 2020. What assumptions do you make about the average rate of change of the population in this prediction?

b. According to the linear model $P(t) = 2.8t + 281$, in what year will the population reach 350 million?

8. The following data was obtained from the 2000 and 2010 censuses for the U.S. population by gender.

YEAR	2000	2010
m, Number of Males (in millions)	138	151.7
f, Number of Females (in millions)	143	156.9

Source: U.S. Census Bureau

a. Write the data for the male population as ordered pairs of the form $(t, m(t))$, where t is the number of years since 2000 and $m(t)$ is the corresponding male population in millions.

b. Assume that the relationship between time and the U.S. male population is linear from 2000 to 2010. Determine the slope of the line containing the two points in part a. What are its units of measure? Interpret the meaning of the slope in the context of this situation.

c. What is the vertical intercept of the line in part b? Interpret the meaning of the vertical intercept in the context of this situation.

d. Write a linear model for $m(t)$ in terms of t.

e. Assume that the relationship between time and the U.S. female population is also linear from 2000 to 2010. Write the data for the female population as ordered pairs of the form $(t, f(t))$, and determine the slope and vertical intercept of the line containing these points.

   f. Write a linear model for $f(t)$ in terms of t.

   g. The population of which gender group, male or female, is growing more rapidly? Explain.

SUMMARY: ACTIVITY 3.7

1. The **error** in a prediction is the difference between the observed value and the predicted value. The **relative error** in prediction is the ratio of the error to the observed value. That is,

$$\text{relative error} = \frac{\text{observed value} - \text{predicted value}}{\text{observed value}} = \frac{\text{error}}{\text{observed value}}.$$

Relative error is usually reported as a percent.

EXERCISES: ACTIVITY 3.7

1. a. Use the linear model $P(t) = 2.8t + 281$ developed in Problem 6 to predict the U.S. population in 2012. What assumptions are you making about the rate of change of the population in this prediction? Recall that t is the number of years since 2000.

   b. The actual U.S. population in 2012 was approximately 314 million. What is the relative error in your prediction?

   c. How accurate was the prediction?

2. a. According to the U.S. Census Bureau, the population of Texas in 2000 was approximately 20.85 million and was increasing at an average rate of approximately 430,000 people per year. Let $P(t)$ represent the Texas population (in millions), and let t represent the number of years since 2000. Assuming that the rate of increase stays constant, complete the following table.

t	$P(t)$ (in millions)
0	
1	
2	

Exercise numbers appearing in color are answered in the Selected Answers appendix.

b. What information in part a indicates that the Texas population growth is linear with respect to time? What are the slope and vertical intercept of the graph of the population data?

c. Write an equation for $P(t)$ in terms of t.

d. Use the linear model in part c to estimate the population of Texas in 2007.

e. Population data is used by Texas state agencies in developing their programs and policies. The population of Texas in 2007 was approximately 23.90 million. Determine the relative error (as a percent) between your prediction in part d and the preceding estimate.

f. Use your linear model to predict the population of Texas in 2015.

g. Do you believe your prediction in part f will be too high, close, or too low? Explain your answer.

3. a. According to the U.S. Census Bureau, the population of Detroit, Michigan, was 936.9 thousand in 2000 and 701.5 thousand in 2012. If t represents the number of years since 2000 and $P(t)$ represents the population (in thousands) at a given time, t, summarize the given information in the accompanying table.

t	$P(t)$

b. Plot the two data points on appropriately scaled and labeled coordinate axes. Draw a line connecting the points.

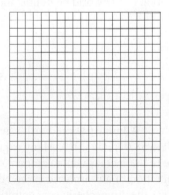

c. What is the slope of the line? What is the practical meaning of the slope in this situation?

d. What is the vertical intercept of the line? What is the practical meaning of the intercept in this situation?

   e. Write an equation to model Detroit's population, $P(t)$, in terms of t.

   f. Use this linear model to predict the population of Detroit in 2025.

4. In each part, determine the equation of the line for the given information.

   a. Two points on the line are (0, 4) and (7, 18). Use the points to determine the slope and y-intercept. Then write the equation of the line.

   b. The graph has y-intercept (0, 6) and contains the point (2, 1).

   c. The graph has y-intercept $(0, -2)$ and x-intercept (5,0).

5. **a.** In 2010, the rate of change of the world population was approximately 0.07792 billion per year (or approximately 1 million people every 5 days). The world population was estimated to be 6.9 billion in 2010. Write an equation to model the population, P (in billions), in terms of t, where t is the number of years since 2010 ($t = 0$ corresponds to 2010).

   b. Use the linear model to predict the world population in 2020.

   c. According to the model, when will the population of the world be double the 2010 population?

6. According to the National Oceanic and Atmospheric Administration, Earth's average surface temperature has increased from 56.58°F in 1900 to 58.09°F in 2013. Many experts believe that this increase in global temperature is largely due to the increase in the amount of carbon dioxide in the atmosphere. The pre-industrial concentration of atmospheric carbon dioxide was 289 parts per million (ppm). The following table shows the atmospheric concentration of carbon dioxide in parts per million, measured at the Mauna Loa Observatory in Hawaii.

	1960	2012
c, Carbon Dioxide Concentration (ppm)	317	394

a. Let t represent the number of years since 1960. Determine the average rate of increase in the atmospheric concentration c of carbon dioxide from 1960 to 2012. Interpret the rate in the context of this situation.

b. What is the slope and c-intercept of the line containing the two data points?

c. Write an equation for c in terms of t.

d. Use the equation in part c to project the atmospheric concentration of carbon dioxide in 2050.

e. Do you have confidence in the projection in part d? Explain your answer.

Activity 3.8

College Tuition

Objectives

1. Determine the equation for a linear function when given two points, neither of which is the vertical intercept.

2. Use the point-slope form, $y - y_1 = m(x - x_1)$, to write the equation of a nonvertical line.

3. Identify lines having zero or undefined slopes.

4. Determine the equation of horizontal and vertical lines.

In Activity 3.7, you were given two points, one of which had an input value of zero (vertical intercept). You determined the slope and produced a linear equation in slope-intercept form. In the next situation, you are given two points, neither of which has an input value of zero, and are asked to determine a linear equation.

1. The cost for a part-time student at the local community college is determined by a fixed activity fee plus a fixed tuition amount per credit. The cost for a student taking 6 credits is $770. The cost for a student taking 9 credits is $1130. The total cost, C, is a function of the number of credits taken, x.

   a. Write two ordered pairs of the form (x, C) for the college cost function.

   b. Use the result from part a to complete the following table.

NUMBER OF CREDITS, x	TOTAL COST, C ($)
6	
9	

   c. Determine the average rate of change for the college cost function.

   d. Interpret the average rate of change in the context of the college costs.

   e. Is the average rate of change constant? Explain.

The cost, C, is a linear function of the number of credits, x. Therefore, the relationship between C and x can be modeled by an equation written in the slope-intercept form, $C = mx + b$.

2. a. What is the value of the slope, m?

   b. Write the equation $C = mx + b$, replacing m with its value.

   c. Use the ordered pair $(6, 770)$ from Problem 1a and rewrite the equation in part b by replacing C and x with the appropriate values in the ordered pair.

   d. Solve the equation in part c for b.

   e. Interpret the value of b in the college cost function.

   f. Finally, rewrite the equation $C = mx + b$, replacing b and m with their respective values.

3. a. In Problem 2c, you can substitute the coordinates of either point into the equation $C = 120x + b$. Explain why this is possible.

b. Use the ordered pair $(9, 1130)$ in Problem 2, parts c and d, to determine the value of b. Compare the result with Problem 2d.

4. Using Problem 2 as a guide, summarize the algebraic procedure for determining an equation of a line from two points, neither of which is the y-intercept. Illustrate your step-by-step procedure using the points $(15, 62)$ and $(21, 80)$.

Point-Slope Form of an Equation of a Line

There is one other very useful way to determine the equation of a line given its slope m and a point other than its y-intercept. This point is customarily denoted (x_1, y_1), where x_1 and y_1 should be understood to represent specific values, unlike x and y, which are understood to be variables.

Definition

The equation of the line with slope m containing point (x_1, y_1) can be written as

$$y - y_1 = m(x - x_1)$$

and is called the **point-slope** form of the equation of a line.

Example 1

a. Use the point-slope form to determine an equation of the line containing the points $(3, 8)$ and $(7, 24)$.

b. Rewrite your equation in slope intercept form.

SOLUTION

a. The slope m of the line containing $(3, 8)$ and $(7, 24)$ is $m = \dfrac{24 - 8}{7 - 3} = \dfrac{16}{4} = 4$.

Choose either given point as (x_1, y_1). Here, $x_1 = 3$ and $y_1 = 8$. The point-slope equation then becomes

$$y - 8 = 4(x - 3).$$

b. This equation can be rewritten in slope intercept form by expanding the expression on the right side and then solving for y.

$$y - 8 = 4(x - 3)$$
$$y - 8 = 4x - 12 \quad \text{using the disributive property}$$
$$y = 4x - 4 \quad \text{adding 8 to both sides}$$

Note that in Problem 1, you determined two ordered pairs, $(6, 770)$ and $(9, 1130)$, for the college cost function and then used the two ordered pairs to determine the slope $m = 120$. Because the input and output variables are x and C, respectively, the point-slope form is $C - C_1 = m(x - x_1)$.

5. a. Use the ordered pair $(6, 770)$ for (x, C_1) and $m = 120$ to write the point-slope form of the college cost function.

  b. Solve the equation for C, and compare the result with the slope-intercept form in Problem 2f.

  c. Repeat parts a and b using the ordered pair $(9, 1130)$ for (x_1, C_1). How does your result compare with the equation obtained in part b?

Problem 5 demonstrates that you obtain the same equation of the line regardless of which point you use for (x_1, C_1).

6. a. Use the point-slop form to determine an equation of the line containing the points $(15, 62)$ and $(21, 80)$.

  b. Rewrite your equation in part a in slope-intercept form and compare with your results from Problem 4.

7. The basal energy requirement is the daily number of calories that a person needs to maintain basic life processes. For a 20-year-old male who weighs 75 kilograms and is 190.5 centimeters tall, the basal energy requirement is 1952 calories. If his weight increases to 95 kilograms, he will require 2226 calories.

The given information is summarized in the following table.

20-YEAR-OLD MALE, 190.5 CENTIMETERS TALL		
w, Weight (kg)	75	95
B, Basal Energy Requirement (cal.)	1952	2226

  a. Assume that the basal energy requirement, B, is a linear function of weight, w, for a 20-year-old male who is 190.5 centimeters tall. Determine the slope of the line containing the two points indicated in the preceding table.

  b. What is the practical meaning of the slope in the context of this situation?

c. Determine a symbolic rule that expresses B in terms of w for a 20-year-old, 190.5-centimeter-tall male.

d. Does the B-intercept have any practical meaning in this situation? Determine the practical domain of the basal energy function.

Horizontal Line

8. a. You found a promotion for unlimited access to the Internet for $20 per month. Complete the following table of values, where t is the number of hours a subscriber spends online during the month and c is the monthly access cost for that subscriber.

t, Time (hr.)	1	2	3	4	5
c, Cost ($)					

b. Sketch a graph of the data points.

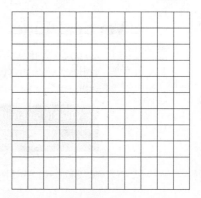

c. What is the slope of the line drawn through the points?

d. What single word best describes a line with zero slope?

e. Determine the horizontal and vertical intercepts (if any).

f. What can you say about the output value of every point on a horizontal line?

Suppose the horizontal and vertical axes are labeled, respectively, as the *x*- and *y*-axes. Then every point on a horizontal line will have the form (x, c), where the *x*-coordinate varies from point to point and the *y*-coordinate always takes the same value, *c*. Because of this, a horizontal line can be completely described by an equation that specifies the constant *y*-value: $y = c$.

> A **horizontal line** has zero slope and is characterized by the fact that *all* its points have identical *y*-coordinates. Therefore, the equation of every horizontal line will have the form $y = c$, where *c* is the constant *y*-coordinate.

Example 2   *A horizontal line contains the point $(3, 7)$. List two additional points on this line, and indicate its equation.*

SOLUTION

A horizontal line is completely described by its *y*-coordinate (in this case, 7). Two additional points could be $(-2, 7)$ and $(11, 7)$. Its equation is $y = 7$.

9. Use function notation to write a symbolic rule that expresses the relationship in Problem 8—the monthly cost for unlimited access to the Internet in terms of the number of hours spent online.

10. The *x*-axis is a horizontal line. List two points on the *x*-axis and indicate its equation.

Vertical Line

11. To cover your weekly expenses while going to school, you work as a part-time aide in your college's health center and earn $100 each week. Complete the following table, where *x* represents your weekly salary and *y* represents your weekly expenses for a typical month.

x, Weekly Salary ($)				
y, Weekly Expenses ($)	50	70	90	60

a. Sketch a graph of the data points. Do the points lie on a line? Explain.

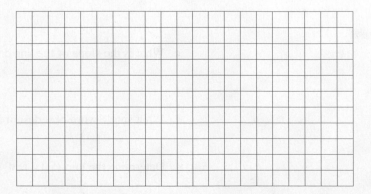

b. Explain what happens if you use the slope formula to determine a numerical value for the slope of the line in part a.

c. Write down the slope formula. What must be true about the change in the input variable for the quotient to be defined?

d. What type of line results whenever the slope is undefined, as in this problem?

e. Determine the vertical and horizontal intercepts (if any).

f. What can you say about the input value of every point on a vertical line?

g. Is y a function of x? Explain.

When the horizontal and vertical axes are labeled, respectively, as the x- and y-axes, then every point on a vertical line will have the form (d, y), where the y-coordinate varies from point to point and the x-coordinate always takes the same value, d. Because of this, the vertical line can be completely described by an equation that specifies the constant x-value: $x = d$.

A **vertical line** has an undefined slope and is characterized by the fact that *all* its points have identical x-coordinates. Therefore, the equation of every vertical line will have the form $x = d$, where d is the constant x-coordinate.

Example 3 *A vertical line contains the point $(3, 7)$. List two additional points on this line, and indicate its equation.*

SOLUTION

A vertical line is completely described by its x-coordinate (in this case, 3). Two additional points could be $(3, -5)$ and $(3, 10)$. Its equation is $x = 3$.

12. a. Write an equation for the situation described in Problem 11.

b. The y-axis is a vertical line. List two points on the y-axis, and indicate its equation.

SUMMARY: ACTIVITY 3.8

1. To determine the equation of a line, $y = mx + b$, given two points on the line:

Step 1. Determine the slope, m.

Step 2. Substitute the value of m into $y = mx + b$, where b is still unknown.

Step 3. Substitute the coordinates of one of the known points for x and y in the equation in step 2.

Step 4. Solve the equation for b to obtain the y-intercept of the line.

Step 5. Substitute the values for m and b into $y = mx + b$.

2. To determine the point-slope equation of a line, $y - y_1 = m(x - x_1)$, given two points on the line:

Step 1. Determine the slope, m.

Step 2. Choose any one of the given points as (x_1, y_1).

Step 3. Replace m, x_1, and y_1 by their numerical values in $y - y_1 = m(x - x_1)$.

3. The slope of a **horizontal line** is zero. Every point on a horizontal line has the same y-coordinate. The equation of a horizontal line is $y = d$, where d is the constant y-coordinate. For example, the graph of $y = 6$ is a horizontal line 6 units above the x-axis as shown below.

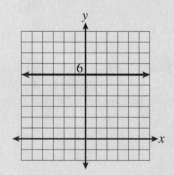

4. The slope of a **vertical line** is undefined; it has no numerical value. Every point on a vertical line has the same x-coordinate. The equation of a vertical line is $x = c$, where c is the constant x-coordinate. For example, the graph of $x = 5$ is a vertical line 5 units to the right of the y-axis as shown below.

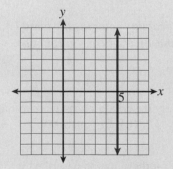

EXERCISES: ACTIVITY 3.8

1. Federal income tax paid by an individual single taxpayer is a function of taxable income. For a recent year, the federal tax for various taxable incomes is given in the following table.

i, Taxable Income ($)	15,000	16,500	18,000	19,500	21,000	22,500	24,000
t, Federal Tax ($)	1,889	2,114	2,339	2,564	2,789	3,014	3,239

    a. Plot the data points, with taxable income *i* as input and tax *t* as output. Scale the input axis from $0 to $24,000 and the output axis from $0 to $4000. Explain why the relationship is linear.

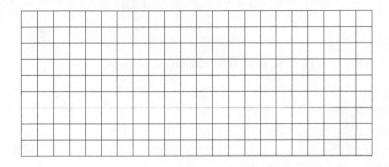

    b. Determine the slope of the line. What is the practical meaning of the slope?

       Write an equation to model this situation. Use the variable *i* to represent the taxable income and the variable *t* to represent the federal tax owed.

    d. What is the *t*-intercept? Does it make sense?

    e. Use the equation from part c to determine the federal tax owed by a college student having a taxable income of $8600.

    f. Use the equation from part c to determine the taxable income of a single person who paid $1686 in federal taxes.

In Exercises 2–8, determine the equation of the line that has the given slope and passes through the given point. Then sketch a graph of the line.

2. $m = 3$, through the point $(2, 6)$

3. $m = -1$, through the point $(5, 0)$

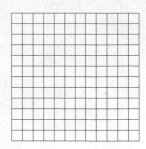
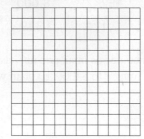

4. $m = 7$, through the point $(-3, -5)$

5. $m = 0.5$, through the point $(8, 0.5)$

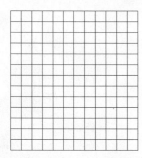
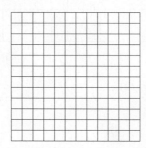

6. $m = 0$, through the point $(5, 2)$

7. $m = -4.2$, through the point $(-4, 6.8)$

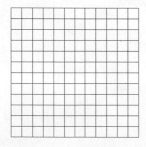
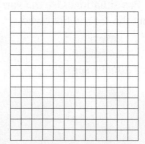
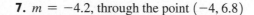

8. $m = -\dfrac{2}{7}$, through the point $\left(5, \dfrac{4}{7}\right)$

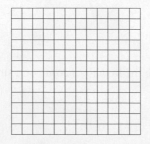

In Exercises 9–14, determine the equation of the line that passes through the given points.

9. $(2, 6)$ and $(4, 16)$

10. $(-5, 10)$ and $(5, -10)$

11. $(3, 18)$ and $(8, 33)$

12. $(0, 6)$ and $(-10, 0)$

13. $(10, 2)$ and $(-3, 2)$

14. $(3.5, 8.2)$ and $(2, 7.3)$

15. A boat departs from a marina and travels so that its distance from the marina is a linear function of time. The table below displays two ordered pairs of this function.

t (hr.)	d (mi.)
2	75
4	145

a. Determine the slope of the line. What is the practical meaning of slope in this situation?

b. Write the equation of the line in slope-intercept form.

16. Straight-line depreciation helps spread the cost of new equipment over a number of years. The value of your company's copy machine after 1 year will be $14,700 and after 4 years will be $4800.

a. Write a linear function that will determine the value of the copy machine for any specified year.

b. The salvage value is the value of the equipment when it gets replaced. What will be the salvage value of the copier if you plan to replace it after 5 years?

17. A horizontal line contains the point $(-3, 7)$. Determine and list three additional points that lie on the line.

18. a. Sketch the graph of the vertical line through $(-2, 3)$.

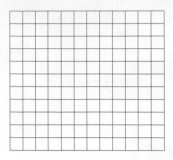

   b. Does the graph represent a function? Explain.

   c. Write the equation of a vertical line through the point $(-2, 3)$.

   d. What is the slope of the line?

19. Explain the difference between a line with a zero slope and a line with an undefined slope.

20. You are retained as a consultant for a major computer company. You receive $2000 per month as a fee no matter how many hours you work.

   a. Using x to represent the number of hours you work each month, write an equation to represent the total amount received from the company each month.

   b. Complete the following table of values.

x, Hours Worked per Month	15	25	35
y, Fee per Month (in $)			

   c. Use your graphing calculator to sketch the graph of this function.

   d. What is the slope of the line? What is the practical meaning of the slope in this situation?

   e. Describe the graph of the function.

Cluster 2   What Have I Learned?

1. A line is given by the equation $y = -4x + 10$.

    a. Determine its x-intercept and y-intercept algebraically from the equation.

    b. Use technology to confirm these intercepts.

2. a. Does the slope of the line having the equation $4x + 2y = 3$ have a value of 4? Why or why not?

    b. Solve the equation in part a for y so that it is in the form $y = mx + b$.

    c. What is the slope of the line?

3. Explain the difference between a line with zero slope and a line with an undefined slope.

4. Describe how you recognize that a function is linear when it is given

    a. graphically   **b.** symbolically

    c. numerically in a table

5. Do vertical lines represent functions? Explain.

6. If you know the slope and the vertical intercept of a line, how do you write the equation of the line? Use an example to demonstrate.

Exercise numbers appearing in color are answered in the Selected Answers appendix.

7. Demonstrate how you would change the equation of a linear function such as $5y - 6x = 3$ into slope-intercept form. Explain your method.

8. What assumption are you making when you say that the cost, c, of a rental car (in dollars) is a linear function of the number, n, of miles driven?

Cluster 2   How Can I Practice?

1. A function is linear if the average rate of change of the output with respect to the input from point to point is constant. Use this idea to determine the missing input (x) and output (y) values in each table, assuming that each table represents a linear function.

a.

x	y
1	4
3	8
5	

b.

x	y
−1	3
0	8
	13
2	

c.

x	y
−3	11
0	8
3	
	2

d. Explain how you used the idea of constant average rate of change to determine the values in the tables.

2. The pitch of a roof is an example of slope in a practical setting. The roof slope is usually expressed as a ratio of rise over run. For example, in the building shown, the pitch is 6 to 24, or in fraction form, $\frac{1}{4}$.

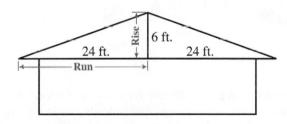

a. If a roof has a pitch of 5 to 16, how high will the roof rise over a 24-foot run?

b. If a roof's slope is 0.25, how high will the roof rise over a 16-foot run?

c. What is the slope of a roof that rises 12 feet over a run of 30 feet?

3. Determine whether any of the following tables contain input and output data that represent a linear function. In each case, give a reason for your answer.

a. You make an investment of $100 at 5% interest compounded semiannually. The following table represents the amount of money you will have at the end of each year.

TIME (yr.)	AMOUNT ($)
1	105.06
2	110.38
3	115.97
4	121.84

b. A cable TV company charges a $45 installation fee and $28 per month for basic cable service. The table values represent the total usage cost since installation.

Number of Months	6	12	18	24	36
Total Cost ($)	213	381	549	717	1053

c. For a fee of $20 a month, you have unlimited video rental. Values in the table represent the relationship between the number of videos you rented each month and the monthly fee.

Number of Rentals	10	15	12	9	2
Cost ($)	20	20	20	20	20

4. After stopping your car at a stop sign, you accelerate at a constant rate for a period of time. The speed of your car is a function of the time since you left the stop sign. The following table shows your speedometer reading each second for the next 7 seconds.

t, TIME (sec.)	0	1	2	3	4	5	6	7
s, SPEED (mph.)	0	11	22	33	44	55	55	55

a. Graph the data by plotting the ordered pairs of the form (t, s) and then connect the points.

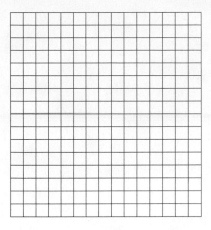

b. For what values of t is the graph increasing?

c. What is the slope of the line segment during the period of acceleration?

d. What is the practical meaning of the slope in this situation?

e. For what values of t is the speed a constant? What is the slope of the line connecting the points of constant speed?

5. a. The three lines shown in the following graphs appear to be different. Calculate the slope of each line.

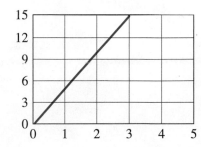 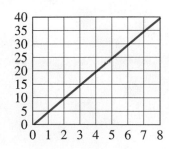 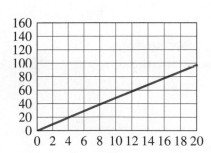

b. Do the three graphs represent the same linear function? Explain.

6. **a.** Determine the slope of the line through the points $(2, -5)$ and $(2, 4)$.

   b. Determine the slope of the line $y = -3x - 2$.

   c. Determine the slope of the line $2x - 4y = 10$.

   d. Determine the slope of the line from the following graph.

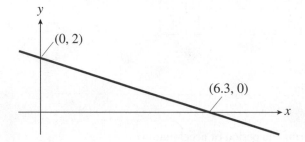

7. Determine the vertical and horizontal intercepts for the graph of each of the following.

   a. $y = 2x - 6$ **b.** $y = -\dfrac{3}{2}x + 10$

   c. $y = 10$

8. Determine the equation of each line.

   a. The line passes through the points $(2, 0)$ and $(0, -5)$.

b. The slope is 7, and the line passes through the point $\left(0, \frac{1}{2}\right)$.

c. The slope is 0, and the line passes through the point $(2, -4)$.

9. Sketch a graph of each of the following. Use technology to verify your graphs.

a. $y = 3x - 6$ **b.** $f(x) = -2x + 10$

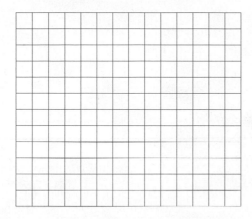 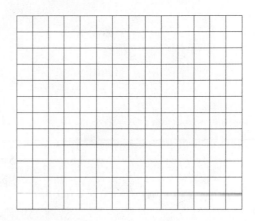

10. Write each equation in slope-intercept form to discover what the graphs have in common. Use technology to verify your graphs.

a. $y = 3x - 4$ **b.** $y - 3x = 6$

c. $3x - y = 0$

11. Write each equation in slope-intercept form to discover what the graphs have in common. Use technology to verify your graphs.

a. $y = -2$ **b.** $y - 3x = -2$

c. $x = y + 2$

12. **a.** Complete the following table by listing four points that are contained on the line $x = 3$.

x	3	3	3	3
y				

b. What is the slope of the line in part a?

    c. Determine the vertical and horizontal intercepts, if any, of the graph of the line in part a.

    d. Does the graph of the line in part a represent a function? Explain.

In Problems 13–20, determine the slope and the intercepts of each line.

13. $y = 2x + 1$

14. $y = 4 - x$

15. $y = -2$

16. $-\dfrac{3}{2}x - 5 = y$

17. $y = \dfrac{x}{5}$

18. $y = 4x + \dfrac{1}{2}$

19. $2x + y = 2$

20. $-3x + 4y = 12$

Determine the equation of each line described in Problems 21–24.

21. The slope is 9, and the *y*-intercept is $(0, -4)$.

22. The line passes through the points $(0, 4)$ and $(-5, 0)$.

23. The slope is $\dfrac{5}{3}$, and the line passes through the point $(0, -2)$.

24. The slope is zero, and the line passes through the origin.

25. Identify the input and output variables, and write a linear function in symbolic form for each of the following situations. Then give the practical meaning of the slope and vertical intercept in each situation.

    a. You make a down payment of $50 and pay $10 per month for your new tablet.

    b. You pay $16,000 for a new car whose value decreases by $1500 each year.

Graph the equations in Problems 26 and 27, and determine the slope and y-intercept of each graph. Describe the similarities and differences in the graphs.

26. a. $y = -x + 2$ **b.** $y = -4x + 2$

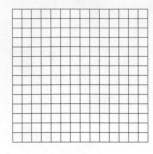

27. a. $y = 3x - 4$ **b.** $y = 3x + 5$

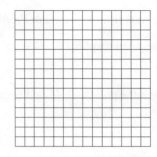

28. Suppose you enter Interstate 90 in Montana and drive at a constant speed of 75 mph.

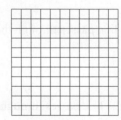

  a. Write a linear function rule that represents the total distance, d, traveled on the highway as a function of time, t, in hours.

  b. Sketch a graph of the function. What are the slope and d-intercept of the line? What is the practical meaning of the slope?

  c. How long would you need to drive 75 miles per hour to travel a total of 400 miles?

d. You start out at 10 A.M. and drive 3 hours at a constant speed of 75 miles per hour. You are hungry and stop for lunch. One hour later, you resume your travel, driving steadily at 60 miles per hour until 6 P.M., when you reach your destination. How far will you have traveled? Sketch a graph that shows the distance traveled as a function of time.

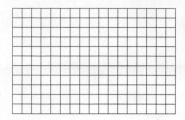

29. Determine the equation of each line in parts a–f.

a. The slope is 3, and the y-intercept is 6.

b. The slope is -4, and the y-intercept is -5.

c. The slope is 2, and the line passes through $(0, 4)$.

d. The slope is 4, and the line passes through $(6, -3)$.

e. The slope is -5, and the line passes through $(4, -7)$.

f. The slope is 2, and the line passes through $(5, -3)$.

30. Determine the equation of the line passing through each pair of points.

a. $(0, 6)$ and $(4, 14)$

b. $(-2, -13)$ and $(0, -5)$

c. $(5, 3)$ and $(-1, 3)$

d. $(-9, -7)$ and $(-7, -3)$

e. $(6, 1)$ and $(6, 7)$

f. $(2, 3)$ and $(2, 7)$

31. Determine the equation of the line shown on the following graph.

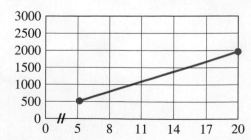

32. Determine the x-intercept of each line having the given equation.

   a. $y = 2x + 4$ **b.** $y = 4x - 27$ **c.** $y = -5x + 13$

33. A camera and lens you want are now available at a cost of $1200. There is a special promotion for students that allows you to make monthly payments for 2 years at 0% interest. You are required to make a 20% down payment, with the balance to be paid in 24 equal monthly payments. This is an opportunity to get the camera and lens you really want, so you investigate to see if you can afford to take advantage of this promotion.

   a. Determine the amount of the down payment required.

   b. Determine the amount owed after the down payment.

   c. What are the monthly payments?

   d. Complete the following table.

PAYMENT NUMBER, n	AMOUNT PAID, A ($)
0	
1	
2	

   e. Is the amount paid, A, a linear function of the payment number, n?

   f. What is the slope, m, of the payment function?

   g. Write the linear equation that gives the amount paid, A, as a function of the number of payments made, n.

h. Graph the equation from part g.

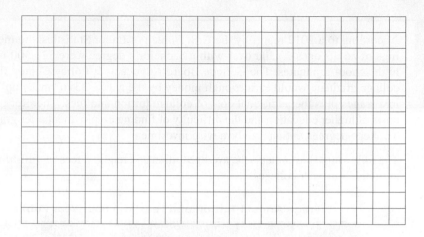

i. What is the practical meaning of the slope in this situation?

j. What is the practical meaning of the vertical intercept in this situation?

34. An architect will charge you a flat fee of $5000 for the plans for your home. The cost of your home is estimated by the square footage. The following table gives the total estimated cost of your home, c, including the architect's fees, as a function of the square footage, h. Assume that the total cost is a linear function of square footage.

TOTAL SQUARE FEET, h	TOTAL COST c ($)
0	5000
3000	380,000

a. What is the vertical intercept of the line containing these points? Explain how you determined this intercept.

b. Using the data in the table and the formula $m = \dfrac{\Delta c}{\Delta h}$, calculate the slope. What is the practical meaning of the slope in this situation?

c. Use the results from parts a and b to write the equation of the line in slope-intercept form that can be used to determine the cost for any given square footage.

d. You decide that you cannot afford a house with 3000 square feet. Using the equation from part c, determine the cost of your home if you decrease its size to 2500 square feet.

Cluster 3   Linear Regression, System, and Inequalities

Activity 3.9

Education Pays

Objectives

1. Recognize when patterns of points in a scatterplot are approximately linear.

2. Estimate and draw a line of best fit through a set of points in a scatterplot.

3. Use technology to determine a line of best fit by the least-squares method.

4. Estimate the error of representing a set of data by a line of best fit.

According to a 2012 report by the U.S. Bureau of Labor Statistics, the median annual earnings of adults ages 25 and over with a bachelor's degree were $21,500 more than those of high school graduates. The College Board Advocacy & Policy Center also reported that a college graduate would earn enough money by the age of 33 to "make up" for the 4 years of lost wages plus the additional cost of college tuition and fees. Therefore, despite the rising costs of higher education and the difficulty of finding employment after graduation, obtaining a college degree continues to be a wise investment.

The following table contains the average tuition and required fees for full-time, in-state students at public four-year colleges.

Average Annual Tuition and Fees at a Four-Year Public College

YEAR	2002	2004	2006	2008	2010	2012
t, Number of Years since 2002	0	2	4	6	8	10
c, Cost in 2012 Dollars	5213	6201	6534	6865	8000	8655

1. Plot the data points from the given table on an appropriately scaled and labeled coordinate axis.

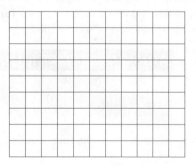

2. Does there appear to be a linear relationship between the years since 2002 and the cost of tuition and fees? Is there an exact linear fit? That is, do all the points lie on the same line?

3. In many situations where the data set is not exactly linear, it can be useful to approximate the data with a linear function whose ordered pairs are reasonable estimates of the original data. Use a straightedge to draw a single line that you believe best represents the linear trend in the data. For the purpose of this activity, draw a line through the points (0, 5500) and (10, 8000). The resulting line is commonly called a **line of best fit**.

Informally drawing a line, called the "eyeball" method, is one way to estimate a line of best fit. This line and its symbolic rule are called **linear models** for the given set of data; and can be used to represent the linear trend in the data.

4. Use the two points in Problem 3 to estimate the slope of the line of best fit. What is the practical meaning of the slope in this situation?

5. What is the vertical intercept of this line? Does this point have any practical meaning in this situation?

6. What is the equation of your linear model?

7. Use the linear model in Problem 6 to predict the average cost of tuition and fees at public four-year colleges in 2015.

Goodness-of-Fit Measure

An estimate of how well a linear model represents a given set of data is called a **goodness-of-fit measure**.

8. Determine a goodness-of-fit measure of the linear model from Problem 6.

Step 1. Use the linear rule you derived in Problem 6 to complete the following table.

t, INPUT	ACTUAL OUTPUT	c, MODEL'S OUTPUT	ACTUAL VALUE − MODEL VALUE	|ACTUAL VALUE − MODEL VALUE|
0	5213			
2	6201			
4	6534			
6	6865			
8	8000			
10	8655			

Step 2. Determine the sum of total the absolute values of the differences in the last column. This sum is called the **error** and is a **goodness-of-fit measure**. The smaller the error, the better the fit.

Regression Line

The method of least squares is a statistical procedure for determining a line of best fit from a set of data pairs. This method produces an equation of a line, called a **regression line**. Appendix E shows you how to use the TI-84 Plus C to determine the equation of a regression line for a set of data pairs.

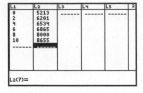

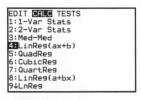

9. Use technology to determine the equation for the regression line in this situation. The final screens for the TI-84 plus C appear in the margin.

10. a. Determine the goodness-of-fit measure for the least-squares regression line in Problem 9.

t, INPUT	ACTUAL OUTPUT	c, MODEL'S OUTPUT	ACTUAL VALUE – MODEL VALUE	|ACTUAL VALUE – MODEL VALUE|
0	5213			
2	6201			
4	6534			
6	6865			
8	8000			
10	8655			

b. Compare the error of your line of best fit with the error of the least-squares regression line.

> **Definition**
>
> Using a regression model to predict an output within the boundaries of the input values of the given data is called **interpolation**. Using a regression model to predict an output outside the boundaries of the input values of the given data is called **extrapolation**. In general, interpolation is more reliable than extrapolation.

11. The number of Internet users in the United States increased steadily from 2008 to 2013, as indicated in the following table.

YEAR	NUMBER OF INTERNET USERS IN U.S. (millions)
2008	203.2
2009	211.7
2010	221.0
2011	229.2
2012	236.9
2013	244.1

a. Plot the data points on an appropriately scaled and labeled coordinate axis. Let x represent the number of years since 2008.

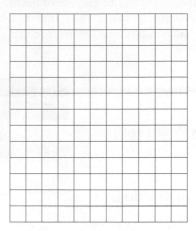

b. Use your graphing calculator's statistics menu (STAT) to determine the equation of the line that best fits the data (the regression line).

c. What is the slope of the line in part b? What is the practical meaning of the slope?

d. Use the linear model from part b to predict when the number of Internet users in the United States will reach 265 million.

SUMMARY: ACTIVITY 3.9

1. A **line of best fit** is used to represent the general linear trend of a set of nonlinear data. This line and its equation form a linear model for the given set of data.

2. A **goodness-of-fit measure** is an estimate of how well a linear model represents a given set of data.

3. Graphing calculators and computer software use the **method of least squares** to determine a particular line of best fit, called a **regression line**.

EXERCISES: ACTIVITY 3.9

1. Over the past quarter century, the number of bachelor's degrees conferred by degree-granting institutions has steadily increased. The following table contains data on the number of bachelor's degrees (in thousands) earned by women in a given year. The input t represents the number of years since 1990.

	YEAR						
	1990	1995	2000	2005	2007	2008	2009
t, Number of Years Since 1990	0	5	10	15	17	18	19
$f(t)$, Number of Degrees in Thousands	558	634	708	838	886	906	928

a. Sketch a scatterplot of the given data.

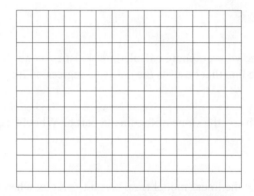

b. Enter the data into your graphing calculator, and determine a linear regression model to represent the data. Write the result here.

c. What is the slope of the line? What is the practical meaning of the slope in this situation?

d. Use the regression model to predict the number of bachelor's degrees that will be granted in 2020.

2. In 1966, the U.S. Surgeon General's health warnings began appearing on cigarette packages. The following data seems to demonstrate that public awareness of the health hazards of smoking has had some effect on consumption of cigarettes.

	YEAR, t							
	1997	1999	2001	2003	2005	2007	2009	2011
% of Total Population 18 and Older Who Smoke, P	24.7	23.5	22.8	21.6	20.9	19.8	20.6	19.0

Source: U.S. National Center for Health Statistics

a. Plot the given data as ordered pairs of the form (t, P), where t is the number of years since 1997 and P is the percent of the total population (18 and older) who smoke. Appropriately scale and label the coordinate axes.

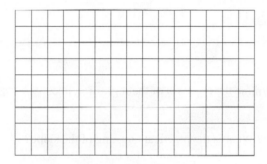

b. Determine the equation of the regression line that best represents the data.

c. Use the equation to predict the percent of the total population 18 and older who will smoke in 2020.

3. Per capita personal income is calculated by taking the total income of a population and dividing it by the total number of people in that population. The per capita income of the United States as reported by the Department of Commerce for the given years is located in the following table.

YEAR	PER CAPITA INCOME
1997	$25,924
1999	$28,546
2001	$30,574
2003	$31,484
2005	$34,757
2007	$38,611
2009	$38,846
2011	$41,633

a. Plot the data points on appropriately scaled and labeled axes. Let *x* represent the number of years since 1997.

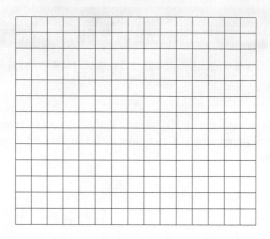

b. Use your graphing calculator to determine the equation of the regression line.

c. What is the slope of the line in part b? What is the practical meaning of the slope?

d. Use the linear model from part b to determine when the per capita income will reach $50,000. How confident are you in the prediction? Explain your answer.

4. As a Fisheries and Wild Life major, you learn that the amount of dissolved oxygen in water is measured in parts per million (ppm). Trout need a minimum of 6 ppm to live. Many variables affect the amount of dissolved oxygen in a stream. One very important variable is the water temperature. To investigate the effect of temperature on dissolved oxygen, you take a water sample from a stream and measure the dissolved oxygen as you heat the water. Your results are as follows:

t, Temperature (°C)	11	16	21	26	31
d, Dissolved Oxygen (ppm)	10.2	8.6	7.7	7.0	6.4

a. Plot the data points as ordered pairs of the form (t, d). Scale your input axis from 0 to 40°C and your output axis from 0 to 15 ppm.

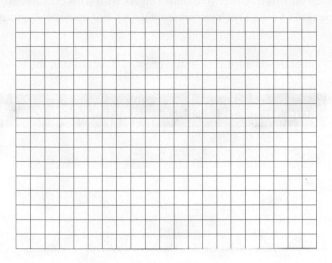

b. Use technology to determine the equation for the regression line in this situation.

c. Use the linear model in part b to approximate the maximum temperature at which trout can survive.

Lab 3.10

Body Parts

Objectives

1. Collect and organize data in a table.

2. Recognize linear patterns in paired data.

Variables arise in many common measurements. Your height is one measurement that has probably been recorded frequently from the day you were born. In this lab, you are asked to pair up and make the following body measurements: height (h); arm span (a), the distance between the tips of your two middle fingers with arms outstretched; wrist circumference (w); foot length (f); and neck circumference (n). For consistency, measure the lengths in inches.

1. Gather the data for your entire class, and record it in the following table:

STUDENT	HEIGHT (h)	ARM SPAN (a)	WRIST (w)	FOOT (f)	NECK (n)	FEMUR (t)

2. What are some relationships you can identify based on a visual inspection of the data? For example, how do the heights relate to the arm spans?

3. Construct a scatterplot for heights versus arm span on the grid below, carefully labeling the axes and marking the scales. Does the scatterplot confirm what you may have guessed in Problem 2?

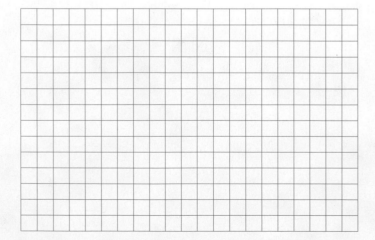

4. Use your calculator to create scatterplots for the following pairs of data, and state whether there appears to be a linear relationship. Comment on how the scatterplots either confirm or go against the observations you made in Problem 2.

VARIABLES	LINEAR RELATIONSHIP?
Height versus Foot Length	
Arm Span versus Wrist Circumference	
Foot Length versus Neck Circumference	

5. Determine a linear regression equation to represent the relationship between the two variables in Problems 3 and 4 that show the strongest linear pattern.

Predicting Height from Bone Length

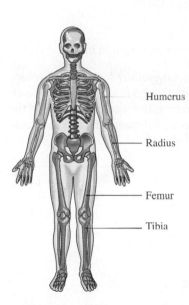

Humerus

Radius

Femur

Tibia

An anthropologist studies human physical traits, place of origin, social structure, and culture. Anthropologists are often searching for the remains of people who lived many years ago. A forensic scientist studies the evidence from a crime scene to help solve a crime. Both of these groups of scientists use various characteristics and measurements of the human skeletal remains to help determine physical traits such as height, as well as racial and gender differences.

In the average person, there is a strong relationship between height and the length of two major arm bones (the humerous and the radius), as well as the length of the two major leg bones (the femur and the tibia).

Anthropologists and forensic scientists can closely estimate a person's height from the length of just one of these major bones.

6. Each member of the class should measure his or her leg from the center of the kneecap to the bone on the outside of the hip. This is the length of the femur. Record the results in the appropriate place in the last column of the table in Problem 1.

   a. If you want to predict height from the length of the femur, which variable should represent the independent variable? Explain.

   b. Make a scatterplot of the data on a carefully scaled and labeled coordinate axes.

   c. Describe any patterns you observe in the scatterplot.

   d. Determine the equation of the regression line for the data.

   e. Use the equation of the regression line in part d to predict the height of a person whose femur measures 17 inches.

   f. Anthropologists have developed the following formula to predict the height of a male based on the length of his femur:

$$h = 1.888L + 32.010,$$

where h represents the height in inches and L represents the length of the femur in inches. Use the formula to determine the height of the person whose femur measures 17 inches.

g. Compare your results from parts e and f. What might explain the difference between the height you obtained using the regression formula in part e and the height using the formula in part f?

h. Determine the regression line equation for femur length vs. height using just the male data from Problem 1. How does this new regression line equation compare with the formula $h = 1.888L + 32.010$ used by anthropologists?

7. a. Determine the linear regression equation (femur length vs. height) for the female data in Problem 1.

   b. Compare your results to the formula used by anthropologists:

$$h = 1.945x + 28.679,$$

where h represents height in inches and x represents femur length in inches.

8. The work of Dr. Mildred Trotter (1899–1991) in skeletal biology led to the development of formulas used to estimate a person's height based on bone length. Her research also led to discoveries about the growth, racial and gender differences, and aging of the human skeleton. Write a brief report on the life and accomplishments of this remarkable scientist.

Activity 3.11

Smartphone Plan Options

Objectives

1. Solve a system of two linear equations numerically.

2. Solve a system of two linear equations graphically.

3. Solve a system of two linear equations using the substitution method.

4. Recognize the connections between the three methods of solution.

5. Interpret the solution to a system of two linear equations in terms of the problem's content.

You have researched the Internet to compare the various types of voice, texting, and data plans that are available for your smartphone. You have decided to select a plan with unlimited text, picture, and video messaging but no voice minutes. You will pay a separate monthly fee for voice minutes used. The following two service providers offer similar plans.

Plan 1: $55.99 per month for Internet access and unlimited text, picture, and video messaging, including 2GB data allowance. There is a voice cost-per-use fee of $0.25 per minute.

Plan 2: $69.99 per month for the same features of Plan 1 but with a voice cost-per-use fee of $0.15 per minute.

Assume that you will not exceed the monthly 2GB data limit.

1. Although the average number of voice minutes used by the average smartphone user is decreasing, in 2013, the average smartphone user used more than 500 voice minutes per month. If you plan on spending about 50 total minutes talking on your smartphone each month, which plan will be more economical?

2. Let m represent the number of voice minutes used in a given month. Write a function rule for the monthly cost, C, in terms of m for plan 1.

3. Write a function rule for the monthly cost C in terms of m for plan 2.

4. **a.** Complete the following table for each plan, showing the monthly cost for 100, 125, 150, 200, 250, 300, and 500 voice minutes. Estimate the number of minutes for which the two plans come closest to being equal in cost. If you have a graphing calculator, use the table feature to complete the table.

Number of Voice Minutes	100	125	150	200	250	300	500
Cost of Plan 1	$80.99	$87.24					
Cost of Plan 2	$84.99	$88.74					

   b. Use the table feature of a graphing calculator to obtain a better estimate. What is that value?

5. a. Graph the cost equation for each plan on the same coordinate axes. Plot the data points, and then use a straightedge to draw a line connecting each set of points. Be sure to properly scale and label the axes.

b. Estimate the coordinates of the point where the lines intersect. What is the significance of this point?

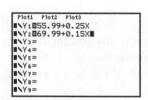

c. Verify your results from part b using technology. Use the trace or intersect feature. See Appendix E for the procedure for the TI-84 Plus *C*. Your final screens should appear as shown in the margin.

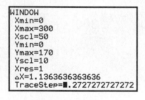

System of Two Linear Equations

Two linear equations that relate the same two variables are called a **system of linear equations**. The two cost equations from Problems 2 and 3 form a system of two linear equations,

$$C = 55.99 + 0.25m$$
$$C = 69.99 + 0.15m.$$

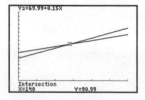

The **solution** of a system is the set of all ordered pairs that satisfy both equations. A system of linear equations generally has only one solution. The solution to the cost system is $(140, 90.99)$. This solution represents the specific number of voice minutes ($x = 140$) that produce identical costs in both accounts (90.99).

In Problem 4, you solved the cost system **numerically** by completing a table and noting the value of the input that resulted in the same output. In Problem 5, you solved the cost system **graphically** by determining the coordinates of the point of intersection.

You can also determine an exact solution by solving the system of equations **algebraically**. In Problem 6, you will explore one method—the **substitution method**—for solving systems of linear equations algebraically.

Substitution Method for Solving a System of Two Linear Equations

Consider the following system of two linear equations.

$$y = 3x - 10$$
$$y = 5x + 14$$

To solve this system, you need to determine for what value of x are the corresponding y-values the same. The idea behind the substitution method is to replace one variable in one of equations with what that variable is equal to in the second equation.

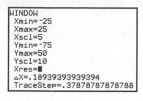

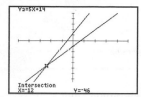

6. a. Use the two linear equations in the preceding system to write a single equation involving just one variable.

b. Solve the equation in part a for the variable.

c. Use the result in part b to determine the corresponding value for the other variable in the system.

d. Write the solution to this system as an ordered pair.

e. Verify this solution numerically as well as graphically using technology.

Procedure

Substitution method

Replace (or *substitute*) the variable y in one equation with its algebraic expression in x from the other equation. Solve for x. Substitute this x-value into either of the original equations to determine the corresponding y-value.

---

7. a. Using the substitution method, solve the following system of smartphone data plan cost functions.

$$C = 55.99 + 0.25m$$
$$C = 69.99 + 0.15m$$

b. Compare your result with the answers obtained using the numerical approach (Problem 4) and the graphing approach (Problem 5).

c. Summarize your results by describing under what circumstances the basic data plan 1 is preferable to plan 2.

8. U.S. online retail sales are forecast to reach 434.2 billion dollars by 2017. Amazon.com is the most popular and well-known example of online shopping. Founded in 1995, the Seattle-based site started out as an online bookstore, but it has since expanded its product line considerably. Amazon began offering a prime membership to its customers, initially priced at $79 per year. One-day shipping on eligible items for Amazon Prime members is $3.99. There is no minimum order size. Without the prime membership, the cost of one-day shipping is $13.99 per shipment. You do occasionally order from Amazon using one-day shipping and wonder if the annual prime membership fee is worthwhile.

a. This situation involves two variables: the number of one-day shipments you expect to order in one year and the total cost of the shipping. Identify the input and output.

b. Determine the total cost of shipping 5 orders in one year:

i. With the prime membership

ii. Without the prime membership

c. Write a function rule for the cost c of making n one-day shipments in a year for each of the following.

i. With Prime Membership

ii. Without Prime Membership

d. Solve the system of two linear equations in part c. What does the solution represent in this situation?

e. When would it be worth purchasing the Amazon Prime membership?

SUMMARY: ACTIVITY 3.11

1. Two equations that relate the same variables are called a **system of equations**. The solution of a system of equations is the set of all ordered pairs that satisfy both equations. The two equations in a system of two linear equations are usually written one below the other as follows:

$$y = ax + b$$
$$y = cx + d.$$

2. There are three standard methods for solving a system of equations:

Numerical method: Make a table of values for both equations. Identify or estimate the input (x-value) that produces the same output (y-value) for both equations.

Graphical method: Graph both equations on the same grid. If the two lines intersect, the coordinates of the point of intersection represent the solution of the system. If the lines are parallel, the system has no solution.

Substitution method: Replace (or *substitute*) the variable y in one equation with its algebraic expression in x from the other equation. Solve for x. Substitute this x-value into either function rule to determine the corresponding y-value. If no value is determined for x, the system has no solution.

3. A solution to a system of two linear equations may be expressed as an ordered pair.

EXERCISES: ACTIVITY 3.11

1. Finals are over, and you are moving back home for the summer. You need to rent a truck to move your possessions from the college residence hall back to your home. You contact two local rental companies and obtain the following information for the 1-day cost of renting a truck.

Company 1: $19.99 per day plus $0.79 per mile

Company 2: $29.99 per day plus $0.59 per mile

Let n represent the total number of miles driven in one day.

a. Write an equation to determine the total cost, C, of renting a truck for 1 day from company 1.

b. Write an equation to determine the total cost, C, of renting a truck for 1 day from company 2.

c. Complete the following table to compare the total cost of renting the vehicle for the day. Verify your results using the table feature of your graphing calculator.

n, NUMBER OF MILES DRIVEN	TOTAL COST, C, COMPANY 1	TOTAL COST, C, COMPANY 2
0		
10		
20		
30		
40		
50		
60		
70		
80		

Exercise numbers appearing in color are answered in the Selected Answers appendix.

d. For what mileage is the 1-day rental cost the same?

e. Which company should you choose if you intend to drive less than 50 miles?

f. Which company should you choose if you intend to drive more than 50 miles?

g. Graph the two cost functions, for *n* between 0 and 90 miles, on the same coordinate axes below.

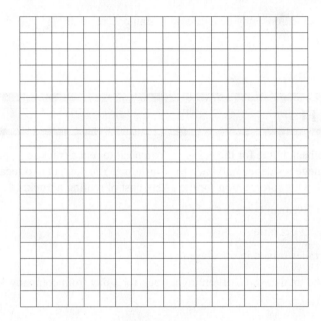

h. Use the table in part g to determine the point where the lines in part g intersect. What is the significance of the point in this situation?

i. Determine the mileage for which the 1-day rental costs are equal by solving the following system for *n* using the substitution method.

$$C = 19.99 + 0.79n$$
$$C = 29.99 + 0.59n$$

2. Your part-time business is growing to a full-time operation. You need to purchase a used car for deliveries.

a. Model 1 costs $13,600 and depreciates $500 a year. Write an equation to determine the resale value, *V*, of the car after *x* years of use.

b. Model 2 costs $16,000 and depreciates $800 a year. Write an equation to determine the resale value, V, of this car after x years of use.

c. Write a system of two linear equations that can be used to determine in how many years both cars will have the same resale value.

d. Solve this system numerically by completing the following table.

NUMBER OF YEARS	VALUE OF MODEL 1 ($)	VALUE OF MODEL 2 ($)
1		
5		
8		
10		
12		

e. Solve the system graphically.

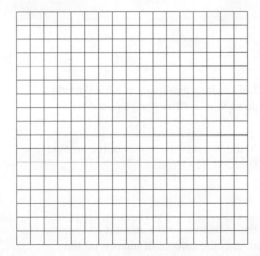

f. Solve the system algebraically using the substitution method.

g. Compare the results in parts d, e, and f.

h. If you plan to keep your car for 5 years, which one would have more value and by how much?

3. Two companies sell software products. In 2015, company A had total sales of $17.2 million. Its marketing department projects that sales will increase $1.5 million per year for the next several years. Company B had total sales of $9.6 million of software products in 2015 and projects that its sales will increase an average of $2.3 million each year.

Let n represent the number of years since 2015.

a. Write an equation that represents the total sales (in millions of dollars), s, of company A since 2015.

b. Write an equation that represents the total sales (in millions of dollars), s, of company B since 2015.

c. The two equations in parts a and b form a system. Solve this system to determine the year in which the total sales of both companies will be the same.

4. Your brother has a total of $10,700 in student loans. Part of the loan was made at a local credit union at 6.50%. The remainder was a Stafford Loan made at 3.86%. After one year, the total amount of interest would accumulate to $531.82.

a. Let c represent the amount borrowed at the local credit union, and let s represent the amount of the Stafford Loan. Does the sum $c + s$ equal 10,700 or 531.82? Explain.

b. Write an algebraic expression that represents the amount of interest on the amount c borrowed at the credit union. Then write an expression that represents the amount of interest on the amount s of the Stafford Loan.

c. Write an equation for the sum of the two amounts in part b.

d. Use the results from parts a and c to write a system of equations involving the variables c and s.

e. Solve the system in part d. What does your solution represent in this situation?

5. You need to replace the heating system in your house. A conventional heating system will cost $5000 with a yearly fuel cost of $5400. A modern heating system will cost $8000 with a yearly fuel cost of $4500.

a. Write an equation for the total cost, C, of a conventional system for t years. The total cost is the cost of the system plus the fuel cost.

b. Write an equation for the total cost, C, of a modern system for t years.

c. Solve the system of equations from parts a and b to find the number of years it takes for the total cost of the conventional system to equal the total cost of the modern system.

Use substitution to determine algebraically the exact solution to each system of equations in Exercises 6–9. Check your solutions numerically and by using the table feature or graphing capability of a graphing calculator or other form of technology.

6. $p = q - 2$

$p = -1.5q + 3$

7. $n = -2m + 9$

$n = 3m - 11$

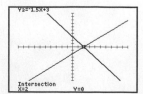

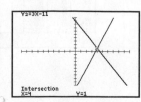

8. $y = 1.5x - 8$

$y = -0.25x + 2.5$

9. $z = 3w - 1$

$z = -3w - 1$

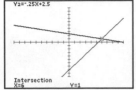

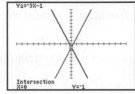

10. You want to hire someone to prune your trees and shrubs. One service you call charges a $15 consultation fee plus $8 an hour for the actual work. A neighborhood gardener says that she does not include a consulting fee, but she charges $10 an hour for her work.

a. Write an equation that describes the pruning service's charge, C, as a function of h, the number of hours worked.

b. Write an equation that describes the local gardener's charge, C, as a function of h, the number of hours worked.

c. Whom would you hire for a 3-hour job?

d. When, if at all, would it be more economical to hire the other service? Set up and solve a system of equations to answer this question.

e. Use technology to verify your results.

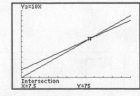

11. You and your friend are going in-line skating at a local park. A 5-mile path along the lake begins at the concession stand. You skate at a rate of 10 miles per hour, and your friend skates at 8 miles per hour. You start skating at the concession stand. Your friend starts farther down the path, 0.5 mile from the concession stand.

   a. Write an equation that models your distance from the concession stand as a function of time. What are the units of the input variable? Output variable? (Recall that distance = rate · time.)

   b. Write an equation that models your friend's distance from the concession stand as a function of time.

   c. How long will it take you to catch up to your friend? In that time, how far will you have skated?

   d. Use technology to verify your results.

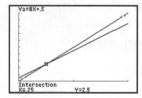
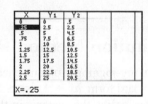

Activity 3.12

Healthy Lifestyle

Objectives

1. Solve a system of two linear equations algebraically using the substitution method.

2. Solve a system of two linear equations algebraically using the addition (or elimination) method.

You are trying to maintain a healthy lifestyle. You eat a well-balanced diet and follow a regular schedule of exercise. One of your favorite exercise activities is a combination of walking and jogging in the nearby park.

On one particular day, it takes you 1 hour and 18 minutes (1.3 hours) to walk and jog a total of 5.5 miles in the park. You are curious about the amount of time you spent walking and the amount of time you spent jogging during the workout.

Let x represent the number of hours (or part of an hour) you walked and y represent the number of hours (or part of an hour) you jogged.

1. Write an equation using x and y that expresses the total time of your workout in the park.

2. a. You walk at a steady speed of 3 miles per hour for x hours. Write an expression that represents the distance you walked.

b. You jog at a constant speed of 5 miles per hour for y hours. Write an expression that represents the distance you jogged.

c. Write an equation for the total distance you walked and jogged in the park.

The situation just described can be represented by this system of linear equations.

$$x + y = 1.3$$
$$3x + 5y = 5.5.$$

Substitution Method Revisited

In the walking-jogging problem, each equation in the system of two linear equations is written in standard form, $ax + by = c$. The solution can be determined by using the substitution method, but unlike the systems in the previous activity, one of the equations must first be solved for one of the variables. The following problem demonstrates this process.

3. Start with the system,

$$x + y = 1.3$$
$$3x + 5y = 5.5.$$

a. First, solve the equation $x + y = 1.3$ for y.

b. Then substitute this expression for y in the equation $3x + 5y = 5.5$ to obtain a single equation involving just x.

c. Solve the equation in part b.

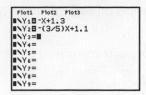

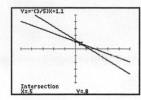

d. What is the solution to the system?

e. Check your answer graphically using your graphing calculator. You may want to use the window Xmin = −2.5, Xmax = 2.5, Ymin = −2.5, Ymax = 2.5.

Addition (or Elimination) Method for Solving a System of Two Linear Equations

Sometimes it is more convenient to leave each of the equations in the linear system in **standard form** ($ax + by = c$) rather than solving for one variable in terms of the other. Look again at the original system.

$$x + y = 1.3$$
$$3x + 5y = 5.5$$

4. Your strategy in solving a linear system is to obtain a single equation involving just one variable. Apply the addition principle of equations by adding the two equations (left side to left side and right side to right side). Do you obtain a single equation containing only one variable?

5. Consider a system that is similar to the one you are trying to solve.

$$-5x - 5y = -6.5$$
$$3x + 5y = 5.5$$

Apply the addition property of equations to system 2. Explain what happens.

6. Compare the two systems under consideration:

$$x + y = 1.3 \qquad\qquad -5x - 5y = -6.5$$
$$3x + 5y = 5.5 \qquad\qquad 3x + 5y = 5.5$$

Do you see any relationship between the two systems? What can you do to system 1 to make it look like system 2?

7. You are now ready to solve system 1 using an algebraic method called the **addition (or elimination) method**.

   a. Multiply the top equation by a factor that will produce opposite coefficients for the variable y.

   b. Apply the addition principle of equations.

   c. Solve the resulting single equation for the variable.

   d. Substitute the result from part c into either one of the equations of the system to determine the corresponding value for y.

To solve a system of two linear equations by the addition (or elimination) method,

Step 1. Line up the like terms in each equation vertically.

Step 2. If necessary, multiply one or both equations by constants so that the coefficients of one of the variables are opposites.

Step 3. Add the corresponding sides of the two equations to obtain a single equation containing one variable.

Step 4. Solve the resulting single equation.

---

8. Solve system 1 again using the addition method. This time multiply the top equation by a number that will eliminate the variable x.

$$x + y = 1.3$$
$$3x + 5y = 5.5$$

You may need to multiply one or both equations by a factor that will produce coefficients of the same variable that are additive inverses (opposites).

9. Solve the following system using the addition method.

$$2x + 3y = 2$$
$$3x + 5y = 4$$

a. Identify which variable you want to eliminate. Multiply each equation by an appropriate factor so that the coefficients of your chosen variable become opposites. Remember to multiply *both* sides of the equation by the factor. Write the two resulting equations.

b. Add the two equations to eliminate the chosen variable.

c. Solve the resulting linear equation.

d. Determine the complete solution. Remember to check by substituting into both of the original equations.

10. a. Solve the following linear system using the substitution method by solving one equation for a chosen variable and then substituting in the remaining equation.

$$x - y = 5$$
$$4x + 5y = -7$$

b. Check your answer in part a by solving the system using the addition method.

SUMMARY: ACTIVITY 3.12

1. There are two common methods for solving a system of two linear equations algebraically:

  a. The **substitution** method

  b. The **addition (or elimination)** method

2. The procedure for solving linear systems by substitution is as follows:

  Step 1. If not already done, solve one equation for one variable (for instance, y) in terms of the other (x).

  Step 2. In the other equation, replace y with the expression in x. This equation should now contain only one variable, x.

  Step 3. Solve this equation for x.

  Step 4. Substitute the x-value you determined in Step 3 into one of the original equations, and solve this equation for y.

3. The procedure for solving linear systems by addition is as follows:

  Step 1. Write each equation in the standard form $ax + by = c$.

  Step 2. Determine which variable you want to eliminate. Multiply one or both equations by the number(s) that will make the coefficients of this variable opposites.

  Step 3. Sum the left and right sides of the two equations, and combine like terms. This step should produce a single equation in one variable.

  Step 4. Solve the resulting equation.

  Step 5. Substitute the value from Step 4 into one of the original equations, and solve for the other variable.

EXERCISES: ACTIVITY 3.12

1. Solve the following systems algebraically using the substitution method. Check your answers numerically and by using your graphing calculator.

  a. $y = 3x + 1$
  $y = 6x - 0.5$

  b. $y = 3x + 7$
  $2x - 5y = 4$

c. $2x + 3y = 5$
$-2x + y = -9$

d. $4x + y = 10$
$2x + 3y = -5$

2. Solve the following systems algebraically using the addition method. Check your answers numerically and by using your graphing calculator.

a. $3x + y = 6$
$2x + y = 8$

b. $-3x + 2y = 7$
$2x + 3y = 17$

c. $7x - 5y = 1$

$3x + y = -\dfrac{1}{5}$

3. Solve the system both graphically and algebraically.

$3x + y = -18$
$5x - 2y = -8$

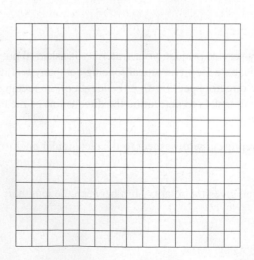

4. A catering service placed an order for eight centerpieces and five glasses, and the bill was $106. For the wedding reception, it was short one centerpiece and six glasses and had to reorder. This order came to $24. Let x represent the cost of one centerpiece, and let y represent the cost of one glass.

   a. Write an equation using x and y that represents the cost of the first order.

   b. Write an equation using x and y that represents the cost of the second order.

   c. The equations in parts a and b form a system of two linear equations that can be used to determine the cost of a single centerpiece and a single glass. Write this system below.

   d. Solve the system using the substitution method.

   e. Use the addition method to check your result in part d.

   f. Use the solution to the system to determine the cost of 15 centerpieces and 10 glasses.

5. As part of a community service project, your fraternity is asked to put up a fence in the playground area at a local daycare facility. A local fencing company will donate 500 feet of fencing. The daycare center director specifies that the length of the rectangular enclosure be 20 feet more than the width. Your task is to determine the dimensions of the enclosed region that meets the director's specifications.

   a. What does the 500 represent with respect to the rectangular enclosure?

   b. Write an equation that gives the relationship between the length, l; width, w; and 500.

c. Write an equation that expresses the length of the rectangle in terms of w.

d. The equations in part b and c form a system of equations involving the variables l and w. Solve this system of equations to determine the dimensions of the enclosed region that satisfies the given conditions.

e. Explain how you know that your result in part d solves the problem you were assigned.

6. The daycare center director would like the playground area in Exercise 5 to be larger than that provided by the 500 feet of fencing. The center's staff also wants the length of the playground to be 30 feet more than the width. Additional donations help the center obtain a total of 620 feet of fencing.

a. Write a system of linear equations involving the length l and the width w.

b. Solve the system in part a to determine the dimensions of the enlarged playground.

c. What is the area of the enlarged playground?

d. Explain how you know your result in part b solves the problem you were assigned.

Project 3.13

Modeling a Business

Objectives

1. Solve a system of two linear equations by any method.

2. Define and interpret the break-even point as the point at which the cost and revenue functions are equal.

3. Determine the break-even point of a linear cost and revenue system algebraically and graphically.

You are employed by a company that manufactures solar collector panels. To remain competitive, the company must consider many variables and make many decisions. Two major concerns are those variables and decisions that affect operating expenses (or costs) of making the product and those that affect the gross income (or revenue) from selling the product.

Costs such as rent, insurance, and utilities for the operation of the company are called *fixed costs*. These costs generally remain constant over a short period of time and must be paid whether or not items are manufactured. Other costs, such as materials and labor, are called *variable costs*. These expenses depend directly on the number of items produced.

1. The records of the company show that fixed costs over the past year have averaged $8000 per month. In addition, each panel manufactured costs the company $95 in materials and $55 in labor. Write a symbolic rule in function notation for the total cost, $C(n)$, of producing n solar collector panels in 1 month.

2. A marketing survey indicates that the company can sell all the panels it produces if the panels are priced at $350 each. The revenue (gross income) is the amount of money collected from the sale of the product. Write a symbolic rule in function notation for the revenue, $R(n)$, from selling n solar collector panels in 1 month.

3. a. Complete the following table.

Number of Solar Panels, n	0	10	20	30	40	50	60
Total Cost ($), $C(n)$							
Total Revenue ($), $R(n)$							

b. Sketch a graph of the cost and revenue functions using the same set of coordinate axes.

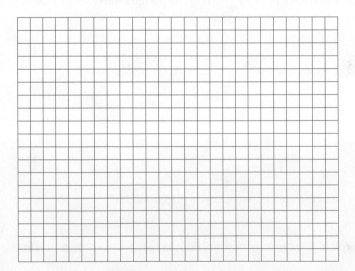

4. The point at which the cost and revenue functions are equal is called the *break-even point*.

    a. Estimate the break-even point on the graph.

    b. What system of equations must be solved to determine the break-even point for your company?

    c. Solve the system algebraically to determine the exact break-even point.

    d. Does your graph confirm the algebraic solution in part c?

5. Revenue exceeds costs when the graph of the revenue function is above the graph of the cost function. For what values of n is $R(n) > C(n)$? What do these values represent in this situation?

6. The break-even point can also be viewed with respect to profit. Profit is defined as the difference between revenue and cost. Symbolically, if $P(n)$ represents profit, then $P(n) = R(n) - C(n)$.

    a. Determine the profit when 25 panels are sold. What does the sign of your answer signify?

    b. Determine the profit when 60 panels are sold.

    c. Use the equations for $C(n)$ and $R(n)$ to write the profit function in terms of n solar panels sold per month. Write the expression in the profit function in simplest terms.

    d. Use your rule in part c to compute the profit for selling 25 and 60 panels. How do your results compare with the answers in parts a and b?

e. What do you expect profit to be at a break-even point?

f. Use the profit equation to confirm that the profit on n panels sold in a month is zero at the break-even point, $n = 40$.

7. a. Graph the profit function $P(n) = R(n) - C(n)$ on the grid provided below.

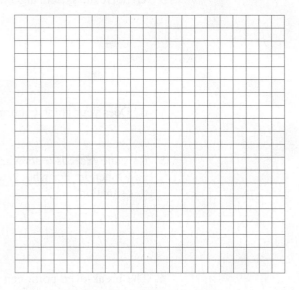

b. Locate and mark the break-even point on the graph in part a. Interpret the meaning of its coordinates.

c. Determine the vertical and horizontal intercepts of the graph. What is the meaning of each intercept in this situation?

d. What is the slope of the line? What is the practical meaning of slope in this situation?

8. Now suppose you are the sales manager of a small company that produces products for home improvement. You sell pavers (interlocking paving pieces for driveways) in bundles of 144 that cost $200 each. The equation $C(x) = 160x + 1000$ represents the cost, $C(x)$, of producing x bundles of pavers.

   a. Write a symbolic rule for the revenue function $R(x)$, in dollars, from the sale of the pavers.

   b. Determine the slope and the vertical intercept of the cost function. Explain the practical meaning of each in this situation.

   c. Determine the slope and the R-intercept of the revenue function. Explain the practical meaning of each in this situation.

   d. Graph the two functions from parts b and c on the same set of axes. Estimate the break-even point from the graph. Express your answer as an ordered pair, giving units. Check your estimate of the break-even point by graphing the two functions on your graphing calculator.

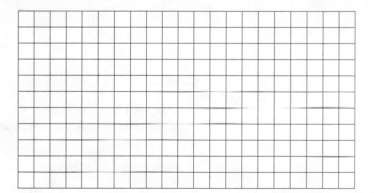

   e. Determine the exact break-even point algebraically. If your algebraic solution does not approximate your answer from the graph in part d, explain why.

f. How many bundles of pavers must the company sell for it to break even?

g. What is the total cost to the company when you break even? Verify that the cost and revenue values are equal at the break-even point.

h. For what values of x will your revenue exceed your cost?

i. As manager, what factors must you consider when deciding how many pavers to make?

j. If you knew you could sell only 30 bundles of pavers, would you make them? Consider how much it would cost you and how much you would make. What if you could sell only 20?

The following are key concepts that were presented in this Project.

1. The profit function P, as a function of number of units n, is the difference between the revenue function R and cost function C: $P(n) = R(n) - C(n)$

2. The break-even point occurs when revenue equals cost (when profit is zero). To determine the break-even point, solve the equation $R(n) = C(n)$.

Activity 3.14

How Long Can You Live?

Objective

1. Use properties of inequalities to solve linear inequalities in one variable algebraically.

We are living longer. Life expectancy in the United States is steadily increasing. The number of Americans aged 100 and older will exceed 850,000 by the middle of this century. Medical advancements have been a primary reason for Americans living longer. Another factor has been the increased awareness of maintaining a healthy lifestyle.

The life expectancy at birth of a male born in or after 1980 can be modeled by the function

$$M(x) = 0.192x + 70.0,$$

where x represents the year of his birth given in number of years since 1980 ($x = 0$ corresponds to a 1980 birth date, $x = 5$ corresponds to a 1985 birth date, etc.), and $M(x)$ represents his life expectancy (predicted age at death).

The life expectancy at birth of a female born in or after 1980 can be modeled by the function

$$W(x) = 0.101x + 77.5,$$

where x represents the year of her birth (given in number of years since 1980) and $W(x)$ represents her life expectancy (predicted age at death).

1. a. Complete the following table.

	YEAR						
	1980	**1985**	**1990**	**1995**	**2000**	**2005**	**2010**
x, Years Since 1980	0	5	10	15	20	25	30
$W(x)$							
$M(x)$							

b. For people born between 1980 and 2010, do men or women have the greater life expectancy?

c. Is the life expectancy of men or women increasing more rapidly? Explain using slope.

You would like to determine when the life expectancy of men will exceed that of women. The phrase *will exceed* means "will be greater than" and indicates a mathematical relationship called an **inequality**. Symbolically, the relationship can be represented by

$$M(x) \qquad > \qquad W(x).$$

life expectancy   life expectancy
for men   for women

2. Use the life expectancy functions to write an inequality involving x to determine after which year a male will have a greater life expectancy than a female born the same year.

Definition

Solving an inequality in one variable is the process of determining the values of the variable that make the inequality a true statement. These values are called the **solutions** of the inequality.

Solving Inequalities in One Variable Algebraically

The process of solving an inequality in one variable algebraically is similar to solving an equation in one variable. Your strategy is to isolate the variable on one side of the inequality symbol. You isolate the variable in an equation by performing the same operations on both sides of the equation so as not to upset the balance. In a similar manner, you isolate the variable in an inequality by performing the same operations on both sides so as not to upset the relative imbalance.

3. a. Write the statement "fifteen is greater than six" as an inequality.

   b. Add 5 to each side of $15 > 6$. Is the resulting inequality a true statement? (That is, is the left side still greater than the right side?)

   c. Subtract 10 from each side of $15 > 6$. Is the resulting inequality a true statement?

   d. Multiply each side of $15 > 6$ by 4. Is the resulting inequality true?

   e. Multiply each side of $15 > 6$ by -2. Is the left side still greater than the right side?

   f. Reverse the direction of the inequality symbol in part e. Is the new inequality a true statement?

Problem 3 demonstrates two very important properties of inequalities.

Property 1. If $a < b$ represents a true inequality, and

  i. if the same quantity is added to or subtracted from both sides,

or

  ii. if both sides are multiplied or divided by the same *positive number*,

then the resulting inequality remains a true statement and the direction of the inequality symbol remains the same.

Example 1   *Because* $-4 < 10$,

  i. $-4 + 5 < 10 + 5$, or $1 < 15$, is true and
$-4 - 3 < 10 - 3$, or $-7 < 7$, is true.

  ii. $-4(6) < 10(6)$, or $-24 < 60$, is true and
$\dfrac{-4}{2} < \dfrac{10}{2}$, or $-2 < 5$, is true.

Property 2. If $a < b$ represents a true inequality and if both sides are multiplied or divided by the same *negative number*, then the inequality symbol in the resulting inequality statement must be reversed ($<$ to $>$ or $>$ to $<$) for the resulting statement to be true.

Example 2   *Because* **−4 < 10,**

$$-4(-5) > 10(-5), \text{ or } 20 > -50.$$

Because **−4 < 10,**

$$\frac{-4}{-2} > \frac{10}{-2}, \text{ or } 2 > -5.$$

Properties 1 and 2 of inequalities are also true if $a < b$ is replaced by $a \leq b$, $a > b$, or $a \geq b$. Note that $a \leq b$, read *a is less than or equal to b*, is true if $a < b$ or $a = b$.

The following example demonstrates how properties of inequalities can be used to solve an inequality algebraically.

Example 3   *Solve* $3(x - 4) > 5(x - 2) - 8.$

SOLUTION

$$3(x - 4) > 5(x - 2) - 8 \qquad \text{Apply the distributive property.}$$
$$3x - 12 > 5x - 10 - 8 \qquad \text{Combine like terms on right side.}$$
$$3x - 12 > 5x - 18$$
$$\underline{-5x \qquad\quad -5x} \qquad\qquad \text{Subtract } 5x \text{ from both sides; the direction of}$$
$$-2x - 12 > -18 \qquad\qquad \text{the inequality symbol remains the same.}$$
$$\underline{+12 \quad +12} \qquad\qquad \text{Add 12 to both sides; the direction of the}$$
$$\frac{-2x}{-2} < \frac{-6}{-2} \qquad\qquad\quad \text{inequality does not change.}$$
$$\qquad\qquad\qquad\qquad \text{Divide both sides by } -2; \text{ reverse the direction!}$$
$$x < 3$$

Therefore, any number less than 3 is a solution of the inequality $3(x - 4) > 5(x - 2) - 8$. The solution set can be represented on a number line by shading all points to the left of 3, as shown.

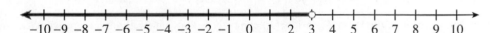

The open circle at 3 indicates that 3 is not a solution. A closed circle indicates that the number is a solution. The arrow shows that the solutions extend indefinitely to the left.

4. Solve the inequality $0.192x + 70.0 > 0.101x + 77.5$ from Problem 2 algebraically to determine after which year a male will have a greater life expectancy than a female born the same year.

Therefore, if the trends given by the equations for $M(x)$ and $W(x)$ were to continue, the approximate solution to the inequality $M(x) > W(x)$ would be $x > 82$. Because x represents the number of years since 1980, $x = 82$ corresponds to the year 2062. If current trends were to continue (possible but unlikely), men born in or after 2062 would have greater life expectancies than women.

5. a. Write an inequality to determine after which year a newborn female is projected to live to an age of 85 or greater.

b. Solve this inequality algebraically.

SUMMARY: ACTIVITY 3.14

1. The **solution set of an inequality** is the set of all values of the variable that satisfy the inequality.

2. The **direction of an inequality** is not changed when

i. the same quantity is added to or subtracted from both sides of the inequality. Stated symbolically, if $a < b$, then $a + c < b + c$ and $a - c < b - c$.

ii. both sides of an inequality are multiplied or divided by the same positive number.

If $a < b$, then $ac < bc$, where $c > 0$, and $\dfrac{a}{c} < \dfrac{b}{c}$, where $c > 0$.

3. The **direction of an inequality** is **reversed** if both sides of an inequality are multiplied or divided by the same negative number. These properties can be written symbolically as

i. if $a < b$ then $ac > bc$, where $c < 0$

ii. if $a < b$ then $\dfrac{a}{c} > \dfrac{b}{c}$, where $c < 0$

Properties 2 and 3 will still be true if $a < b$ is replaced by $a \le b, a > b$, or $a \ge b$.

4. Inequalities of the form $f(x) < g(x)$ can be solved using an **algebraic approach**, in which the properties of inequalities (items 2 and 3 above) are used to isolate the variable.

Similar statements can be made for solving inequalities of the form $f(x) \le g(x)$, $f(x) > g(x)$, and $f(x) \ge g(x)$.

EXERCISES: ACTIVITY 3.14

In Exercises 1–3, translate the given statement into an algebraic inequality.

1. The sum of the length, l, width, w, and depth, d, of a piece of luggage to be checked on a commercial airline cannot exceed 61 inches without incurring an additional charge.

2. A PG-13 movie rating means that your age, a, must be at least 13 years for you to view the movie.

3. The cost, $C(A)$, of renting a car from company A is less expensive than the cost, $C(B)$, of renting from company B.

Solve Exercises 4–11 algebraically.

4. $3x > -6$

5. $3 - 2x \leq 5$

6. $x + 2 > 3x - 8$

7. $5x - 1 < 2x + 11$

8. $8 - x \geq 5(8 - x)$

9. $5 - x < 2(x - 3) + 5$

10. $\dfrac{x}{2} + 1 \leq 3x + 2$

11. $0.5x + 3 \geq 2x - 1.5$

12. You contacted two local rental companies and obtained the following information for the 1-day cost of renting an SUV.

Company 1 charges $60.00 per day plus $0.75 per mile, and company 2 charges $30.00 per day plus $1.00 per mile.

Let n represent the total number of miles driven in 1 day.

a. Write an equation to determine the total cost, C, of renting an SUV for a day from company 1.

b. Write an equation to determine the total cost, C, of renting an SUV for a day from company 2.

c. Use the results in parts a and b to write an inequality that can be used to determine for what number of miles it is less expensive to rent the SUV from company 2.

d. Solve the inequality in part c.

13. The sign on the elevator in a seven-story building states that the maximum weight it can carry is 1200 pounds. You need to move a large shipment of books to the sixth floor. Each box weighs 60 pounds.

    a. Let n represent the number of boxes placed in the elevator. Assuming that you weigh 150 pounds, write an expression that represents the total weight in the elevator. Assume that only you and the boxes are in the elevator.

    b. Using the expression in part a, write an inequality that can be used to determine the maximum number of boxes you can place in the elevator at one time.

    c. Solve the inequality.

14. The local credit union is offering a special student checking account. The monthly cost of the account is $15. The first 10 checks are free, and each additional check costs $0.75. You search the Internet and find a bank that offers a student checking account with no monthly charge. The first 10 checks are free, but each additional check costs $2.50.

    a. Assume that you will be writing more than 10 checks a month. Let n represent the number of checks written in a month. Write a function rule for the cost c of each account in terms of n.

    b. Write an inequality to determine what number of checks in the bank account would be more expensive than the credit union account.

    c. Solve the inequality in part b.

    d. Which student checking account would you choose? Explain.

15. Let x represent the number of years since 1990. The male population of the United States can be modeled by the linear function defined by

$$m(x) = 1.6x + 121,$$

where $m(x)$ represents the male population in millions. The female population of the United States can be modeled by the linear function defined by

$$f(x) = 1.54x + 127,$$

where $f(x)$ represents the female population in millions.

a. Determine the slope and vertical intercept of each linear model. Interpret the slope and intercept of each line. Interpret the slope and intercept in the context of this situation.

b. Will the male population exceed the female population? Explain.

c. Write an inequality in terms of x to determine after which year the male population will be greater than the female population.

d. Solve this inequality.

Cluster 3   What Have I Learned?

1. When a scatterplot of input-output values from a data set suggests a linear relationship, you can determine a line of best fit. Why might this line be useful in your analysis of the data?

2. Explain how you would determine a line of best fit for a set of data. How would you estimate the slope and y-intercept?

3. Suppose a set of data pairs suggests a linear trend. The input values range from a low of 10 to a high of 40. You use your graphing calculator to calculate the regression equation in the form $y = ax + b$.

   a. Do you think that the equation will provide a good prediction of the output value for an input value of $x = 20$? Explain.

   b. Do you think that the equation will provide a good prediction of the output value for an input value of $x = 60$? Explain.

4. What is meant by a *solution to a system of linear equations*? How is a solution represented graphically?

5. Briefly describe three methods for solving a system of linear equations.

6. Describe a situation in which you would be interested in determining a break-even point. Explain how you would determine the break-even point mathematically. Interpret the break-even point with respect to the situation you describe.

7. Typically, a linear system of equations has one unique solution. Under what conditions would a linear system not have a solution. Give an example.

Cluster 3 | How Can I Practice?

1. What input value results in the same output value for $y_1 = 2x - 3$ and $y_2 = 5x + 3$?

2. What point on the line given by $4x - 5y = 20$ is also on the line $y = x + 5$?

3. Solve the following system of two linear equations.

$y = 2x + 8$

$y = -3x + 3$

4. You sell centerpieces online for $19.50 each. Your fixed costs are $500 per month, and each centerpiece costs $8 to produce.

   a. Let n represent the number of centerpieces you sell. Write an equation to determine the cost, C.

   b. Write an equation to determine the revenue, R.

   c. What is your break-even point?

5. Suppose you want to break even (in Exercise 4) by selling only 25 centerpieces. At what price would you need to sell each centerpiece?

Exercise numbers appearing in color are answered in the Selected Answers appendix.

6. a. Solve the system algebraically.

$$y = -25x + 250$$
$$y = 25x + 300$$

b. Use the addition method to solve the following system.

$$4m - 3n = -7$$
$$2m + 3n = 37$$

7. Solve this system algebraically and graphically.

$$u = 3v - 17$$
$$u = -4v + 11$$

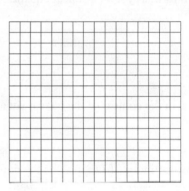

8. Use technology to estimate the solution to the following system. (**Note:** Both equations first must be written in the form $y = mx + b$.)

$$342x - 167y = 418$$
$$-162x + 103y = -575$$

Use the window Xmin $= -10$, Xmax $= 0$,
Xscl $= 2$, Ymin $= -28$, Ymax $= 0$, Yscl $= 2$.

9. The formula $A = P + Prt$ represents the value, A, of an investment of P dollars at a yearly simple interest rate, r, for t years.

a. Write an equation to determine the value, A, of an investment of \$100 at 8% for t years.

b. Write an equation to determine the value, A, of an investment of \$120 at 5% for t years.

c. The equations in parts a and b form a system of two linear equations. Solve this system using an algebraic approach.

d. Interpret your answer in part c.

10. When warehouse workers use hand trucks to load a boxcar, it costs management $40 for labor for each boxcar. After management purchases a forklift for $2000, it costs only $15 for labor to load each boxcar.

a. Write a symbolic rule that describes the cost, $C(n)$, of loading n boxcars with a hand truck.

b. Write a symbolic rule that describes the total cost, $C(n)$, including the purchase price of the forklift, of loading n boxcars with the forklift.

c. The equations in parts a and b form a system of linear equations. Solve the system using an algebraic approach.

d. Interpret your answer in part c in the context of the boxcar situation.

11. The consumption of cigarettes is declining. If t represents the number of years since 1990, then the consumption, C, is modeled by

$$C = -14.25t + 598.69.$$

where C represents the number of billions of cigarettes smoked per year.

a. Write an inequality that can be used to determine the first year in which cigarette consumption is less than 200 billion cigarettes per year.

b. Solve the inequality in part a using an algebraic as well as a graphical approach.

12. On an average winter day, the Auto Club receives 125 calls from people who need help starting their cars. The number of calls varies, however, depending on the temperature. Here is some data giving the number of calls as a function of the temperature (in degrees Celsius).

Temperature (°C)	−12	−6	0	4	9
Number of Auto Club Service Calls	250	190	140	125	100

a. Sketch the given data on appropriately scaled and labeled coordinate axes.

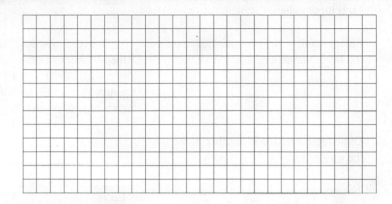

b. Use technology to determine the equation of the regression line for the data in the preceding table.

c. Use the regression equation from part b, $y = -7.11x + 153.9$, to determine how many service calls the Auto Club can expect if the temperature drops to −20°C.

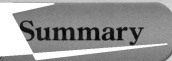

Summary

The bracketed numbers following each concept indicate the activity in which the concept is discussed.

CONCEPT/SKILL	DESCRIPTION	EXAMPLE
Function [3.1]	A function is a correspondence relating an input variable and an output variable in a way that assigns a single output value to each input value.	*(see table below)*

x	2	4	6	8	10
y	−1	1	2	3	4

CONCEPT/SKILL	DESCRIPTION	EXAMPLE
Independent variable [3.1]	Independent variable is another name for the input variable of a function.	Your weekly earnings depend on the number of hours you work. The two variables in this relationship are hours worked (independent) and total earnings (dependent).
Dependent variable [3.1]	Dependent variable is another name for the output variable of a function.	
Domain of a function [3.1]	The domain of a function is the collection of all input values.	*(see table below)*

x	0	1	3	7	12
y	5	3	8	3	−2

The domain is $\{0, 1, 3, 7, 12\}$.

CONCEPT/SKILL	DESCRIPTION	EXAMPLE
Range of a function [3.1]	The range of a function is the collection of all possible output values.	*(see table below)*

x	0	1	3	7	12
y	5	3	8	3	−2

The range $= \{5, 3, 8, -2\}$

CONCEPT/SKILL	DESCRIPTION	EXAMPLE
Function notation [3.1]	The notation "$f(x)$" is a symbol that represents the **output** of function f that is associated with input x. The input variable, or a replacement value for the input, is placed inside the parentheses. The parentheses do not indicate multiplication.	$f(10)$ represents the output of function f that is associated with input replacement value 10. The statement $f(10) = 7$ indicates that the output value is 7, corresponding to the input-output pair $(10, 7)$.
Quadrants [3.1]	Two perpendicular coordinate axes divide the plane into four quadrants, labeled counterclockwise with Quadrant I being the upper-right quadrant.	*(see figure below)*

Quadrant II, Quadrant I, Quadrant III, Quadrant IV (coordinate plane with axes from −5 to 5)

CONCEPT/SKILL	DESCRIPTION	EXAMPLE

Ways to represent a function [3.1], [3.3], [3.4]

A function can be represented numerically by a table, graphically by a curve or scatterplot, verbally by stating how the output value is obtained for a given input, and symbolically using an algebraic rule.

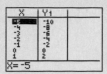

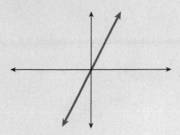

Delta notation for change [3.2]

Let y_1 and y_2 represent the output values corresponding to inputs x_1 and x_2, respectively. As the variable x changes in value from x_1 to x_2, the change in input is represented by $\Delta x = x_2 - x_1$ and the change in output is represented by $\Delta y = y_2 - y_1$. The order in which the output pairs are subtracted must match the order in which the inputs are subtracted.

Given the input/output pairs $(4, 14)$ and $(10, 32)$, $\Delta x = 10 - 4 = 6$ and $\Delta y = 32 - 14 = 18$.

Average rate of change over an interval [3.2]

The quotient

$$\frac{\Delta y}{\Delta x} = \frac{y_2 - y_1}{x_2 - x_1}$$

is called the average rate of change of y (output) with respect to x (input) over the x-interval from x_1 to x_2.

x	−3	4	7	10
y	0	14	27	32

The average rate of change over the interval from $x = 4$ to $x = 10$ is

$$\frac{\Delta y}{\Delta x} = \frac{32 - 14}{10 - 4} = \frac{18}{6} = 3.$$

Practical domain and practical range [3.4]

The practical domain and range are determined by the context of the problem.

$C(x) = 35x + 15$ represents the cost to rent a car for x days. The practical domain $= \{$whole numbers $\geq 1\}$, and the practical range $= \{50, 85, 120, 155, \dots\}$.

Linear function [3.5]

A linear function is a function whose average rate of change of output with respect to input from any one data point to any other data point is always the same (constant) value.

x	1	2	3	4
f(x)	10	15	20	25

The average rate of change between any two of these points is 5.

CONCEPT/SKILL	DESCRIPTION	EXAMPLE
Graph of a linear function [3.5]	The graph of every linear function is a line.	
Slope of a line [3.5]	The slope of the line that contains the two points (x_1, y_1) and (x_2, y_2) is denoted by $$m; m = \frac{\Delta y}{\Delta x} = \frac{y_2 - y_1}{x_2 - x_1}, (x_1 \neq x_2)$$	The slope of the line containing the two points $(2, 7)$ and $(5, 11)$ is $$\frac{11 - 7}{5 - 2} = \frac{4}{3}.$$
Positive slope [3.5]	The graph of every linear function with positive slope is a line rising to the right. The output values increase as the input values increase. A linear function is increasing if its slope is positive.	
Negative slope [3.5]	The graph of every linear function with negative slope is a line falling to the right. The output values decrease as the input values increase. A linear function is decreasing if its slope is negative.	
Horizontal intercept of a graph [3.5]	The horizontal intercept is the point at which the graph crosses the horizontal axis. Its ordered-pair notation is $(a, 0)$; that is, the second (y-) coordinate is equal to zero.	A line with horizontal intercept $(-3, 0)$ crosses the (input) x-axis 3 units to the left of the origin.
Vertical intercept of a graph [3.5]	The vertical intercept is the point at which the graph crosses the vertical axis. Its ordered-pair notation is $(0, b)$; that is, the first (x-) coordinate is equal to zero.	A line with vertical intercept $(0, 5)$ crosses the (output) y-axis 5 units above the origin.
Slope-intercept form of the equation of a line [3.6]	Represent the input variable with x and the output variable with y. Denote the slope of the line with m and the y-intercept with $(0, b)$. Then the coordinate pair, (x, y), of *every* point on the line satisfies the equation $y = mx + b$.	The line with equation $y = 3x + 4$ has a slope of 3 and y-intercept $(0, 4)$. The point $(2, 10)$ is on the line because its coordinates satisfy the equation $$10 = 3(2) + 4.$$

CONCEPT/SKILL	DESCRIPTION	EXAMPLE
Identifying the slope and y-intercept of a line defined by a symbolic rule of the form $y = mx + b$. [3.6], [3.7]	The slope can be identified as the coefficient of the x-term. The y-intercept can be identified as the constant term.	$y = \dfrac{5}{2}x - 3$ The slope is $\dfrac{5}{2}$. The y-intercept is $(0, -3)$.
Using the slope and y-intercept to write an equation of the line [3.6], [3.7]	Given a line whose slope is m and y-intercept is $(0, b)$, an equation of the line is $$y = mx + b.$$	An equation of the line with slope $\dfrac{2}{3}$ and y-intercept $(0, -6)$ is $$y = \dfrac{2}{3}x - 6.$$
Determining the horizontal intercept of a line given its equation [3.6]	Because the horizontal intercept is the point whose y-coordinate is 0, set $y = 0$ in the equation and solve for x.	Given the line with equation $$y = 2x + 6,$$ set $y = 0$ to obtain equation $$0 = 2x + 6$$ $$-6 = 2x$$ $$-3 = x.$$ The x-intercept is $(-3, 0)$.
Determining the point-slope equation of a line, $y - y_1 = m(x - x_1)$, given two points on the line. [3.8]	1. Determine the slope m. 2. Choose either given point as (x_1, y_1). 3. Replace m, x_1, and y_1 with their numerical values in $y - y_1 = m(x - x_1)$.	Given $(2, 1)$ and $(5, 10)$, the slope of this line is $m = \dfrac{10 - 1}{5 - 2} = 3$. The equation is $$y - 1 = 3(x - 2)$$ $$y - 1 = 3x - 6$$ $$y = 3x - 5.$$
Zero slope [3.8]	The graph of every linear function with zero slope is a horizontal line. Every point on a horizontal line has the same output value. A linear function is constant if its slope is zero.	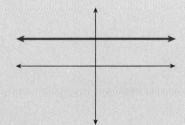
Undefined slope [3.8]	A line whose slope is not defined (because its denominator is zero) is a vertical line. A vertical line is the only line that does not represent a function. Every point on a vertical line has the same input value.	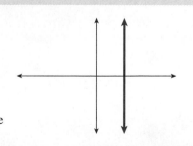
Equation of a horizontal line [3.8]	The slope, m, of a horizontal line is 0, and the output of each of its points is the same constant value, c. Its equation is $y = c$.	An equation of the horizontal line through the point $(-2, 3)$ is $y = 3$.

CONCEPT/SKILL	DESCRIPTION	EXAMPLE
Equation of a vertical line [3.8]	The input of each point on a vertical line is the same constant value, d. Its equation is $x = d$.	An equation of the vertical line through the point $(-2, 3)$ is $x = -2$.
Regression line [3.9], [3.10]	The regression line is a line that approximates a set of data points, even if all the points do not lie on a single line and is determined by the method of least squares.	Input the data pairs into a graphing calculator or a spreadsheet computer application whose technology is pre-programmed to apply the "method of least squares" to determine the equation of the regression line.
System of equations [3.11]	Two equations that relate the same variables are called a system of equations. The solution of the system is the ordered pair(s) that satisfies the two equations.	For the system of equations $$2x + 3y = 1$$ $$x - y = 3,$$ the ordered pair $(2, -1)$ is a solution of the system because $x = 2, y = -1$ satisfy both equations.
Graphical method for solving a system of linear equations [3.11]	Graph both equations on the same coordinate system. If the two lines intersect, then the solution to the system is given by the coordinates of the point of intersection.	
Substitution method for solving a system of linear equations [3.11] and [3.12]	• Solve one equation for either variable (e.g., y) as an expression in x. • In the other equation, replace y with the expression in x. (This equation should now contain only the variable x.) • Solve this equation for x. • Use this value of x to determine y.	For the system of equations $$2x + 3y = 11$$ $$y = x - 3:$$ • In the second equation, y is already written as an expression in x. • Rewrite the first equation $2x + 3(y) = 11$ as $$2x + 3(x - 3) = 11.$$ • Solve this equation for x: $$2x + 3x - 9 = 11$$ $$5x = 20$$ $$x = 4$$ • Solve for y: $$y = x - 3, \text{ so } y = 4 - 3 = 1$$

CONCEPT/SKILL	DESCRIPTION	EXAMPLE
Addition (or elimination) method for solving a system of linear equations [3.12]	Used when both equations are written in the form $ax + by = c$. • Multiply each term in one equation (or both equations) by the appropriate constant that will make the coefficients of one variable (e.g., y) opposites. • Add the like terms in the two equations so that the y terms cancel. • Solve the remaining equation for x. • Use this value of x to determine y.	For the system of equations $$3x + 2y = 5$$ $$5x + 4y = 7:$$ • Multiply *each* term in the top equation by -2 so that its y-coefficient will become -4: $$-6x - 4y = -10$$ • Add like terms: $$\begin{array}{r} -6x - 4y = -10 \\ 5x + 4y = 7 \\ \hline -x \quad\quad = -3 \end{array}$$ • Multiply each side by -1 to solve for x: $x = 3$. • Substitute $x = 3$ into either original equation to determine y: $$3(3) + 2y = 5$$ $$9 + 2y = 5$$ $$2y = -4; y = -2$$
Solution set of an inequality [3.14]	The solution set of an inequality is the set of all values of the variable that satisfy the inequality.	The solution set to the inequality $2x + 3 \leq 9$ is all real numbers x which are less than or equal to 3.
Solving inequalities: direction of inequality sign [3.14]	**Property 1.** The direction of an inequality is not changed when **i.** the same quantity is added to or subtracted from both sides of the inequality. Stated algebraically, if $a < b$, then $a + c < b + c$ and $a - c < b - c$. **ii.** both sides of an inequality are multiplied or divided by the same positive number. If $a < b$, then $ac < bc$, where $c > 0$ and $\dfrac{a}{c} < \dfrac{b}{c}$, where $c > 0$. **Property 2.** The direction of an inequality is reversed if both sides of an inequality are multiplied or divided by the same negative number. These properties can be written symbolically as **i.** if $a < b$, then $ac > bc$, where $c < 0$ **ii.** if $a < b$, then $\dfrac{a}{c} > \dfrac{b}{c}$, where $c < 0$ Properties 1 and 2 will still be true if $a < b$ is replaced by $a \leq b, a > b$, or $a \geq b$.	For example, because $-4 < 10$, **i.** $-4 + 5 < 10 + 5$, or $1 < 15$, is true; $-4 - 3 < 10 - 3$, or $-7 < 7$, is true. **ii.** $-4(6) < 10(6)$, or $-24 < 60$, is true; $\dfrac{-4}{2} < \dfrac{10}{2}$, or $-2 < 5$, is true. Because $-4 < 10$, $-4(-5) > 10(-5)$, or $20 > -50$. Because $-4 < 10$, $\dfrac{-4}{-2} > \dfrac{10}{-2}$, or $2 > -5$. These properties will be true if $a < b$ is replaced by $a \leq b, a > b$, or $a \geq b$.

CONCEPT/SKILL	DESCRIPTION	EXAMPLE

Solving inequalities of the form

$f(x) < g(x)$

[3.14]

Inequalities of the form $f(x) < g(x)$ can be solved using two different methods.

Method 1 is a numerical approach in which a table of input-output pairs is used to determine values of x for which $f(x) < g(x)$.

Method 2 is an algebraic approach in which the two properties of inequalities are used to isolate the variable.

Solve $5x + 1 \geq 3x - 7$.

$$5x + 1 \geq 3x - 7$$
$$2x \geq -8$$
$$x \geq -4$$

1. Determine whether each of the following is a function.

   a. The area of a circle is a function of its radius.

   b. $\{(2, 9), (3, 10), (2, -9)\}$

2. For an average yard, the fertilizer costs $39. You charge $12 per hour to do yard work. If x represents the number of hours worked on the yard and y represents the total cost, including fertilizer, complete the following table.

x	0	2	3	5	7
y					

   a. Is the total cost a function of the hours worked? Explain.

   b. What is the independent variable?

   c. Which is the dependent variable?

   d. Which value(s) of the domain would not be realistic for this situation? Explain.

3. Your college service organization has volunteered to help with Spring Cleanup Day at a youth summer camp. You have been assigned the job of supplying paint for the exterior of the bunkhouses. You discover that 1 gallon of paint will cover 400 square feet of flat surface.

   a. If n represents the number of gallons of paint you supply and s represents the number of square feet you can cover with the paint, complete the following table.

n, Number of Gallons of Paint	1	2	4	6
s, Square Feet Covered by the Paint	400	800		

   b. Plot the ordered pairs you determined in part a on an appropriately scaled and labeled set of axes.

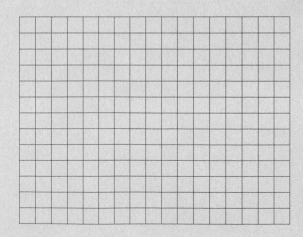

c. Let s be represented by $f(n)$, where f is the name of the function. Determine $f(6)$.

d. Write a sentence explaining the meaning of $f(4) = 1600$.

4. Determine the slope and y-intercept of the line whose equation is $4y + 10x - 16 = 0$.

5. Determine the equation of the line through the points $(1, 0)$ and $(-2, 6)$.

6. Determine the slope of the following line, and use the slope to determine an equation for the line.

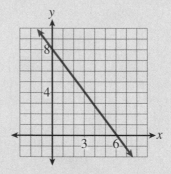

7. Match the graphs with the given equations. Assume that x is the input variable and y is the output variable. Each tick mark on the axes represents 1 unit.

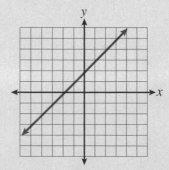 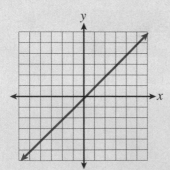 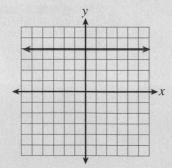

a. $y = x - 6$ **b.** $y = 4$ **c.** $y = x + 2$

d. $y = -3x - 5$ **e.** $y = x$ **f.** $y = 4 - 3x$

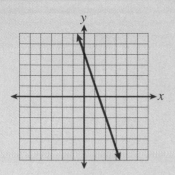 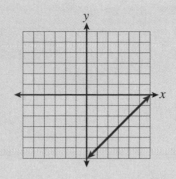 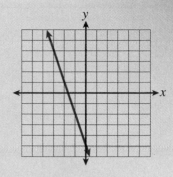

a. $y = x - 6$

d. $y = -3x - 5$

b. $y = 4$

e. $y = x$

c. $y = x + 2$

f. $y = 4 - 3x$

8. What is the equation of the line passing through the points $(0, 5)$ and $(2, 11)$? Write the final result in slope-intercept form, $y = mx + b$.

9. The equation of a line is $5x - 10y = 20$. Determine the horizontal and vertical intercepts.

10. Determine the horizontal and vertical intercepts of the line whose equation is $-3x + 4y = 12$. Use the intercepts to graph the line.

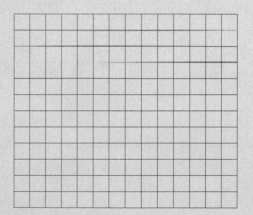

11. Determine the equation of the line that passes through the point $(0, -4)$ and has slope $\dfrac{5}{3}$.

12. Consider the lines represented by the following pair of equations: $3x + 2y = 10$ and $x - 2y = -2$.

a. Solve the system algebraically.

b. Determine the point of intersection of the lines by solving the system graphically.

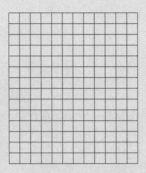

c. Determine the point of intersection of the lines by solving the system using the tables feature of your calculator, if available.

13. Solve the system of equations:

$b = 2a + 8$

$b = -3a + 3$

14. Solve the system of equations:

$x = 10 - 4y$

$5x - 2y = 6$

15. Corresponding values for p and q are given in the following table.

p	4	8	12	16
q	95	90	85	80

Write a symbolic rule for q as a linear function of p.

16. The following table was generated by a linear function. Determine the symbolic rule that defines this function.

x	95	90	85	80	75
y	55.6	58.4	61.2	64.0	66.8

17. Consider a graph of Fahrenheit temperature, °F, as a linear function of Celsius temperature, °C. You know that 212°F and 100°C represent the temperature at which water boils. You also know the input is °C, the output is °F, and 32°F and 0°C each represent the freezing point of water.

a. What is the slope of the line through the points representing the boiling and freezing points of water?

b. What is the equation of the line?

c. Use the equation to determine the Fahrenheit temperature corresponding to 40°C.

d. Use the equation to determine the number of degrees Celsius corresponding to 170°F.

18. Your power company currently charges $20.63 per month as a fixed basic service charge plus a variable amount based on the number of kilowatt-hours used. The following table displays several sample monthly usage amounts and the corresponding bill for each.

a. Complete the table of electricity cost, rounding to the nearest cent.

KILOWATT-HOURS, k (kWh)	COST, C(k) ($)
600	84.35
650	89.66
700	
750	100.28
800	105.59
850	
900	
950	121.52
1000	

b. Sketch a graph of the electricity cost function, labeling the axes.

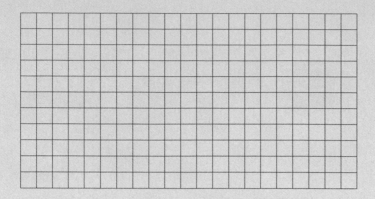

c. Write an equation for your total electric bill, $C(k)$, as a function of the number of kilowatt hours used, k.

d. Determine the total cost if 875 kilowatt hours of electricity are used.

e. Approximately how much electricity did you use if your total monthly bill was $150?

19. For tax purposes, you may have to report the value of assets, such as a computer. The value of some assets depreciates, or drops, over time. For example, a computer that originally cost $3000 may be worth only $1500 a few years later. The simplest way to calculate the value of an asset is by using *straight-line depreciation*, which assumes that the value is a linear function of time. If a $950 refrigerator depreciates completely (to zero value) in 10 years, determine a formula for its value, $V(t)$, as a function of time, t.

20. A coach of a local basketball team decides to analyze the relationship between his players' heights and their weights.

PLAYER	HEIGHT (in.)	WEIGHT (lb.)
1	68	170
2	72	210
3	78	235
4	75	220
5	71	175

a. Use technology to determine the equation of the regression line for this data.

b. Player 6 is 80 inches tall. Predict his weight.

c. If a player weighs 190 pounds, how tall would you expect him to be?

21. Use the substitution method to solve the following system.

$$5x + 2y = 18$$
$$y = x + 2$$

22. Use the addition method to solve the following system.

$$3x + 5y = 27$$
$$x - y = 1$$

23. Solve the following inequalities algebraically.

a. $2x + 3 \geq 5$ **b.** $3x - 1 \leq 4x + 5$

24. The weekly revenue, $R(x)$, and cost, $C(x)$, generated by a product can be modeled by the following equations.

$$R(x) = 75x$$
$$C(x) = 50x + 2500$$

Determine when $R(x) \geq C(x)$ numerically, algebraically, and graphically.

x	R(x)	C(x)

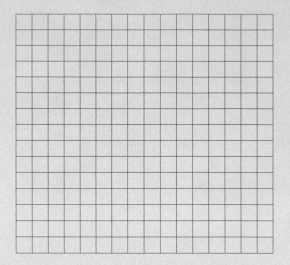

An Introduction to Nonlinear Problem Solving

Cluster 1   Mathematical Modeling Involving Polynomials

Activity 4.1

Fatal Crashes

Objectives

1. Identify polynomials and polynomial functions.

2. Classify a polynomial as a monomial, binomial, or trinomial.

3. Determine the degree of a polynomial.

4. Simplify a polynomial by identifying and combining like terms.

5. Add and subtract polynomials.

6. Evaluate and interpret polynomials.

In a recent year, approximately 43,000 people died in motor vehicle accidents in the United States. Does a driver's age have any relationship to his or her chances of being involved in a fatal car crash?

1. To compare the fatalities of one age group with another, you must use relative data (in the form of number of fatal crashes per 100 million miles driven). The number of fatal vehicle crashes, per 100 million miles driven, in a recent year can be approximated by the function

$$F(x) = 0.013x^2 - 1.19x + 28.24,$$

where x is the age of the driver, $x \geq 16$.

   a. Use the given function rule to complete the following table. Round your answers to the nearest tenth.

x, AGE OF DRIVER	F(x), FATAL CRASHES PER 100 MILLION MILES DRIVEN
16	
18	
20	
25	
35	
45	
55	
65	
75	
79	

   b. Interpret the value $F(20)$ in the context of this problem.

c. Which age group is involved in

    i. the highest number of fatal car crashes per 100 million miles driven?

    ii. the lowest number of fatal crashes per 100 million miles driven?

Polynomials

The expression $0.013x^2 - 1.19x + 28.24$ in Problem 1 of this activity is a special type of algebraic expression called a **polynomial**. Polynomials are formed by adding and subtracting terms containing positive integer powers of a variable such as x and possibly a constant term. Most of the algebraic expressions you have encountered in this text have been polynomials.

The function $F(x) = 0.013x^2 - 1.19x + 28.24$ is an example of a **polynomial function**.

The polynomial $0.013x^2 - 1.19x + 28.24$ contains three terms: an x^2 term, with its coefficient 0.013; an x term, with its coefficient -1.19; and a constant term, 28.24. Although the second term is being subtracted, the negative sign is assigned to its coefficient. Thus, it can be written as a sum: $0.013x^2 + (-1.19)x + 28.24$.

Definition

In general, a **polynomial in one variable** is an algebraic expression consisting of a sum of terms, each term being a constant or a variable raised to a positive integer power and multiplied by a coefficient. The polynomial is in **standard form** when the terms are written with the powers of x in decreasing order.

Example 1 *Consider the polynomial $2x^4 - 5x^3 + 6x + 7$. The exponent on x in each of the first three terms is a positive integer. This polynomial is in standard form because it is written with the powers of x in decreasing order. The coefficient of x^4 is 2, the coefficient of x^3 is -5, the coefficient of x is 6, and the constant term is 7. Notice that the x^2 term is missing. However, you can think of it as being there, with coefficient 0.*

Example 2

POLYNOMIALS	EXPRESSIONS THAT ARE NOT POLYNOMIALS
$10x$	$\dfrac{1}{x}$
$3x^2 + 5$	$\sqrt{x}$
$2x^3 - 3x^2 + \dfrac{4}{3}x - \dfrac{1}{2}$	$10x^2 + \dfrac{1}{x}$
$-10x^4$	$10x\sqrt{x^2 - 1}$

Note that in a polynomial expression, the variable cannot appear in a denominator or inside a radical.

2. Write yes if the expression is a polynomial. Write no if it is not a polynomial. In each case, give a reason for your answer.

   a. $3x + \dfrac{2}{5}$

   b. $-10x$

   c. $\dfrac{1}{3}x^3 + 2x - 1$

   d. $5x + \dfrac{10}{x}$

   e. $\dfrac{x + 1}{x - 2}$

   f. $\sqrt{3x + 1}$

Terminology

Polynomials with one, two, or three terms can be classified by the number of terms they contain.

> ### Definition
>
> **1.** A **monomial** is a polynomial containing a single term. It may be a number (constant) or a constant times a variable raised to a positive integer exponent.
>
> Examples: -3, $2x^4$, $\dfrac{1}{2}s^2$
>
> **2.** A **binomial** is a polynomial that has exactly two terms.
>
> Examples: $4x^3 + 2x$, $3t - 4$, $5w^6 + 4w^2$
>
> **3.** A **trinomial** is a polynomial that has exactly three terms.
>
> Examples: $3x^4 - 5x^2 + 10$, $5x^2 + 3x - 4$

3. Identify each of the following polynomials as a monomial, binomial, or trinomial.

   a. $15x$ _____

   b. $10x^3$ _____

   c. $3x^2 + 10$ _____

   d. $5x^4 + 2x^3 - 1$ _____

The **degree of a monomial** is defined as the exponent on its variable. The monomial $4x^5$ has degree 5. The monomial $2x$ has degree 1. The monomial consisting of a constant, such as 8, is said to have degree 0. The **degree of a polynomial** is the highest degree among all its terms. The degree of $2x^3 + 3x^7 - 5x^2 + 10$ is 7. When a polynomial is written in standard form, its degree is the degree of its first term.

4. Determine the degree of each polynomial. Then write the polynomial in standard form.

   a. $5 + 7x$

   b. $31 - 6x^3 + 11x^2$

   c. $-3x + 7x^2 - 8$

   d. $15 + x^5 - 2x^2 + 9x$

   e. 5

5. A polynomial function of degree one should look familiar to you. In general, it takes the form $f(x) = ax + b$. What type of function is this?

The fatal-crash function defined by $F(x) = 0.013x^2 - 1.19x + 28.24$ is a second-degree polynomial function. Such functions are called **quadratic functions** and are good models for a variety of situations. The graphs of quadratic functions are U-shaped curves called **parabolas**. The properties of these graphs and the algebra of quadratic (second-degree) polynomials are explored in Cluster 2.

Simplifying, Adding, and Subtracting Polynomials

Monomials containing common variable factors with identical degrees are **like terms**. For example, the monomials $5x^3$, $-2x^3$, and $3x^3$ are like terms with common variable factor x^3. The terms differ only in their coefficients. Recall from Chapter 2 that the distributive property provides a way to simplify expressions containing like terms by combining the like terms into a single term as follows:

$$5x^3 - 2x^3 + 3x^3 = (5 - 2 + 3)x^3 = 6x^3$$

The algebra of polynomials is an extension of the procedures you already know.

Example 3　*Simplify the polynomial $5x^2 - 7 + 2(3x^2 - 4x + 1)$.*

SOLUTION

$5x^2 - 7 + 2(3x^2 - 4x + 1)$	Remove the parentheses by applying the distributive property.
$= 5x^2 - 7 + 6x^2 - 8x + 2$	Combine like terms.
$= 11x^2 - 8x - 5$	

6. Simplify the following expressions.

    a. $-2(x + 3) + 5(3x - 1) + 3x$

    b. $(2x^2 - 7) - (4x - 3)$

    c. $2(4x^2 + 3x - 1) + 3(-2x^2 + 1) - 5x^2$

7. In tennis, the length of a singles court is 3 feet less than 3 times its width.

    a. Let w represent the width of a singles court. Express the length of the court, l, as a function of w.

    b. Express the perimeter, P, as a function of w.

    c. Simplify the function in part b.

The approach used to simplify polynomials can also be used to add and subtract polynomials.

Example 4   *Add the polynomials* $3x^3 - 5x^2 + 7x - 1$ *and* $9x^2 - 7x + 2$.

SOLUTION

Often it is easiest to match up like terms by writing one polynomial above the other, with like terms lined up in columns.

$$3x^3 - 5x^2 + 7x - 1$$
$$\underline{9x^2 - 7x + 2} \quad \text{Sum like terms in columns.}$$
$$3x^3 + 4x^2 + 1 \quad \text{The sum is } 3x^3 + 4x^2 + 1.$$

8. Add the following polynomials.

    a. $-2x^2 + 7x - 4$ and $4x^2 - 3x + 2$

    b. $4x^3 + 15x - 4$ and $4x^2 - 3x + 20$

c. $125x^3 - 14x^2 - 13$ and $-100x^3 + 25x^2 - x + 11$

d. $3x^2 + 5x - 7$, $-7x^2 + 2x - 2$, and $x^2 - 3x + 2$

Example 5 *Subtract* $2x^3 - 9x^2 - 7x + 2$ *from* $3x^3 - 5x^2 + 7x - 1$.

SOLUTION

Recall that subtracting $2x^3 - 9x^2 - 7x + 2$ is the same as adding its opposite. The opposite of $2x^3 - 9x^2 - 7x + 2$ is

$$-1(2x^3 - 9x^2 - 7x + 2) = -2x^3 + 9x^2 + 7x - 2.$$

Notice that the coefficient of each term in the opposite polynomial is the opposite of each coefficient in the original polynomial. Once again, you can write the polynomials in a vertical format and combine like terms:

$$
\begin{array}{l}
3x^3 - 5x^2 + 7x - 1 \\
\underline{-2x^3 + 9x^2 + 7x - 2} \qquad \text{Add the opposite polynomial.} \\
x^3 + 4x^2 + 14x - 3
\end{array}
$$

---

9. Perform the indicated operations, and express your answer in simplest form.

a. $(10x^2 + 2x - 4) - (6x^2 - 3x + 2)$

b. $(4x^3 + 13x - 4) - (4x^2 - 3x + 10)$

c. $(3x^2 + 2x - 3) - (8x^2 + x - 2) + (x^2 - 3x + 2)$

Evaluating Polynomial Functions

The stopping distance in feet, $D(v)$, for a car moving at a velocity (speed) of v miles per hour is given by the following function.

$$D(v) = 0.04v^2 + 1.1v$$

Note that the function rule for $D(v)$ is a binomial in the variable v.

10. **a.** Determine the value $D(55)$.

b. Interpret the meaning of this value.

c. Determine and interpret $D(80)$.

11. Evaluate each polynomial function for the given input value.

a. $P(a) = 15 - 2a$ Determine $P(8)$.

b. $g(x) = 2x^2 - 3x + 4$ Determine $g(-2)$.

c. $F(x) = -x^2 + 3x - 5$ Determine $F(-1)$.

SUMMARY: ACTIVITY 4.1

1. A **monomial** is a single term consisting of either a number (constant) or a variable raised to a positive integer exponent along with its coefficient.

2. A **polynomial** is an algebraic expression formed by adding and subtracting monomials.

3. Polynomials can be categorized by the number of terms they contain: A **monomial** is a polynomial with one term; a **binomial** is a polynomial with exactly two terms, and a **trinomial** is a polynomial with three terms.

4. A polynomial is in **standard form** when its terms are written with the powers of x in decreasing order.

5. The **degree of a monomial** is defined as the exponent on its variable. The **degree of a polynomial** is the highest degree among all its terms.

6. A **polynomial function** is defined by $y = P(x)$, where $P(x)$ is a polynomial expression.

7. To simplify a polynomial expression, use the distributive property to expand and then combine like terms.

8. Polynomials are added and subtracted using the same techniques for simplifying polynomials.

EXERCISES: ACTIVITY 4.1

1. a. Write yes if the expression is a polynomial. Write no if the expression is not a polynomial. In each case, give a reason for your answer.

i. $4x - 10$

ii. $3x^2 + 2x - 1$

iii. $\dfrac{2}{3}a$

iv. $\dfrac{1}{2}t^2 + \dfrac{4}{\sqrt{t}}$

v. $\dfrac{2}{w} - 8$

vi. $-x^2 + 3x^4 - 9$

b. Classify each expression as a monomial, binomial, or trinomial.

i. $4x - 10$ _____

ii. $3x^2 + 2x - 1$ _____

iii. $\dfrac{2}{3}a$ _____

iv. $-4x^3$ _____

c. How many terms are in the expression $2x^2 + 3x - x - 3$ as written?

d. How many terms are in the simplified expression?

2. Determine the degree of each polynomial. Write the polynomial in standard form.

a. $12 - 5x^4 + 3x^2$

b. $9 + 2x$

c. $5x - 15 + 8x^2$

d. 27

In Exercises 3–12, perform the indicated operations and express your answer in simplest form.

3. $3x + 2(5x - 4)$

4. $(3x - 1) + 4x$

5. $3.1(a + b) + 8.7a$

6. $6x + 2(x - y) - 5y$

7. $(3 - x) + (2x - 1)$

8. $(5x^2 + 4) - (3x^2 - 2)$

9. $9.5 - (3.5 - x)$

10. $4(x^3 + 2x) + 5(x^2 + 2x - 1)$

11. $3(2 - x) - 4(2x + 1)$

12. $10(0.3x + 1) - (0.2x + 3)$

In Exercises 13–18, evaluate the polynomial function at the given value.

13. $f(x) = 3x^2 + 2x - 1; f(4)$

14. $p(l) = 2(l + 2.8); p(3.5)$

15. $V(r) = \dfrac{4\pi}{3}r^3; V(6)$

16. $S(d) = 0.3d^2 + 4d; S(2.1)$

17. $C(x) = 4x^3 + 15; C(10)$

18. $D(m) = 3(m - 20); D(-5)$

19. One side of a square is increased by 5 units, and the other side is decreased by 3 units. If x represents a side of the original square, write a polynomial expression that represents the perimeter of the newly formed rectangle.

20. You want to boast to a friend about the stock you own without telling him how much money you originally invested in the stock. Let your original stock be valued at x dollars. You watch the market once a month for 4 months and record the following.

Month	1	2	3	4
Stock Value	increased $10	doubled	decreased $4	increased by 50%

a. Use x as a variable to represent the value of your original investment, and write a polynomial expression to represent the value of your stock after the first month.

b. Use the result from part a to write a polynomial expression to determine the value at the end of the second month, simplifying when possible. Continue until you determine the value of your stock at the end of the fourth month.

c. Do you have good news to tell your friend? Explain to him what happened to the value of your stock over 4 months.

d. Instead of simplifying after each step, write a single expression that represents the stock's value at the end of the 4-month period.

e. Simplify the expression in part d. How does the simplified expression compare with the result in part b?

21. Add the following polynomials.

a. $-5x^2 + 2x - 4$ and $4x^2 - 3x + 6$

b. $7x^3 + 12x - 9$ and $-4x^2 - 3x + 20$

c. $45x^3 - 14x^2 - 13$ and $-50x^3 + 20x^2 - x + 13$

d. $x^2 + x - 7$, $-9x^2 + 2x - 3$, and $5x^2 - x + 2$

22. Perform the indicated operations, and express your answer in simplest form.

a. $(13x^2 + 12x - 5) - (6x^2 - 3x + 2)$

b. $(3x^3 + x - 4) - (3x^2 - 3x + 11)$

c. $(6x^2 + 2x - 3) - (7x^2 + x - 2) + (x^2 - 3x + 1)$

d. $(-5x^3 - 4x^2 - 7x - 8) - (-5x^3 - 4x^2 - 7x + 8)$

23. According to statistics from the U.S. Department of Commerce, the per capita personal income (i.e., the average annual income) of each resident of the United States from 1960 to 2012 can be modeled by the polynomial function

$$P(t) = 10.4t^2 + 263.0t + 1347.8,$$

where t equals the number of years since 1960 and $P(t)$ represents the per capita income.

a. Use $P(t)$ to estimate the per capita personal income in 1989.

b. Determine the value of $P(56)$. Interpret the meaning of the answer in the context of this situation.

24. Smartphones are very popular and are becoming more powerful each year. The number of smartphone users in the United States starting in 2010 and projected through 2017 can be modeled by the polynomial function

$$S(t) = -1.3t^2 + 29.7t + 63.8,$$

where S represents the number of U.S. smartphone users in millions and t represents the number of years since 2010.

a. Determine the number of smartphone users in 2013.

b. Determine and interpret $S(5)$.

c. According to the model, what is the projected number of smartphone users in 2017?

Activity 4.2

Volume of a Storage Box

Objectives

1. Use properties of exponents to simplify expressions and combine powers that have the same base.

2. Use the distributive property and properties of exponents to write expressions in expanded form.

You are building a wooden storage box to hold your vinyl record collection. The box has three dimensions: length, l, width, w, and height, h, as shown in the following figure.

Recall that the formula for the volume V of a box (also called a *rectangular solid*) is $V = lwh$. The units of volume are cubic units, such as cubic inches, cubic feet, and cubic meters.

1. Determine the volume of a storage box whose length is 3 feet, width is 2 feet, and height is 1.5 feet.

You decide to design a storage box with a square base.

2. What is the relationship between the length and width of a box with a square base?

3. Write the volume formula for this square-based box in terms of l and h in two ways.

   a. Using three factors (including repeated factors)

   b. Using two factors (including exponents)

Now suppose the box is to be a cube.

4. What is the relationship between the length, width, and height of a cube?

5. Write the volume formula for a cube in terms of l in three ways.

   a. Using three factors (including repeated factors)

   b. Using two factors (including exponents)

   c. Using one factor (including exponents)

6. Suppose the cube has length of 4 ft. Evaluate each of the expressions in Problem 5 to verify that all three expressions are equivalent.

The three expressions lll, $l^2 l$, and l^3 are equivalent expressions that all represent the volume of a cube with side of length l.

Property 1 of Exponents: Multiplying Powers Having the Same Base

One of the most important skills of algebra is the ability to rewrite a given expression into an equivalent expression. In the case of the three expressions for volume of a cube, it is helpful to note that the symbol l itself can be written in exponential form as l^1. In this way, you can see that

$$l^1 l^1 l^1 = l^2 l^1 = l^3.$$

This equivalence illustrates the first Property of Exponents: When multiplying exponential factors with identical bases, you retain the base and add the exponents.

> **Property 1 of Exponents: Multiplying Powers with the Same Base**
>
> For any base b and positive integers m and n, $b^m b^n = b^{m+n}$.

7. Simplify the following products by rewriting each as a single power.

   a. $x^2 x^5$ **b.** $ww^3 w^4$ **c.** $a^3 a^6$ **d.** $h^4 h$

When the exponential factors include coefficients, you can reorder the factors by first grouping the coefficients and then combining the exponential factors using Property 1.

8. Simplify the following products.

   a. $2x^3 x^2$ **b.** $4w^3 3w^4$ **c.** $a^3 5a$ **d.** $6h^4 4h2h^2$

9. A box is constructed so that its length and height are each 3 times its width, w.

   a. Represent the length and height in terms of w.

   b. Determine the volume of the box in terms of w.

10. Show how you would evaluate the expression $2^4 \cdot 3^2$. Are you able to add the exponents first? Explain.

The results from Problem 10 illustrate that when the bases are different, the powers cannot be combined. For example, $x^4 y^2$ cannot be simplified. However, a product such as $(x^3 y^2)(x^2 y^5)$ can be rewritten as $x^3 x^2 y^2 y^5$, which can be simplified to $x^5 y^7$.

> **Procedure**
>
> **Multiplying a Series of Monomial Factors**
>
> **1.** Multiply the numerical coefficients.
>
> **2.** Use Property 1 of Exponents to simplify the product of the variable factors that have the same base. That is, add their exponents and keep the base the same.

11. Multiply the following.

   a. $(-3x^2)(4x^3)$ **b.** $(5a^3)(3a^5)$

   c. $(a^3 b^2)(ab^3)(b)$ **d.** $(3.5x)(-0.1x^4)$

   e. $(r^2)(4.2r)$ **f.** $(4x^3 y^6)(2x^3 y)(3y^2)$

12. Use the distributive property and Property 1 of Exponents to write each of the following expressions in expanded, standard polynomial form.

a. $x^3(x^2 + 3x - 2)$ **b.** $-2x(x^2 - 3x + 4)$

c. $2a^3(a^3 + 2a^2 - a + 4)$

13. Rewrite each of the following in factored form.

a. $x^5 + 3x^4 - 2x^3$ **b.** $-2x^3 + 6x^2 - 8x$ **c.** $2a^6 + 4a^5 - 2a^4 + 8a^3$

Property 2 of Exponents:
Dividing Powers Having the Same Base $b, b \neq 0$

Related to Property 1 is Property 2 of Exponents involving quotients, such as $\dfrac{x^5}{x^2}$, that have exponential factors with identical bases.

14. a. Complete the following table for the given values of x.

x	$\dfrac{x^5}{x^2}$	x^3
2		
3		
4		

b. How does the table demonstrate that $\dfrac{x^5}{x^2}$ is equivalent to x^3?

If you write both the numerator and denominator of $\dfrac{x^5}{x^2}$ as a chain of multiplicative factors, you obtain

$$\frac{x^5}{x^2} = \frac{x \cdot x \cdot x \cdot x \cdot x}{x \cdot x}.$$

You can then divide out the two factors of x in the denominator and the numerator to obtain x^3.

This suggests that when dividing exponential factors with identical bases, you retain the base and subtract the exponents.

> **Property 2 of Exponents: Dividing Powers with the Same Base**
>
> For any base $b \neq 0$ and positive integers m and n, $\dfrac{b^m}{b^n} = b^{m-n}$.

15. Simplify the following products and quotients. Assume none of the bases is equal to 0.

    a. $\dfrac{x^9}{x^2}$

    b. $\dfrac{10x^8}{2x^5}$

    c. $\dfrac{c^3 2c^5}{3c^6}$

    d. $\dfrac{24z^{10}}{3z^3 4z}$

    e. $\dfrac{15d^4 2d}{3d^5}$

Property 3 of Exponents: Raising a Product or Quotient to a Power

Consider a cube whose side measures s inches. The surface of the cube consists of six squares—bottom and top, front, back, and left and right sides.

16. a. Determine a formula for the volume, V, of a cube whose side measures s inches.

    b. Use the formula to calculate the volume of a cube whose side measures 5 inches. Include appropriate units of measure.

Suppose you construct a new cube whose side is twice as long as the cube in Problem 16a. The side of the new cube is represented by $2s$. The volume of this new cube can be determined using the formula $V = s^3$, replacing s with $2s$.

$$V = (2s)^3$$

This new formula can be simplified as follows:

$$V = (2s)^3 = 2s \cdot 2s \cdot 2s = 2 \cdot 2 \cdot 2 \cdot sss = 2^3 s^3$$

Therefore, the expression $(2s)^3$ is equivalent to $2^3 s^3$, or $8s^3$. In a similar fashion, the fourth power of the product xy, $(xy)^4$, is written as an equivalent expression in Problem 17.

17. a. Use the definition of an exponent to write the expression $(xy)^4$ as a repeated multiplication.

    b. How many factors of x are in your expanded expression?

    c. How many factors of y are in your expanded expression?

d. Regroup the factors in the expanded expression so that the x factors are written first, followed by the y factors.

e. Rewrite the expression in part d using exponential notation.

The preceding results illustrate Property 3 of Exponents: When raising a product (or quotient) to a power, you raise each of its factors to that power.

> **Property 3 of Exponents: Raising a Product or Quotient to a Power**
>
> For any numbers a and b and positive integer n,
>
> $$(ab)^n = a^n b^n \quad \text{and} \quad \left(\frac{a}{b}\right)^n = \frac{a^n}{b^n}, \ b \neq 0.$$

18. Use Property 3 to simplify the following expressions.

a. $(2a)^3$ **b.** $(3mn)^2$ **c.** $\left(\dfrac{10}{p}\right)^6$ **d.** $\left(\dfrac{x}{2}\right)^3$

19. a. The **surface area**, A, of a cube is the sum of the areas of its six sides (squares). Determine a formula for the surface area, A, of a cube whose side measures s inches.

b. Use your formula to calculate the surface area of a cube whose side measures 5 inches. Include appropriate units of measure.

c. If you double the length of the side of the cube whose side measures s units, write a new formula, in expanded and simplified form, for the surface area.

20. Determine which of the following expressions are equivalent.

$$2(xy)^3 \qquad (2xy)^3 \qquad 2xy^3 \qquad 8x^3y^3$$

Property 4 of Exponents: Raising a Power to a Power

The final Property of Exponents is a combination of Property 1 and the definition of exponential notation itself. Consider the expression

$$a^2 \cdot a^2 \cdot a^2 \cdot a^2 \cdot a^2.$$

Because this expression consists of a^2 written as a factor 5 times, you can use exponential notation to rewrite this product as $(a^2)^5$.

However, Property 1 indicates that $a^2 \cdot a^2 \cdot a^2 \cdot a^2 \cdot a^2$ can be rewritten as $a^{2+2+2+2+2}$, which simplifies to a^{10}.

Therefore, $(a^2)^5 = a^{10}$, illustrating Property 4 of Exponents: When raising a power to a power, you multiply the exponents.

> **Property 4 of Exponents: Raising a Power to a Power**
>
> For any number b and positive integers m and n, $(b^m)^n = b^{mn}$.

21. Use the properties of exponents to simplify each of the following.

   a. $(t^3)^5$ **b.** $(y^2)^4$ **c.** $(3^2)^4$

   d. $2(a^5)^3$ **e.** $x(x^2)^3$ **f.** $-3(t^2)^4$

   g. $(5xy^2)(3x^4y^5)$ **h.** $r\left(\dfrac{r}{4}\right)^3$

Defining Zero and Negative Integer Exponents

22. a. Determine the numerical value of the expression $\dfrac{2^3}{2^3}$.

   b. Apply Property 2 of Exponents, $\dfrac{b^n}{b^m} = b^{n-m}$, to simplify $\dfrac{2^3}{2^3}$.

   c. What do your results from parts a and b suggest about the numerical value of 2^0?

> **Definition**
>
> **Zero Exponent**
>
> For any nonzero number b, $b^0 = 1$.

23. a. Evaluate the expression $\dfrac{2^3}{2^4}$, and write your result as a fraction in simplest form.

   b. Use Property 2 of Exponents to simplify $\dfrac{2^3}{2^4}$.

c. What do your results from parts a and b suggest about the numerical value of 2^{-1}?

24. a. Simplify the expression $\dfrac{b^3}{b^5}$ by dividing out common factors.

b. Use Property 2 of Exponents to simplify $\dfrac{b^3}{b^5}$.

c. Use the result of part a to explain the meaning of the expression b^{-2}.

Problems 23 and 24 illustrate the meaning of negative exponents: Negative exponents indicate taking reciprocals.

Definition

Negative Exponents

For any nonzero number b and positive integer n, $b^{-n} = \dfrac{1}{b^n}$.

---

25. Simplify each of the following expressions. Write the result without negative exponents and in simplest fraction from $\dfrac{a}{b}$. Assume all variables are nonzero.

a. 5^{-2} **b.** $\dfrac{b^0}{b^3}$ **c.** $\dfrac{x^5}{x^9}$

d. $\dfrac{12x^2}{3x^4}$ **e.** $\left(\dfrac{5}{2}\right)^{-1}$ **f.** $\left(\dfrac{2a}{5}\right)^{-3}$

SUMMARY: ACTIVITY 4.2

Properties of Exponents

1. Multiplying Powers Having the Same Base

For any number b, if m and n represent positive integers, then $b^m \cdot b^n = b^{m+n}$.

2. Dividing Powers Having the Same Base

For any nonzero number b, if m and n represent positive integers, then $\dfrac{b^m}{b^n} = b^{m-n}$.

3. **Raising a Product or a Quotient to a Power**

   For any numbers a and b, if n represents a positive integer, then

   $$(a \cdot b)^n = a^n \cdot b^n \quad \text{and} \quad \left(\frac{a}{b}\right)^n = \frac{a^n}{b^n}, \quad b \neq 0.$$

4. **Raising a Power to a Power**

   For any number b, if m and n represent positive integers, then $(b^m)^n = b^{mn}$.

5. **Defining the Zero Exponent**

   For any nonzero number b, $b^0 = 1$.

6. **Defining Negative Exponents**

   For any nonzero number b, if n represents a positive integer, then $b^{-n} = \frac{1}{b^n}$.

EXERCISES: ACTIVITY 4.2

1. Refer to the properties and definitions of exponents, and explain how to simplify each of the following. (All variables are nonzero.)

   a. $x \cdot x \cdot x \cdot x \cdot x$

   b. $x^3 \cdot x^7 \cdot x$

   c. $\dfrac{x^6}{x^4}$

   d. $(x^4)^3$

   e. x^{-3}

   f. y^0

   g. $\dfrac{r^3}{r^8}$

   h. $(5w)^3$

In Exercises 2–23, use the properties of exponents to simplify the expressions. Write the result without negative exponents. (All variables are nonzero.)

2. aa^3

3. $3xx^4$

4. $y^2y^3y^4$

5. $3t^45t^2$

6. $-3w^24w^5$

7. $3.4b^51.05b^3$

8. $(a^5)^3$

9. $4(x^2)^4$

10. $(-x^{10})^5$

11. $(-3x^2)(-4x^7)(2x)$

12. $(-5x^3)(0.5x^6)(2.1y^2)$

13. $(-2s^2t)(t^2)^3(s^4t)$

14. $\dfrac{16x^7}{24x^3}$

15. $\dfrac{10x^4y^6}{5x^3y^5}$

16. $\dfrac{x^4}{x^7}$

17. $\dfrac{9a^4b^2}{18a^3b^2}$

18. $\dfrac{4w^2z^2}{10w^2z^4}$

19. $32\left(\dfrac{x}{4}\right)^4$

20. $11x^2(y^2)^0$

21. $5a(4a^3)^2$

22. $\left(\dfrac{3a^2}{2b}\right)^4$

23. $18t^2\left(\dfrac{t}{3}\right)^3$

In Exercises 24–31, use the distributive property and the properties of exponents to expand the algebraic expression and write it as a polynomial in standard form.

24. $2x(x+3)$

25. $y(3y-1)$

26. $x^2(2x^2+3x-1)$

27. $2a(a^3+4a-5)$

28. $5x^3(2x-10)$

29. $r^4(3.5r-1.6)$

30. $3t^2(6t^4-2t^2-1.5)$

31. $1.3x^7(-2x^3-6x+1)$

32. Simplify the expressions $4(a+b+c)$ and $4(abc)$. Are the results the same? Explain.

33. a. You are drawing up plans to enlarge your square patio. You want to triple the length of one side and double the length of the other side. If x represents a side of your square patio, write a formula for the area, A, of the new patio in terms of x.

b. You discover from the plan that after doubling the one side of the patio, you must cut off 3 feet from that side to clear a bush. Write an expression in terms of x that represents the length of this side.

c. Use the result from part b to write a formula without parentheses to represent the area of the patio in part b. Remember that the length of the other side of the original square patio was tripled.

34. A rectangular bin has the following dimensions.

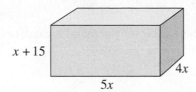

a. Write a formula that represents the area, A, of the base of the bin.

b. Using the result from part a, write a formula that represents the volume, V, of the bin.

35. A cube measures b^4 units on a side. Write a formula in terms of b that represents the volume of the cube.

36. A square measures $3xy^2$ units on each side. Write an expression that represents the area of the square.

37. A car travels 4 hours at an average speed of $2a - 4$ miles per hour. Write an expression, without parentheses, that represents the distance traveled.

38. The radius of a cylindrical container is 3 times its height. Write a formula for the volume of the cylinder in terms of its height. Remember to indicate what each variable in your formula represents.

39. A tuna fish company wants to make a tuna fish container whose radius is equal to its height. Write a formula for the volume of the desired cylinder in terms of its height. Be sure to indicate what each variable in your formula represents. Write the formula in its simplest form.

40. A manufacturing company says that the size of a can is the most important factor in determining profit. The company prefers the height of a can to be 4 inches greater than the radius. Write a formula for the volume of this desired cylindrical can in terms of its radius. Be sure to indicate what each variable in your formula represents. Write the formula in expanded form as a standard polynomial.

41. In the cylinder shown here, the relationship between the height and radius is expressed in terms of x. Write a formula, in expanded form, for the volume of the cylinder.

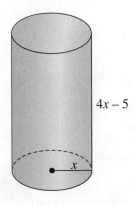

$4x - 5$

x

42. Determine the greatest common factor of each polynomial, and rewrite the expression in completely factored form. See Activity 2.9, page 235, for a review of greatest common factor.

a. $18x^7 + 27x^3 - 15x^2$ **b.** $14y^8 - 21y^5$

c. $6x^5 - 12x^4 + 9x^3$ **d.** $15t^9 - 25t^6 + 20t^2$

43. Simplify each of the following expressions. Write the result without negative exponents and in simplified fraction form $\dfrac{a}{b}$. Assume $x \neq 0$.

a. 4^{-1} **b.** $\left(\dfrac{2}{3}\right)^{-1}$

c. $\left(\dfrac{3}{4}\right)^{-2}$ **d.** $2 \cdot \left(\dfrac{4}{3}\right)^{-1}$

e. $\dfrac{3}{2} \cdot \left(\dfrac{5x}{6}\right)^{-1}$ **f.** $\dfrac{8x}{9} \cdot \left(\dfrac{2x}{3}\right)^{-2}$

Activity 4.3

Room for Work

Objectives

1. Expand and simplify the product of two binomials.

2. Expand and simplify the product of any two polynomials.

3. Recognize and expand the product of conjugate binomials: difference of squares.

4. Recognize and expand the product of identical binomials: perfect-square trinomials.

The basement of your family home is only partially finished. Currently, there is a 9- by 12-foot room that is being used for storage. You want to convert this room to a home office but want more space than is now available. You decide to knock down two walls and enlarge the room.

1. What is the current area of the storage room?

2. If you extend the long side of the room 4 feet and the short side 2 feet, what will be the area of the new room?

3. Start with a diagram of the current room (9 by 12 feet), and extend the length 4 feet and the width 2 feet to obtain the following representation of the new room.

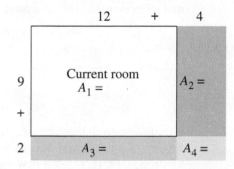

a. Calculate the area of each section, and record the results in the appropriate places on the diagram.

b. Determine the total area of the new room by summing the areas of the four sections.

c. Compare your answer from part a with the area you calculated in Problem 2.

You are not sure how much larger a room you want. However, you decide to extend the length and the width each by the same number of feet, represented by x.

4. Starting with a geometric representation of the current room (9 by 12 feet), you can extend the length and the width x feet to obtain the following diagram of the new room.

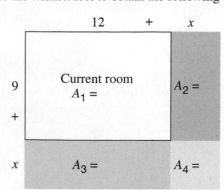

a. Determine the area of each section in your geometric model, and record your answer in the appropriate place. Note that the area in several sections will be an expression in x.

b. Write a formula for the total area by summing the areas of the four sections. Combine like terms, and write your answer as a polynomial in standard form.

You can represent the area of the new room in an equivalent way by multiplying the expression for the new length by the expression for the new width. The new length of the room is represented by $12 + x$, and the new width is represented by $9 + x$; so the area of the room can be expressed as the product $(12 + x)(9 + x)$.

This product can be expanded using the distributive property twice. Each term of the first binomial is distributed over the entire second binomial. This means that each term of the first binomial multiplies every term in the second. This process results in a polynomial with four terms.

$$(12 + x)(9 + x) = \underbrace{12(9 + x)}_{\substack{\text{first column of} \\ \text{geometric} \\ \text{representation}}} \quad + \quad \underbrace{x(9 + x)}_{\substack{\text{second column} \\ \text{of geometric} \\ \text{representation}}}$$

$$= 12 \cdot 9 + 12 \cdot x + x \cdot 9 + x \cdot x$$

This four-step multiplication process is often referred to as the FOIL (**F**irst + **O**uter + **I**nner + **L**ast) Method, in which each letter refers to the position of the terms in the two binomials. FOIL is not a new method—it just describes a double use of the distributive property. However, it is a helpful reminder to perform all four multiplications when you are expanding the product of two binomial factors.

$$(12 + x)(9 + x) = \underbrace{12 \cdot 9}_{F} + \underbrace{12 \cdot x}_{O} + \underbrace{x \cdot 9}_{I} + \underbrace{x \cdot x}_{L}$$

Notice that $12 \cdot x$ and $x \cdot 9$ are like terms that can be combined as $21x$ so that the simplified product becomes $x^2 + 21x + 108$.

Compare the final expression with your result in Problem 4b. They should be the same.

5. The expressions $(12 + x)(9 + x)$ and $x^2 + 21x + 108$ are equivalent expressions that represent the area of a 9- by 12-foot room whose length and width have each been extended x feet.

Use each of these expressions to determine the area of the remodeled room when $x = 3$ feet. What are the dimensions of the room? Remember to include the units.

6. Write $(x + 4)(x - 5)$ in expanded form, and simplify by combining like terms.

7. Write $(3x + 1)(x + 2)$ in expanded form, and simplify by combining like terms.

Product of Any Two Polynomials

A similar extension of the distributive property is used to expand the product of any two polynomials. Here again, each term of the first polynomial multiplies every term of the second.

Example 1 *Expand and simplify the product* $(2x + 3)(x^2 + 3x - 2)$.

SOLUTION

$$(2x + 3)(x^2 + 3x - 2)$$
$$= 2x \cdot x^2 + 2x \cdot 3x + 2x(-2) + 3 \cdot x^2 + 3 \cdot 3x + 3(-2)$$
$$= 2x^3 + 6x^2 - 4x + 3x^2 + 9x - 6$$
$$= 2x^3 + 9x^2 + 5x - 6$$

8. Expand and simplify the product $(x - 4)(3x^2 - 2x + 5)$.

Difference of Squares

You are drawing up plans to change the dimensions of your square patio. You want to increase the length by 2 feet and reduce the width by 2 feet. This new design will change the shape of your patio, but will it change the size?

9. a. Determine the current area and new area of a square patio whose side originally measured

    i. 10 ft.   **ii.** 12 ft.   **iii.** 15 ft.

b. In all three cases, how does the proposed new area compare with the current area?

You can examine why your observation in Problem 9b is always true regardless of the size of the original square patio. If x represents the length of a side of the square patio, the new area is represented by the product

$$(x + 2)(x - 2).$$

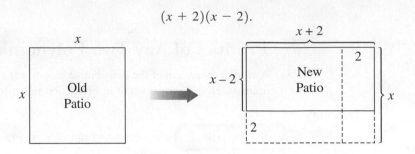

10. a. Represent the area of the old patio in terms of x.

   b. Represent the area of the new patio in terms of x. Expand and simplify this expression.

   c. How does the new area compare with the current area?

11. Expand and simplify the following binomial products using the FOIL Method. What do you notice about their expanded forms?

   a. $(x + 5)(x - 5)$ **b.** $(3x + 4)(3x - 4)$

Look closely at the binomial factors in each product of Problem 11. In each case, the two binomial factors are **conjugate binomials**. That is, the first terms of the binomials are identical and the second terms differ only in sign (for example, $x + 5$ and $x - 5$).

> When two conjugate binomials are multiplied, the like terms (corresponding to the outer and inner products) are opposites. Because opposites always sum to zero, the only terms remaining in the expanded product are the first and last products. This can be expressed symbolically as
>
> $$(A + B)(A - B) = A^2 - B^2.$$
>
> The expanded expression $A^2 - B^2$ is commonly called a **difference of squares**.

12. Any time two conjugate binomials are multiplied and expanded, the result is a difference of squares. Use this pattern to write (with minimal calculation) the expanded forms of the following products.

a. $(x + 3)(x - 3)$ **b.** $(x - 10)(x + 10)$

c. $(2x + 7)(2x - 7)$ **d.** $(4x - 9)(4x + 9)$

e. $(5x + 1)(5x - 1)$ **f.** $(3x + 2)(3x - 2)$

Perfect-Square Trinomials

You rethink your plans to change the size of your square patio. You decide to extend the length of each side by 3 feet. By how many square feet will this new design enlarge your current patio?

13. a. Determine the current area and new area of a square patio if the side originally measured

    i. 7 feet   **ii.** 10 feet   **iii.** 15 feet

    b. Determine the increase in patio area for each original patio dimension.

    c. Do the results in part b indicate that the new design will enlarge your current patio by a constant amount? Explain.

You can verify your observations in Problem 13c algebraically. If x represents the length of a side of the square patio, the new area is represented by the product

$$(x + 3)(x + 3).$$

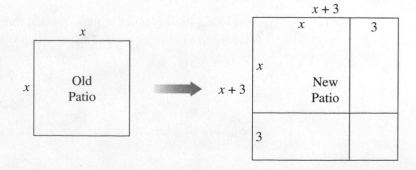

14. a. Represent the area of the current patio in terms of x.

    b. Represent the area of the new patio in terms of x. Expand and simplify this expression.

    c. Represent in terms of x the number of square feet by which the patio has been enlarged?

    d. Is your result in part c consistent with the results from Problem 13c?

15. Expand and simplify the following binomial products.

    a. $(x + 5)(x + 5)$ **b.** $(x - 4)(x - 4)$

In each product of Problem 15, the binomial factors are identical. Note that the product $(x + 3)(x + 3)$ can be written as $(x + 3)^2$.

16. Rewrite each product of Problem 15 using exponential notation.

    a.   **b.**

17. Look back at your solutions to Problem 15 for a relationship between the repeated binomial, $(x + a)^2$, and its expanded form.

    a. What is the relationship between the constant in the repeated binomial and the constant term in the expanded form?

    b. What is the relationship between the coefficient of x in the repeated binomial and the coefficient of the x term in the expanded form?

When a binomial is squared, the expanded form contains the square of each term *plus* twice the product of its terms. This is written symbolically as

$$(x + a)^2 = (x + a)(x + a) = x^2 + 2ax + a^2$$

$$\text{or} \quad (x - a)^2 = (x - a)(x - a) = x^2 - 2ax + a^2.$$

The trinomials $x^2 + 2ax + a^2$ and $x^2 - 2ax + a^2$ are called **perfect-square trinomials**.

Example 2   *Write the expanded form of* $(x - 7)^2$.

SOLUTION

$$\underbrace{x^2}_{\substack{\text{square} \\ \text{of the} \\ \text{first term}}} + \underbrace{2 \cdot (-7x)}_{\substack{\text{twice the} \\ \text{product of the} \\ \text{two terms}}} + \underbrace{49}_{\substack{\text{square of} \\ \text{the second} \\ \text{term}}} = x^2 - 14x + 49$$

18. Write the expanded form of each of the following using the pattern illustrated in Example 2.

a. $(x + 8)^2$ **b.** $(x + 10)^2$

c. $(x - 6)^2$ **d.** $(x - 1)^2$

e. $(x - 4)(x - 4)$ **f.** $(x + 9)(x + 9)$

19. Identify which of the following are perfect-square trinomials. Write each of the perfect-square trinomials as a square of the appropriate binomial.

a. $x^2 - 12x + 24$ **b.** $x^2 - 4x + 4$ **c.** $x^2 - 8x - 16$

d. $x^2 + 2x + 1$ **e.** $x^2 + 13x + 36$

SUMMARY: ACTIVITY 4.3

1. The product of two polynomials can be expanded using an extension of the distributive property. Each term of the first polynomial multiplies every term in the second.

2. **Conjugate binomials** are binomials whose first terms are identical and whose second terms differ only in sign: for example, $2x + 5$ and $2x - 5$.

3. When two conjugate binomials are multiplied, the expanded form is called a **difference of squares**.

$$(A + B)(A - B) = \underbrace{A^2 - B^2}_{\text{difference of squares}}$$

4. When a binomial is squared, the expanded form is called a **perfect-square trinomial**.

$$(x + a)^2 = (x + a)(x + a) = \underbrace{x^2 + 2ax + a^2}_{\text{perfect-square trinomial}}$$

$$(x - a)^2 = (x - a)(x - a) = x^2 - 2ax + a^2$$

EXERCISES: ACTIVITY 4.3

1. Multiply each of the following pairs of binomial expressions. Remember to simplify by combining like terms.

  a. $(x + 1)(x + 7)$

  b. $(w - 5)(w - 2)$

  c. $(x + 3)(x - 2)$

  d. $(x - 6)(x + 3)$

  e. $(5 + 2c)(2 + c)$

  f. $(x + 3)(x - 3)$

  g. $(3x + 2)(2x + 1)$

  h. $(5x - 1)(2x + 7)$

  i. $(6w + 5)(2w - 1)$

  j. $(4a - 5)(8a + 3)$

  k. $(x + 3y)(x - 2y)$

  l. $(3x - 1)(x - 4)$

  m. $(2a - b)(a - 2b)$

  n. $(4c + d)(4c - d)$

2. Multiply the following. Write your answer in simplest form.

  a. $(x - 1)(3x - 2w + 5)$

  b. $(x + 2)(x^2 - 3x + 5)$

  c. $(x - 4)(3x^2 + x - 2)$

  d. $(a + b)(a^2 - 2b - 1)$

  e. $(x - 3)(x^2 + 3x + 9)$

  f. $(x - 2y)(3x + xy - 2y)$

3. Expand and simplify the following binomial products.

  a. $(x - 4)^2$

  b. $(x + 4)^2$

  c. $(x - 4)(x + 4)$

  d. $(2x - 3)(2x + 3)$

  e. $(x - 20)^2$

  f. $(x + 12)^2$

4. a. You have a circular patio that you want to enlarge. The radius of the existing patio is 10 feet. Extend the radius by x feet, and express the area of the new patio in terms of x. Leave your answer in terms of π. Do not expand.

b. Use the FOIL Method to express the factor $(10 + x)^2$ in the formula $A = \pi(10 + x)^2$ in expanded form.

c. You decide to increase the radius 3 feet. Use the formula for area you obtained in part a to determine the area of the new patio.

d. Now use the formula for area you obtained in part b to determine the area of the patio. Compare your results with your answer in part c.

5. a. You have an old rectangular birdhouse just the right size for wrens. The birdhouse measures 7 inches long, 4 inches wide, and 8 inches high. What is the volume of the birdhouse?

b. You want to use the old birdhouse as a model to build a new birdhouse to accommodate larger birds. You will increase the length and width the same amount, x, and leave the height unchanged. Express the volume of the birdhouse in factored form in terms of x.

c. Use the FOIL Method to expand the expression you obtained in part b.

d. If you increase the length and width 2 inches each, determine the volume of the new birdhouse. How much have you increased the volume? By what percent have you increased the volume?

e. Instead of increasing both the length and width of the house, you choose to decrease the length and increase the width the same amount, x. Express the formula for the volume of the new birdhouse in factored form in terms of x.

f. Use the FOIL Method to expand the expression you obtained in part e.

g. Use the result from part f to determine the volume of the birdhouse if the length is decreased 2 inches and the width is increased 2 inches. Check your result using the volume formula you obtained in part e.

Cluster 1 | What Have I Learned?

1. Determine numerically and algebraically which of the following four algebraic expressions are equivalent. Select any three input values, and then complete the table. Describe your findings in a few sentences.

x	$(2x^2)^3$	$6x^6$	$8x^5$	$8x^6$

2. Explain the difference between the two expressions $-x^2$ and $(-x)^2$.

3. a. Write an algebraic expression that does not represent a polynomial, and explain why the expression is *not* a polynomial.

b. Write a polynomial that has three terms.

4. Does the distributive property apply when simplifying the expression $3(2xy)$? Explain.

5. What effect does the negative sign to the left of the parentheses have in simplifying the expression $-(x - y)$?

6. Simplify the product $x^3 \cdot x^3$. Explain how you obtained your answer.

7. Is there a difference between $(3x)^2$ and $3x^2$? Explain.

Exercise numbers appearing in color are answered in the Selected Answers appendix.

8. As you simplify the following expression, list each mathematical principle that you use.

$$(x - 3)(2x + 4) - 3(x + 7) - (3x - 2) + 2x^3x^5 - 7x^8 + (2x^2)^4 + (-x)^2$$

9. Is $(x + 3)^2$ equivalent to $x^2 + 9$? Explain.

Cluster 1   How Can I Practice?

1. Select any three input values, and complete the following table. Then determine numerically and algebraically which of the following expressions are equivalent.

   a. $4x^2$

   b. $-2x^2$

   c. $(-2x)^2$

x	$4x^2$	$-2x^2$	$(-2x)^2$

2. Given that none of the variables are zero, use the properties of exponents to simplify the following. Write the result using non-negative exponents.

   a. $3x^3 \cdot x$

   b. $-x^2 \cdot x^5 \cdot x^7$

   c. $(2x)(-6y^2)(x^3)$

   d. $8(2x^2)(3xy^4)$

   e. $(p^4)^5 \cdot (p^3)^2$

   f. $(3x^2y^3)(4xy^4)(x^7y)$

   g. $\dfrac{w^9}{w^5}$

   h. $\dfrac{z^5}{z^5}$

   i. $\dfrac{15y^{12}}{5y^5}$

   j. $\dfrac{x^8y^{11}}{x^3y^5}$

   k. $\dfrac{21xy^7z}{7xy^5}$

   l. $(a^4)^3$

   m. $3(x^3)^2$

   n. $(-x^3)^5$

   o. $(-2x^3)(-3x^5)(4x)$

   p. $(-3s^2t)(t^2)^3(s^3t^5)$

   q. $\left(\dfrac{2x}{3}\right)^3$

   r. $\dfrac{a^8b^{10}}{a^9b^{12}}$

   s. $\dfrac{6x^5z}{14x^5z^2}$

   t. $\left(\dfrac{x}{3}\right)^{-2}$

   u. $x^2\left(\dfrac{2x}{3}\right)^4$

3. Multiply each of the following and write your answers in standard form.

   a. $3(x - 5)$ **b.** $2x(x + 7)$ **c.** $-(3x - 2)$

   d. $-3.1(x + 0.8)$ **e.** $3x(2x^4 - 5x^2 - 1)$ **f.** $x^2(x^5 - x^3 + x)$

4. For each of the following expressions, list the specific operations indicated in the order in which they are to be performed. Your sequence of operations should begin as "Start with input x, then. . . ."

   a. $12 + 5(x - 2)$ **b.** $(x + 3)^2 - 12$

   c. $5(3x - 2)^2 + 1$ **d.** $-x^2 + 1$

   e. $(-x)^2 + 1$

5. Simplify the following algebraic expressions.

   a. $7 - (2x - 3) + 9x$ **b.** $4x^2 - 3(4x^2 - 7) + 4$

   c. $x(2x - 1) + 2x(x - 3)$ **d.** $3x(x^2 - 2) - 5(x^3 + x^2) + 2x^3$

   e. $3[7 - 5(a - b) + 7a] - 5b$ **f.** $4 - [2x + 3(x + 5) - 2] + 7x$

6. Determine the product for each of the following, and write the final result in simplest form.

  a. $(x + 2)(x - 3)$ **b.** $(a - b)(a + b)$ **c.** $(x + 3)(x + 3)$

  d. $(x - 2y)(x + 4y)$ **e.** $(2y + 1)(3y - 2)$ **f.** $(4x - 1)(3x - 1)$

  g. $(x + 3)(x^2 - x + 3)$ **h.** $(x - 1)(x^2 + 2x - 1)$

  i. $(a - b)(a^2 - 3ab + b^2)$ **j.** $(x^2 + 3x - 1)(2x^2 - x + 3)$

7. Write an expression in simplest form that represents the area of the shaded region.

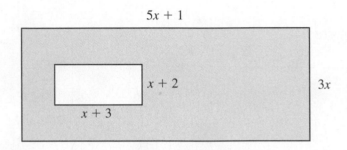

8. A volatile stock began the last week of the year worth x dollars per share. The table shows the changes during that week. If you own 30 shares, express in symbolic form the total value of your stock at the end of the year.

Day	1	2	3	4	5
Change in Value/Share	doubled	lost 10	tripled	gained 12	lost half its value

9. a. You are drawing up plans to enlarge your square garden. You want to triple the length of one side and quadruple the length of the other side. If x represents a side of your square garden, draw a plan and write a formula for the new area in terms of x.

b. You discover from the plan that after tripling the one side of your garden, you must cut off 5 feet from that side to clear a bush. Draw this revised plan and express the area of your new garden in factored form in terms of x. Then write your result as a polynomial in expanded form.

c. You also realize that if you add 5 feet to the side that you quadrupled, you can extend the garden to the walkway. What would be the new area of your garden with the changes to both sides incorporated into your plan? Draw this revision, express this area in factored form in terms of x, and expand the result.

Cluster 2

Problem Solving with Quadratic Equations and Functions

Activity 4.4

The Amazing Property of Gravity

Objectives

1. Evaluate quadratic functions of the form $y = ax^2$.

2. Graph quadratic functions of the form $y = ax^2$.

3. Interpret the coordinates of points on the graph of $y = ax^2$ in context.

4. Solve a quadratic equation of the form $ax^2 = c$ graphically.

5. Solve a quadratic equation of the form $ax^2 = c$ algebraically by taking square roots.

6. Solve a quadratic equation of the form $(x \pm a)^2 = c$ algebraically by taking square roots.

Note: $a \neq 0$ in Objectives 1–5.

In the sixteenth century, scientists such as Galileo were experimenting with the physical laws of gravity. In a remarkable discovery, they learned that if the effects of air resistance are neglected, any two objects dropped from a height above Earth will fall at exactly the same speed. That is, if you drop a feather and a brick down a tube whose air has been removed, the feather and brick will fall at the same speed. Surprisingly, the function that describes the distance that such an object falls in terms of elapsed time is a very simple one:

$$s = 16t^2,$$

where t represents the number of seconds elapsed and s represents the distance (in feet) the object has fallen.

The algebraic rule $s = 16t^2$ indicates a sequence of two mathematical operations:

Start with a value
for input t → *square the value* → *multiply by 16* → to obtain output s.

1. Use the algebraic rule to complete the table.

t (sec.)	s (ft.)
0	
1	
2	
3	

2. **a.** How many feet does the object fall during the first second after being dropped?

   b. How many feet does the object fall during the first two seconds after being dropped?

3. **a.** Determine the average rate of change of distance fallen from time $t = 0$ to $t = 1$.

   b. What are the units of measurement of the average rate of change?

   c. Explain what the average rate of change indicates about the falling object.

4. Determine and interpret the average rate of change of distance fallen from time $t = 1$ to $t = 2$.

5. Is the function $s = 16t^2$ a linear function? Explain your answer.

6. a. If the object hits the ground after 5 seconds, determine the practical domain of the distance function.

b. On the following grid, plot the points in the table in Problem 1 and sketch a curve representing the distance function over its practical domain.

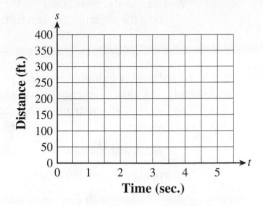

c. Use technology to verify the graph in part b in the window $0 \leq x \leq 5$, $0 \leq y \leq 400$. Your graph should resemble the one below.

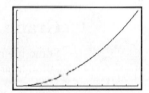

It is important to understand that the graph of the function $s = 16t^2$ does *not* represent a picture of the trajectory (path) of the object. The trajectory is vertical. The object is dropped from some large height (think of the edge of a high cliff) and falls straight down. The function values describe the vertical distance fallen (feet) at a given instant of time (seconds).

7. a. Does the point $(2.5, 100)$ lie on the graph? What do the coordinates of this point indicate about the falling object?

b. Does the point $(4.5, 324)$ lie on the graph? What do the coordinates of this point indicate about the falling object?

8. Use the graph in Problem 6b to estimate the amount of time it takes the object to fall 256 feet.

Solving Equations of the Form $ax^2 = c$, $a \neq 0$

9. Use the algebraic rule $s = 16t^2$ to write an equation to determine the amount of time it takes the object to fall 256 feet.

To solve this equation, reverse the order of operations indicated by the function rule, replacing each operation by its inverse. Here, one of the operations is "square a number." The inverse of squaring is to take a square root, denoted by the symbol $\sqrt{\ }$, called a **radical sign**.

Start with a value for s → *divide by 16* → *take its square root* → to obtain t.

In particular, if $s = 256$ feet:

Start with 256 → *divide by 16* → *take its square root* → to obtain t.
$$256 \div 16 = 16 \qquad \sqrt{16} = 4 \qquad t = 4$$

Therefore, it takes 4 seconds for the object to fall 256 feet.

Graph of a Quadratic Function

Some interesting properties of the function defined by $s = 16t^2$ arise when you ignore the falling object context and consider just the algebraic rule itself.

Replace t with x and s with y in $s = 16t^2$, and consider the general equation $y = 16x^2$. First, by ignoring the context, you can allow x to take on a negative, positive, or zero value. For example, suppose $x = -5$; then

$$y = 16(-5)^2 = 16 \cdot 25 = 400.$$

10. a. Use the algebraic rule $y = 16x^2$ to complete the table. You already calculated the entries for $t = 0$, 1, 2, and 3 when you answered Problem 1; so there is no need to recalculate those values. Just copy them into the appropriate boxes.

x	−5	−4	−3	−2.5	−2	−1.5	−1	−0.5	0	0.5	1	1.5	2	2.5	3	4	5
y																	

b. What pattern (symmetry) do you notice from the table?

11. a. Sketch the graph of $y = 16x^2$, using the coordinates listed in the table in Problem 10. Scale the axes appropriately to plot the points, and then draw a curve through them.

b. Use technology to produce a graph of this function in the window $-5 \leq x \leq 5$, $-100 \leq y \leq 400$. Your graph should resemble the following.

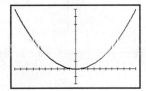

> Recall from Activity 4.1 that second-degree polynomial functions are called *quadratic functions*. Their graphs are U-shaped curves called **parabolas**. The U-shaped graph of the quadratic function defined by $y = 16x^2$ is an example of a *parabola*.

Solving Equations of the Form $ax^2 = c, a \neq 0$: A Second Look

12. a. In the table in Problem 10a, how many points on the graph of $y = 16x^2$ lie 256 units above the x-axis?

b. Identify the points. What are their coordinates?

The x-values of the points on the graph of $y = 16x^2$ that lie 256 units above the x-axis can be determined algebraically by solving the quadratic equation $256 = 16x^2$. Example 1 demonstrates a systematic procedure to solve this equation.

Example 1   *Solve the equation* $16x^2 = 256$ *algebraically.*

SOLUTION

As discussed after Problem 9 (page 488), you can solve the equation algebraically by performing the operations in reverse order: Divide by 16 and then take the square root of both sides.

Step 1. Divide both sides by 16:

$$\frac{16x^2}{16} = \frac{256}{16} \text{ to obtain } x^2 = 16.$$

Step 2. Calculate the square root.

When x is no longer restricted to positive (or nonnegative) values only, there are *two* square roots of 16, one denoted by $\sqrt{16}$ and the other denoted by $-\sqrt{16}$. These two square roots are often written in condensed form as $\pm\sqrt{16}$. Because $\sqrt{16} = 4$ and $-\sqrt{16} = -4$, the solutions to the equation $x^2 = 16$ are $x = \pm 4$.

The solution to $16x^2 = 256$ can be written in the following systematic manner.

$$16x^2 = 256$$

$$\frac{16x^2}{16} = \frac{256}{16} \qquad \textbf{Divide both sides by 16 and simplify.}$$

$$x^2 = 16 \qquad \textbf{Take the positive and negative square roots.}$$
$$x = \pm\sqrt{16}$$

$$x = \pm 4$$

13. a. How many points on the graph in Problem 11 lie 400 units above the x-axis?

   b. Identify the points. What are their coordinates?

   c. Set up the appropriate equation to determine the values of x for which $y = 400$, and solve it algebraically.

14. Solve the following quadratic equations algebraically. Write decimal results to the nearest hundredth.

   a. $x^2 = 36$ **b.** $2x^2 = 98$

c. $3x^2 = 375$ **d.** $5x^2 = 50$

15. a. Refer to the graph in Problem 11a to determine how many points on the graph of $y = 16x^2$ lie 16 units *below* the x-axis.

b. Set up an equation that corresponds to the question in part a, and solve it algebraically.

c. How many solutions does this equation have? Explain.

16. What does the graph of $y = 16x^2$ (Problem 11a) indicate about the number of solutions to the following equations? (You do not need to solve these equations.)

a. $16x^2 = 100$ **b.** $16x^2 = 0$ **c.** $16x^2 = -96$

17. Solve the following equations.

a. $5x^2 = 20$ **b.** $4x^2 = 0$ **c.** $3x^2 = -12$

Solving Equations of the Form $(x \pm a)^2 = c$

The following example demonstrates a procedure for solving a slightly more complex equation, such as $(x - 4)^2 = 9$.

Example 2   *Solve the equation* $(x - 4)^2 = 9$.

SOLUTION

The sequence of operations indicated by the expression on the left is

Start with input x → **subtract 4** → **square** → *to obtain output value* 9.

To solve the equation, reverse these steps by first taking the square roots and then adding 4 as follows:

$$(x - 4)^2 = 9$$ Take the square roots.

$$x - 4 = \pm\sqrt{9}$$

$$x - 4 = \pm 3$$ Rewrite as two separate equations.

$$x - 4 = 3 \quad \text{or} \quad x - 4 = -3$$ Solve each of these equations by adding 4.

$$x = 7 \quad \text{or} \quad x = 1$$

Therefore, $x = 7$ and $x = 1$ are solutions to the original equation $(x - 4)^2 = 9$.

18. Solve the following equations.

a. $(x + 4)^2 = 9$ **b.** $(x - 1)^2 = 25$

c. $(x - 10)^2 = 70$

19. Suppose you have a 10- by 10-foot square garden plot. This year you want to increase the area of your plot and still keep its square shape.

a. Use the inner square of the sketch to represent your existing plot and the known dimensions. Use the variable x to represent the increase in length to each side. Write an expression for the length of a side of the larger square.

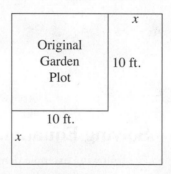

b. Express the area, A, of your new larger plot as the square of a binomial in function notation.

c. Suppose you decide to double the original area. Use the area expression in part b to write an equation to determine the required increase in the length of each side.

d. Solve the equation in part c. Round the answer to hundredths.

e. What is the length of each side of your new garden, to the nearest hundredth of a foot? Check your answer by calculating the area of this new plot.

SUMMARY: ACTIVITY 4.4

1. The graph of a function of the form $y = ax^2$, $a \neq 0$, is a U-shaped curve and is called a **parabola**. Such functions are called **quadratic functions**.

2. An equation of the form $ax^2 = c$, $a \neq 0$, is one form of a **quadratic equation**. Such equations can be solved algebraically by dividing both sides of the equation by a and then determining the positive and negative square roots of both sides.

3. An equation of the form $(x \pm a)^2 = c$ is solved algebraically by determining the positive and negative square roots of c and then solving the two resulting linear equations.

4. Every positive number a has two square roots, one positive and one negative. The square roots are equal in magnitude. The positive, or principal, square root of a is denoted by $\sqrt{a}$. The negative square root of a is denoted by $-\sqrt{a}$. The symbol $\sqrt{}$ is called the **radical sign**. For example, the square roots of 25 are $\sqrt{25} = 5$ and $-\sqrt{25} = -5$. The square roots of 25 can be written in a condensed form as $\pm\sqrt{25} = \pm 5$.

EXERCISES: ACTIVITY 4.4

1. On Earth's Moon, gravity is only one-sixth as strong as it is on Earth; so an object on the Moon will fall one-sixth the distance it would fall on Earth in the same time. This means that the distance function for a falling object on the Moon is

$$s = \left(\frac{16}{6}\right)t^2, \text{ or } s = \frac{8}{3}t^2,$$

where t represents time since the object is released, in seconds, and s is the distance fallen, in feet.

a. How far does an object on the Moon fall in 3 seconds?

b. How long does it take an object on the Moon to fall 96 feet?

c. Graph the Moon's gravity function on a properly scaled and labeled coordinate axis or on a graphing calculator for $t = 0$ to $t = 5$.

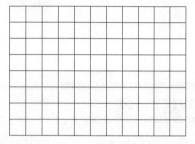

d. Use the graph to estimate how long it takes an object on the Moon to fall 35 feet.

e. Write an equation to determine the required time in part d. Solve the equation algebraically.

f. How does your answer in part e compare with your estimate in part d?

2. Solve the following equations. Write decimal results to the nearest hundredth.

a. $5x^2 = 45$ **b.** $9x^2 = 0$ **c.** $-25x^2 = 100$

d. $x^2 = 5$ **e.** $2x^2 = 20$ **f.** $\dfrac{x^2}{2} = 32$

3. Solve the following by first writing the equation in the form $x^2 = c$.

    a. $t^2 - 49 = 0$ **b.** $15 + c^2 = 96$ **c.** $3a^2 - 21 = 27$

4. Solve the following equations.

    a. $(x + 1)^2 = 16$ **b.** $(x - 1)^2 = 4$

    c. $(x + 7)^2 = 64$ **d.** $(x - 9)^2 = 81$

    e. $3(x - 2)^2 = 75$ **f.** $5(x + 3)^2 = 80$

    g. $(2x - 5)^2 = 81$ **h.** $(4 + x)^2 - 20 = 124$

In a right triangle, as shown in the diagram to the right, the side c opposite the right angle is called the *hypotenuse* and the other two sides a and b are called *legs*. The Pythagorean theorem states that in any right triangle, the lengths of the three sides are related by the equation $c^2 = a^2 + b^2$.

Use the Pythagorean theorem to answer Exercises 5–7.

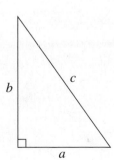

5. Determine the length of the hypotenuse in a right triangle with legs 5 inches and 12 inches.

6. One leg of a right triangle measures 8 inches, and the hypotenuse 17 inches. Determine the length of the other leg.

7. A 16-foot ladder is leaning against a wall so that the bottom of the ladder is 5 feet from the wall (see diagram). You want to determine how much farther from the wall the bottom of the ladder must be moved so that the top of the ladder is exactly 10 feet above the floor.

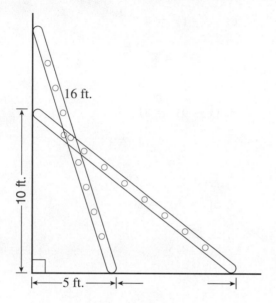

a. The new ladder position forms another right triangle in which x represents the additional distance the foot of the ladder has moved from the wall. Use the Pythagorean theorem to write an equation relating the base, height, and hypotenuse of this new right triangle.

b. Solve the equation from part a to determine x.

Activity 4.5

What Goes Up, Comes Down

Objectives

1. Evaluate quadratic functions of the form $y = ax^2 + bx, a \neq 0$.

2. Graph quadratic functions of the form $y = ax^2 + bx, a \neq 0$.

3. Identify the x-intercepts of the graph of $y = ax^2 + bx$ graphically and algebraically.

4. Interpret the x-intercepts of a quadratic function in context.

5. Factor a binomial of the form $ax^2 + bx$.

6. Solve an equation of the form $ax^2 + bx = 0$ using the zero-product property.

When a golf ball is hit, its height is a function of the time it has been in flight. The approximate height of the ball above the ground is modeled by

$$h(t) = -16t^2 + 80t,$$

where h is the height (in feet) of the ball and t is the time (in seconds) that the ball has been in flight.

Because the expression $-16t^2 + 80t$ is a second-degree polynomial, the height function is a quadratic function.

1. Use the function rule to complete the table.

t (sec.)	0	0.5	1	1.5	2	2.5	3	3.5	4	4.5	5
$h(t)$ (ft.)											

2. a. What does $h(t) = 0$ signify?

b. What do you notice about the values of $h(1)$ and $h(4)$? Explain why you might expect this.

c. Use the table values to help determine the practical domain of the height function.

3. Explain why the heights increase and then decrease.

4. a. Sketch the graph of $h(t) = -16t^2 + 80t$ using the table in Problem 1. Plot the points, and then draw a curve through them. Scale the axes appropriately.

b. Use technology to produce a graph of this function in the window $0 \leq x \leq 5, 0 \leq y \leq 120$. Your graph should resemble the one below.

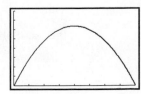

Once again, it is important to understand that the graph of the height function $h(t) = -16t^2 + 80t$ does *not* represent a picture of the trajectory (path) of the object. The function values describe the height of the golf ball (feet) at a specific time (seconds). The horizontal axis variable represents time, not distance.

Solving Equations of the Form $ax^2 + bx = 0, a \neq 0$

The graph of $h(t) = -16t^2 + 80t$ is a parabola whose graph rises and then falls.

5. a. How many horizontal intercepts (points at which the graph intersects the *t*-axis) does this graph have?

   b. What are their coordinates?

   c. What do these horizontal intercepts signify in terms of the golf ball?

The **horizontal intercepts** have an algebraic interpretation as well. When the golf ball is at ground level, its height, $h(t)$, is zero. Substituting 0 for $h(t)$ in the formula $h(t) = -16t^2 + 80t$, results in a quadratic equation:

$$0 = -16t^2 + 80t, \text{ or equivalently } -16t^2 + 80t = 0.$$

Note that you *cannot* solve this equation by reversing the sequence of operations. The difficulty here is that the input variable, *t*, occurs twice, in two distinct terms, $-16t^2$ and $80t$. These two terms are not like terms, so they cannot be combined. However, both terms share common factors: Their coefficients are both multiples of 16, and they each have at least one factor of *t*.

> A common factor is called a *greatest common factor* (GCF) if there are no additional factors common to the terms in the expression.

6. Determine the greatest common factor of $-16t^2$ and $80t$.

7. Use the greatest common factor to write the expression on the left side of the equation $-16t^2 + 80t = 0$ in factored form.

The solution of this equation depends on a unique property of the zero on the right side.

> **Zero-Product Property**
>
> If the product of two factors is zero, then at least one of the factors must also be zero. Written symbolically, if $a \cdot b = 0$, then either $a = 0$ or $b = 0$.

Because the right side of $16t(-t + 5) = 0$ is zero, the zero-product property guarantees that either $16t = 0$ or $-t + 5 = 0$.

8. a. Solve $16t = 0$ for *t*.   **b.** Solve $-t + 5 = 0$ for *t*.

c. Verify that each solution in parts a and b satisfies the original equation $-16t^2 + 80t = 0$.

9. Use the zero-product property to solve the following equations.

a. $x(x - 3) = 0$

b. $x(x + 5) = 0$

c. $(x - 2)(x + 2) = 0$

10. In each equation, use the greatest common factor to rewrite the expression in factored form; then apply the zero-product property to determine the solutions.

a. $x^2 - 10x = 0$

b. $x^2 + 7x = 0$

c. $2x^2 - 12x = 0$

d. $4x^2 - 6x = 0$

11. To apply the zero-product property, one side of the equation must be zero. In the following equations, first add or subtract an appropriate term to both sides of the equation to get zero on one side. Then write in factored form and apply the zero-product property.

a. $2x^2 = 8x$

b. $6x^2 = -4x$

In an equation such as $3x^2 - 15x = 0$, it is *always* permissible to divide *every* term of the equation (both sides) by a nonzero number. Here, you see that dividing by 3 simplifies the original equation:

$$\frac{3x^2}{3} - \frac{15x}{3} = \frac{0}{3} \text{ simplifies to the equivalent equation } x^2 - 5x = 0.$$

Factoring and applying the zero-product property leads to $x(x - 5) = 0$, whose solutions are $x = 0$ and $x = 5$. However, note that it is *never* permissible to divide through by a variable quantity such as x because that variable may equal 0 and division by 0 is undefined. If you were to divide both sides of $3x^2 - 15x = 0$ by x, you would obtain

$$\frac{3x^2}{x} - \frac{15x}{x} = \frac{0}{x},$$

which simplifies to $3x - 15 = 0$. Solving the resulting equation, $3x - 15 = 0$, you would obtain a single solution, $x = 5$. You would have lost the other solution, $x = 0$.

12. A golf ball is hit with an initial velocity of 164 feet per second at an inclination of 45° to the horizontal. Using physics, the **actual path** of the golf ball is modeled by the quadratic function

$$h(x) = -0.0012x^2 + x,$$

where x is the horizontal distance in feet that the golf ball has traveled.

a. What is the height of the ball after it traveled 200 feet?

b. Determine the x-intercept of the graph of the height function.

c. Interpret the x-intercepts in the context of this situation.

d. Use technology to graph the path of the golf ball. Use the results in part b to help you determine the window.

SUMMARY: ACTIVITY 4.5

1. The **x-intercepts** on the graph of $y = ax^2 + bx$, $a \neq 0$, are the points at which the graph intersects the x-axis. (At these points, the y-coordinate is zero.) The x-intercepts can be determined algebraically by setting $y = 0$ and solving the associated equation, $ax^2 + bx = 0$.

2. Zero-product property: If the product of two factors is zero, then at least one of the factors must also be zero. Stated symbolically, if $a \cdot b = 0$, then either $a = 0$ or $b = 0$.

3. A quadratic equation of the form $ax^2 + bx = 0$, $a \neq 0$, is solved algebraically by factoring the equation's left side and then using the zero-product property.

$$3x^2 + 6x = 0$$
$$3x(x + 2) = 0$$
$$3x = 0 \quad \text{or} \quad x + 2 = 0$$
$$x = 0 \quad \text{or} \qquad x = -2$$

4. The solutions of the equation $ax^2 + bx = 0$ correspond precisely to the x-intercepts of the graph of $y = ax^2 + bx$. For example, the x-intercepts of the graph of $y = 3x^2 + 6x$ are $(-2, 0)$ and $(0, 0)$.

5. An equation can *always* be simplified by dividing each term on *both* sides by a *nonzero* number. The original and simplified equations will have identical solutions.

6. An equation may *never* be simplified by dividing both sides by a variable quantity. The resulting equation will have lost one or more solutions.

EXERCISES: ACTIVITY 4.5

1. You are relaxing on a float in your in-ground pool casually tossing a ball straight upward into the air. The height of the ball above the water is given by the formula $s = -16t^2 + 40t$, where t represents the time since the ball was tossed in the air.

a. Use the formula to determine the height of the ball for the times given in the following table.

t (sec.)	0	0.5	1	1.5	2	2.5
s (ft.)						

b. Set up and solve an equation to determine when the ball will hit the water.

c. Use the result in parts a and b to help you determine a practical domain.

d. Graph the height function over its practical domain. Label and scale the axes appropriately.

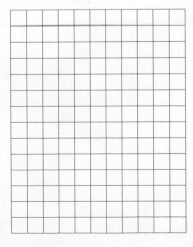

2. Use the zero-product property to solve the following equations.

a. $x(x - 2) = 0$

b. $x(x + 7) = 0$

c. $(x - 3)(x + 2) = 0$

d. $(2x - 1)(x + 1) = 0$

3. In each equation, use the greatest common factor to rewrite the expression in factored form; then apply the zero-product property to determine the solutions.

a. $x^2 - 8x = 0$

b. $x^2 + 5x = 0$

c. $3x^2 - 12x = 0$

d. $4x^2 - 2x = 0$

e. $x^2 = 5x$

f. $3x^2 = 48x$

4. Your model rocket is popular with the neighborhood children. Its powerful booster propels the rocket straight upward so that its height is described by the function $H(t) = -16t^2 + 480t$, where t represents the number of seconds after launch and $H(t)$ the height of the rocket in feet.

a. Write an equation that can be used to determine the horizontal intercepts of the graph of $H(t)$.

b. Solve the equation obtained in part a by factoring.

c. What do the solutions signify in terms of the rocket's flight?

d. What is the practical domain of this function?

e. Use the practical domain to help you set a window for your graphing calculator. Use technology to produce a graph of this function in the window $0 \leq x \leq 30, 0 \leq y \leq 4000$.

f. Use the graph to estimate how high the model rocket will go and how long after launch it will reach its maximum height.

5. Use the zero-product property to determine the solutions of each equation. Remember, to use the zero-product property, one side of the equation must be zero. Check your solutions by substituting back into the original equation.

a. $x^2 + 10x = 0$

b. $2x^2 = 25x$

c. $15y^2 + 75y = 0$

d. $2(x - 5)(x + 1) = 0$

e. $24t^2 - 36t = 0$

f. $12x = x^2$

g. $2w^2 - 3w = 5w$

h. $14p = 3p - 5p^2$

Activity 4.6

How High Did It Go?

Objectives

1. Recognize and write a quadratic equation in standard form, $ax^2 + bx + c = 0$, $a \neq 0$.

2. Factor trinomials of the form $x^2 + bx + c$.

3. Solve a factorable quadratic equation of the form $x^2 + bx + c = 0$ using the zero-product property.

4. Identify a quadratic function from its algebraic form.

Suppose a soccer ball is kicked directly upward at a speed of 96 feet per second. The algebraic rule that describes the height of the soccer ball as a function of time is

$$h(t) = -16t^2 + 96t + 2,$$

where t represents the number of seconds since the kick and $h(t)$ represents the ball's height (in feet) above the ground.

1. a. Complete the table, and sketch the graph of $h(t) = -16t^2 + 96t + 2$. Scale the axes appropriately.

t	0	0.5	1	1.5	2	2.5	3	3.5	4	4.5	5	5.5	6
$h(t)$													

b. Use technology to graph the function $h(t) = -16t^2 + 96t + 2$ in the window $0 \leq x \leq 10$, $-10 \leq y \leq 150$. Your graph should resemble the one here.

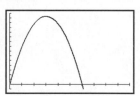

Solving Equations of the Form $ax^2 + bx + c = 0, a \neq 0$

In this activity, you will explore additional questions of interest about the flight of an object. For instance, when will the soccer ball reach a height of 82 feet?

2. Use the table or the graph in Problem 1a to estimate the time(s) at which the soccer ball is 82 feet above the ground. How often during its flight does this occur?

The answers to Problem 2 have an algebraic interpretation as well. When the soccer ball is at a height of 82 feet, $h(t) = 82$. Substituting 82 for $h(t)$ in the formula $h(t) = -16t^2 + 96t + 2$, you have

$$82 = -16t^2 + 96t + 2, \quad \text{or equivalently} \quad -16t^2 + 96t + 2 = 82.$$

3. The equation $-16t^2 + 96t + 2 = 82$ can be written as $-16t^2 + 96t = 80$. Explain why you *cannot* solve the equation $-16t^2 + 96t = 80$ by factoring the expression on the left side and applying the zero-product rule.

To use the zero-product property to solve an equation, one side of the equation *must* be zero. If necessary, you can force one side of an equation to be zero by subtracting appropriate terms from both sides of the equation.

4. Rewrite the equation $-16t^2 + 96t = 80$ as an equivalent equation whose right side is zero.

5. a. Identify the greatest common factor of the trinomial on the left side of the equation.

   b. Is it permissible to divide each term of the equation by this common factor? If so, do it. If not, explain the problem.

If you completed Problems 4 and 5 correctly, the original equation $-16t^2 + 96t = 80$ is most likely rewritten in equivalent form $-t^2 + 6t - 5 = 0$.

6. What is the coefficient of the t^2 term in the equation $-t^2 + 6t - 5 = 0$?

To simplify the equation $-t^2 + 6t - 5 = 0$ once more, multiply each term in the equation by -1. The effect of multiplication by -1 is always a reversal of the signs of each term in the equation. You should now obtain the equation

$$t^2 - 6t + 5 = 0.$$

Note here that there are no longer any common factors among the three terms on the left side. However, the expression $t^2 - 6t + 5$ can still be factored. The key to factoring $t^2 - 6t + 5$ lies in your recall of binomial multiplication from Activity 4.3.

When two binomials such as $x + a$ and $x + b$ are multiplied, the coefficients in the expanded form are closely related to the constant terms a and b in the original factored form.

$$(x + a)(x + b) = x^2 + bx + ax + ab$$
$$= x^2 + (a + b)x + ab$$

The product is a trinomial in which

i. the constant term, ab, in the expanded form on the right side is the *product* of the constants a and b

ii. the coefficient, $a + b$, of the x term in the expanded form on the right side is the *sum* of the constants a and b

Therefore, factoring a trinomial such as $t^2 - 6t + 5$ is accomplished by determining the two constants a and b whose product is 5 and whose sum is -6.

To determine the constants a and b, it is preferable to start by first considering the product. In this case, the product is 5, a prime number with only two possible pairs of factors, 5 and 1 or -5 and -1. Only the second pair has the property that its sum is -6.

7. Complete the factoring of $t^2 - 6t + 5$.

Note that the order in which you write these factors does not matter because multiplication is commutative.

$$(t - 5)(t - 1) \text{ is exactly the same product as } (t - 1)(t - 5).$$

8. a. Use the zero-product property to solve the factored form of the equation $t^2 - 6t + 5 = 0$.

b. How many solutions did you obtain? Interpret these solutions in the context of the soccer ball situation.

The solution of $-16t^2 + 96t = 80$ that was completed in Problems 4–8 is summarized as follows:

$$-16t^2 + 96t = 80$$

$$\underline{ - 80 - 80}$$ **Subtract 80 from each side to obtain 0 on one side.**

$$\frac{-16t^2}{16} + \frac{96t}{16} - \frac{80}{16} = \frac{0}{16}$$ **Divide each side by the greatest common factor, 16.**

$$-1(-t^2 + 6t - 5) = -1(0)$$ **Multiply each side by -1.**

$$t^2 - 6t + 5 = 0$$ **Factor the trinomial.**

$$(t - 5)(t - 1) = 0$$ **Apply the zero-product property.**

$$t - 5 = 0 \quad t - 1 = 0$$

$$t = 5 \quad\quad\quad t = 1$$

The equation $-16t^2 + 96t = 80$ is a **quadratic equation**. When the equation is written with zero on one side and the polynomial is written in standard form on the other side, then the quadratic equation $-16t^2 + 96t - 80 = 0$ is said to be written in **standard form**.

Definition

Any equation that can be written in the general form $ax^2 + bx + c = 0$ is called a **quadratic equation**. Here, a, b, and c are understood to be arbitrary constants, with the single restriction that $a \neq 0$.

A quadratic equation is said to be in **standard form** when the terms on the left side are written in decreasing degree order: the x^2 term first, followed by the x term, and trailed by the constant term. The right side in the standard form of the equation is *always* zero.

9. Write the quadratic equations in standard form, and determine the constants, a, b, and c.

a. $3x^2 + 5x + 8 = 0$

b. $x^2 - 2x + 6 = 0$

c. $x^2 + 6x = 10$

d. $x^2 - 4x = 0$

e. $x^2 = 9$

f. $5x^2 = 2x + 6$

Procedure

Solving a Quadratic Equation by Factoring

1. Write the quadratic equation in standard form, $ax^2 + bx + c = 0$, $a \neq 0$.

2. When appropriate, divide each term in the equation by the greatest common numerical factor.

3. Factor the nonzero side.

4. Apply the zero-product property to obtain two linear equations.

5. Solve each linear equation.

6. Check your solutions in the original equation.

---

10. If not already done, rewrite each quadratic equation in standard form. Then factor the nonzero side, and use the zero-product property to solve the equation. Check your solutions in the original equation.

a. $x^2 + 5x + 6 = 0$

b. $x^2 + 8 = 7x - 4$

c. $x^2 + 5x - 24 = 0$

d. $-3x^2 + 6x = -24$

Return once again to the function defined by $h(t) = -16t^2 + 96t + 2$ that describes the height (in feet) of a soccer ball as a function of time (in seconds).

11. a. Set up and solve the quadratic equation that describes the time when the ball is 130 feet above the ground.

b. How many solutions does this equation have?

c. Interpret the solution(s) in the soccer ball context.

12. a. Set up and solve the quadratic equation that describes the time when the ball is 146 feet above the ground.

b. How many solutions does this equation have?

c. Interpret the solution(s) in the soccer ball context.

SUMMARY: ACTIVITY 4.6

1. Any function that is defined by an equation of the form $y = ax^2 + bx + c, a \neq 0$, is called a **quadratic function**. Here, a, b, and c are understood to be arbitrary constants, with the single restriction that $a \neq 0$.

2. Algebraically, an associated **quadratic equation** arises when y is assigned a value and you need to solve for x. A quadratic equation, written as $ax^2 + bx + c = 0$, is said to be in **standard form**.

3. If the nonzero side of a quadratic equation written in standard form can be factored, then the equation can be solved using the zero-product property.

EXERCISES: ACTIVITY 4.6

1. Solve the following quadratic equations by factoring.

a. $x^2 + 7x + 6 = 0$

b. $x^2 - 10x - 24 = 0$

c. $y^2 + 11y = -28$

d. $5x^2 - 75x + 180 = 0$

e. $x^2 + 9x + 18 = 0$

f. $x^2 - 9x = 36$

g. $x^2 - 9x + 20 = 0$

h. $2x^2 - 12 = 2x$

i. $x^2 - 5x = 3x - 15$

j. $2x^2 - 3x = x^2 + 10$

2. Your friend's model rocket has a powerful engine. Its height (in feet) is described by the function $h(t) = -16t^2 + 432t$, where t represents the number of seconds since launch. Set up and solve a quadratic equation to answer each of the parts a–e, and record your results in the following table.

	HEIGHT h (ft.) OF ROCKET FROM GROUND	FIRST TIME t (sec.) THAT HEIGHT WAS REACHED	SECOND TIME t (sec.) THAT HEIGHT WAS REACHED	AVERAGE OF FIRST TIME AND SECOND TIME HEIGHT WAS REACHED
a.	0			
b.	800			
c.	1760			
d.	2240			
e.	2720			

a. At what times does your friend's rocket touch the ground?

b. When is the rocket 800 feet high?

c. When is the rocket 1760 feet high?

d. When is the rocket 2240 feet high?

e. When is the rocket 2720 feet high?

f. For each given height in parts a–e, determine the average of the first and second times the height was reached and record in the last column of the table.

g. Use the results in the table to determine the maximum height the rocket reached. To verify your answer, graph the solutions from parts a–e and draw a curve through the points or use technology.

3. Solve $(x - 1)(x + 2) = 10$. (*Hint:* First expand the expression and write the equation in standard form.)

4. A fastball is hit straight up (vertically) over home plate. The ball's height, h (in feet), from the ground is modeled by

$$h = -16t^2 + 80t + 5,$$

where t is measured in seconds.

a. Write the equation you would need to determine when the ball is 101 feet above the ground.

b. Solve the equation you determined in part a algebraically to determine the time it will take the ball to reach a height of 101 feet.

5. You are on the roof of your apartment building with a friend. You lean over the edge of the building and toss an apple straight up into the air. The height h, of the apple above the ground can be determined by the function defined by

$$h = -16t^2 + 16t + 96,$$

where t is time measured in seconds. Note that $t = 0$ corresponds to the time the apple leaves your hand.

a. Determine the height of the apple above the ground at the time you release it.

b. Write an equation that can be used to determine the time the apple strikes the ground.

c. Solve the equation in part b by the factoring method.

d. What is the practical domain of the height function?

e. Graph the height function on the following grid, or use a graphing calculator. Use the results from part d to help you determine a scale for the axes or set the proper window.

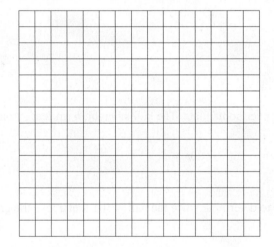

f. Use the graph to estimate the coordinates of the maximum point of the parabola.

g. What do the coordinates of the point in part f indicate about the position of the apple with respect to the ground?

Activity 4.7

More Ups and Downs

Objectives

1. Use the quadratic formula to solve quadratic equations.

2. Identify the solutions of a quadratic equation with points on the corresponding graph.

Suppose a soccer goalie punted the ball in such a way as to kick the ball as far as possible. The height of the ball above the field as a function of the distance travelled horizontally can be approximated by

$$h(x) = -0.017x^2 + 0.98x + 0.33,$$

where $h(x)$ represents the height of the ball (in yards) and x represents the horizontal distance (in yards) from where the goalie kicked the ball.

In this situation, the graph of $h(x) = -0.017x^2 + 0.98x + 0.33$ is the actual path of the flight of the soccer ball. The graph of this quadratic function appears below.

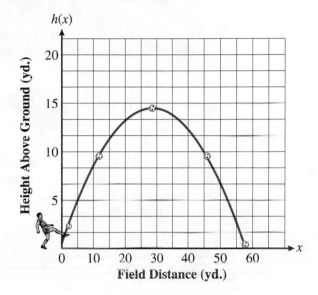

1. Use the graph to estimate how far from the point of contact the soccer ball is 10 yards above the ground. How often during its flight does this occur?

2. **a.** Use the function rule $h(x) = -0.017x^2 + 0.98x + 0.33$ to write the quadratic equation that arises in determining when the ball is 10 yards above the ground.

   b. Rewrite the equation in standard form and simplify if possible.

   c. What does the graph tell you about the number of solutions to this equation?

In the previous activity, you solved the quadratic equation $-16t^2 + 96t + 2 = 82$ by simplifying its standard form to $t^2 - 6t + 5 = 0$, factoring, and applying the zero-product property. Suppose that instead, the constant term was 6 and the equation to solve was $t^2 - 6t + 6 = 0$. The quadratic trinomial in this equation cannot be factored.

You have just encountered one of the main difficulties in solving quadratic equations algebraically—most quadratic equations cannot be solved by factoring. However, this does not imply that most quadratic equations are not solvable. Algebraic manipulations can be applied to the equation $ax^2 + bx + c = 0, a \neq 0$, to develop a general formula, called the quadratic formula, that produces solutions to the equation. The solutions will depend on the specific values of the coefficients, a, b, and c.

> **The Quadratic Formula**
>
> For a quadratic equation in standard form, $ax^2 + bx + c = 0, a \neq 0$, the solutions are given by the *quadratic formula*:
>
> $$x = \frac{-b \pm \sqrt{b^2 - 4ac}}{2a}.$$
>
> The $\pm$ in the formula indicates that there are two solutions, one in which the terms in the numerator are added and another in which the terms are subtracted. The two solutions can be written separately as follows:
>
> $$x_1 = \frac{-b + \sqrt{b^2 - 4ac}}{2a} \quad \text{and} \quad x_2 = \frac{-b - \sqrt{b^2 - 4ac}}{2a}$$

When you use this formula, the quadratic equation *must* be in standard form so that the values (and signs) of the three coefficients a, b, and c are correct. Be careful when you key the expression into your calculator. The square root and quotient computations require some thought; you often must include appropriate sets of parentheses.

Example 1   *Use the quadratic formula to solve the quadratic equation $6x^2 - x = 2$.*

SOLUTION

$6x^2 - x - 2 = 0$ Write the equation in standard form, $ax^2 + bx + c = 0$.

$a = 6, b = -1, c = -2$ Identify a, b, and c.

$x = \dfrac{-(-1) \pm \sqrt{(-1)^2 - 4(6)(-2)}}{2(6)}$ Substitute for a, b, and c in $x = \dfrac{-b \pm \sqrt{b^2 - 4ac}}{2a}$.

$x = \dfrac{1 \pm \sqrt{1 + 48}}{12} = \dfrac{1 \pm 7}{12}$

$x = \dfrac{1 + 7}{12} = \dfrac{8}{12} = \dfrac{2}{3}, \quad x = \dfrac{1 - 7}{12} = \dfrac{-6}{12} = -\dfrac{1}{2}$

3. Use the quadratic formula to solve the quadratic equation in Problem 2b.

4. Solve $3x^2 + 20x + 7 = 0$ using the quadratic formula.

SUMMARY: ACTIVITY 4.7

The solutions of a quadratic equation in standard form, $ax^2 + bx + c = 0, a \neq 0$, are

$$x = \frac{-b \pm \sqrt{b^2 - 4ac}}{2a}.$$

The $\pm$ symbol in the formula indicates that there are two solutions—one in which the terms in the numerator are added and one in which the terms are subtracted.

EXERCISES: ACTIVITY 4.7

Use the quadratic formula to solve the equations in Exercises 1–5.

1. $x^2 + 2x - 15 = 0$

2. $4x^2 + 32x + 15 = 0$

3. $-2x^2 + x + 1 = 0$

4. $2x^2 + 7x = 0$ **5.** $-x^2 + 10x + 9 = 0$

6. Solve $x^2 - 2x - 15 = 0$ in the following two ways.

    a. Factor and then use the zero-product property.   **b.** Use the quadratic formula.

    c. Which process seems more efficient to you?

7.   **a.** Determine the x-intercepts, if any, of the graph of $y = x^2 - 2x + 5$.

    b. Sketch the graph of $y = x^2 - 2x + 5$ using technology, and verify your answer to part a.

8. The following data from the National Health and Nutrition Examination Survey indicates that the number of American adults who are overweight or obese is increasing.

Years since 1960, t	1	12	18	31	39	44	46	50
Percent of American Adults Who Are Overweight or Obese, $P(t)$	45	47	47	56	64.5	66.3	66.9	69.2

This data can be modeled by the equation
$P(t) = 0.0076t^2 + 0.1549t + 44.015$.

a. Determine the percentage of Americans expected to be overweight or obese in 2020.

b. Use the quadratic formula to determine the year when the model predicts that the percentage of overweight Americans will be 75%.

9. The height of a bridge arch located in the Thousand Islands is modeled by the function $y = -0.04x^2 + 28$, where x is the distance, in feet, from the center of the arch and y is the height of the arch above the water.

a. Sketch a picture of this arch on a grid, using the vertical axis as the center of the arch.

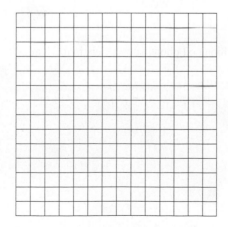

b. Determine the y-intercept. What is the practical meaning of this intercept in this situation?

c. Determine the x-intercepts algebraically using the quadratic formula.

d. If the arch straddles the river exactly, how wide is the river?

e. A sailboat is approaching the bridge. The top of the mast measures 30 feet. Will the boat clear the bridge? Explain.

f. You want to install a flagpole on the bridge at an arch height of 20 feet. Write the equation you must solve to determine how far to the right of center the arch height is 20 feet.

g. Solve the equation in part f using the quadratic formula. Use technology to check your result.

10. The number n (in millions) of cell phone subscribers in the United States from 2000 to 2012 is given in the following table.

Year	2000	2002	2004	2006	2007	2011	2012
Number of Subscribers (millions)	109.5	149.8	182.1	233.0	255.4	316.0	326.4

This data can be approximated by the quadratic model

$$n(t) = -0.3553t^2 + 22.84t + 106.3,$$

where $t = 0$ corresponds to the year 2000.

a. Estimate the year in which there were 300 million cell phone subscribers in the United States.

b. How confident are you in the solutions in part a? Explain your answer.

Cluster 2 | What Have I Learned?

1. Write (but do not solve) a quadratic equation that arises in each of the following situations.

   a. In determining the x-intercepts of the graph of $y = x^2 - 4x - 12$.

   b. In determining the point(s) on the graph of $y = 100 - x^2$ whose y-coordinate is 40.

   c. In determining the point(s) at which the graph of $y = 3x^2 - 75$ crosses the x-axis.

2. Describe a reasonable process for solving each of the following quadratic equations.

   a. $2x^2 = 50$

   b. $3x^2 - 12x = 0$

   c. $x^2 + 2x - 15 = 0$

   d. $(x - 4)^2 = 9$

   e. $x^2 + 3x - 1 = 0$

3. Solve each of the quadratic equations in Exercise 2 using the process you described. Remember to check your answers in the original equations.

   a. $2x^2 = 50$ **b.** $3x^2 - 12x = 0$

   c. $x^2 + 2x - 15 = 0$ **d.** $(x - 4)^2 = 9$

   e. $x^2 + 3x - 1 = 0$

Exercise numbers appearing in color are answered in the Selected Answers appendix.

4. Which of the following equations are equivalent (have identical solutions) to $4x^2 - 12x = 0$? Explain how to rewrite $4x^2 - 12x = 0$ into each equivalent equation.

$$2x^2 - 6x = 0 \qquad 4x - 12 = 0 \qquad x^2 - 3x = 0$$

5. What is the graph of a quadratic function called?

6. Does the graph of every parabola have x-intercepts? Use the graph of $y = x^2 + 5$ and $y = -x^2 - 4$ to explain.

7. How many y-intercepts are possible for a quadratic function, $y = ax^2 + bx + c$? Explain.

Cluster 2   How Can I Practice?

1. Given $f(x) = 2x^2 - 8x$.

a. Complete the table of values.

x	−4	−2	−1	0	2	3	5	6
f(x)								

b. Use the points in the table in part a to sketch a graph of $f(x)$. Scale the axes appropriately.

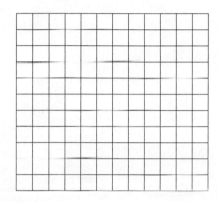

c.

c. Verify the graph using a graphing calculator, if available.

d. Determine the x-intercepts algebraically, and verify your answer graphically.

2. Solve the following equations *by factoring* if possible. Check your solutions in the original equation.

a. $x^2 - 9x + 20 = 0$ **b.** $x^2 - 9x + 14 = 0$

c. $x^2 - 11x = -24$ **d.** $m^2 + m = 6$

    e. $2x^2 - 4x - 6 = 0$ **f.** $3t^2 + 21t - 18 = 0$

    g. $-7x + 10 = -x^2$ **h.** $-9a - 12 = -3a^2$

    i. $2x^2 + 3x = 0$ **j.** $2x^2 = 12x$

    k. $v^2 + 8v = 0$

3. a. Determine the x-intercepts of $f(x) = x^2 + 5x - 6$.

    b. Verify your answer to part a using technology.

4. Use the quadratic formula to solve the following equations.

    a. $x^2 = -5x + 6$ **b.** $6x^2 + x = 15$

c. $3x^2 - 8x + 4 = 0$

d. $4x^2 + 28x - 32 = 0$

e. $2x^2 + 4x - 3 = 0$

f. $x^2 + 7 = 6x$

g. $x^2 + 2 = 4x$

h. $2x - 6 = -x^2$

5. The stopping distance of a moving vehicle is the distance required to stop the vehicle from a given speed. The stopping distance, d (in feet), of a car moving at a speed of v miles per hour is a quadratic function defined by the equation

$$d = 0.04v^2 + 1.1v.$$

a. Determine the stopping distance of a car traveling 40 miles per hour.

b. How fast is a car traveling if it needs 307.5 feet to stop?

6. The area of a circular region is given by the formula $A = \dfrac{\pi d^2}{4}$, where d is its diameter.

   a. Determine the area of a circular region whose diameter is 8 feet.

   b. Determine the area of a circular region whose radius is 15 feet.

   c. Determine the diameter of a circular region whose area is 25π square feet.

   d. Determine to the nearest inch the diameter of a circle whose area is 300 square inches.

7. An object is thrown upward with an initial velocity, v (measured in feet per second), from an initial height above the ground, h (in feet). The object's height above the ground, s (in feet), in t seconds after it is thrown, is given by the general formula

$$s = h + vt - 16t^2.$$

A ball is tossed upward with an initial velocity of 20 feet per second from a height of 4 feet. So the general formula becomes $s = 4 + 20t - 16t^2$.

   a. How high is the ball after half a second?

   b. How high is it after 1 second?

c. How high is the ball after 3 seconds?

d. Complete the following table to determine the height of the ball for the given number of seconds. Use the data in the table to determine the maximum height attained by the ball. After how many seconds does the ball reach this height? Explain your reasoning.

Number of Seconds	0.6	0.61	0.62	0.625	0.63	0.64	0.65
Height (ft.)							

8. Since 1988, the number of voters who identify themselves as Hispanic has been increasing. The following table shows the number of Hispanic voters for each election since 1988.

Year	1988	1992	1996	2000	2004	2008	2012
Number of Hispanic Voters in the Presidential Election, V (millions)	3.7	4.1	5.0	5.9	7.6	9.7	11.2

If t represents the number of years since 1988, then $V(t)$ can be modeled by the following quadratic function defined by

$$V(t) = 0.00975t^2 + 0.0902t + 3.63.$$

a. Identify the V-intercept from the quadratic function. What does it indicate? How does it compare to the data?

b. Use the quadratic function to predict the number of Hispanic voters in 2024.

c. Use the function to predict when the number of Hispanic voters in presidential elections will first equal 25 million.

Cluster 3   Other Nonlinear Functions

Activity 4.8

Inflation

Objectives

1. Recognize an exponential function as a rule for applying a growth factor or a decay factor.

2. Graph exponential functions from numerical data.

3. Recognize exponential functions from equations.

4. Graph exponential functions from equations.

Exponential Growth

Inflation means that a current dollar will buy less in the future. According to the U.S. Consumer Price Index, the inflation rate for the 12 months from July 2012 to July 2013 was 2%. This means that a pound of apples that cost a dollar in July 2012 cost $1.02 in July 2013. The change in price is usually expressed as an annual percentage rate, known as the **inflation rate**.

1. a. In 2012, a pair of athletic shoes cost $80. At the current inflation rate of 2%, how much will the $80 pair of athletic shoes cost in 2013?

b. Assume that the rate of inflation remains at 2% for two consecutive years. How much will the shoes cost in 2014?

2. If you assume that the inflation rate increases somewhat and remains at 4% per year for the next decade, you can calculate the cost of a currently priced $12 pizza for each of the next 10 years. Complete the following table. Round to the nearest cent.

Years from Now, t	0	1	2	3	4	5	6	7	8	9	10
Cost of Pizza, $c(t)$, $											

3. a. Determine the average rate of change of the cost of the pizza in the first year (from $t = 0$ to $t = 1$), the fifth year (from $t = 4$ to $t = 5$), and the tenth year (from $t = 9$ to $t = 10$). Explain what these results mean for the cost of pizza over the next 10 years.

b. Is the function linear? Explain.

To complete the table in Problem 2, you could calculate each cost value by multiplying the previous output by 1.04, the inflation's growth factor. Thus, to obtain the cost after 10 years, you would multiply the original cost, $12, by 1.04, a total of 10 times. Symbolically, you write $12(1.04)^{10}$. Therefore, you can model the cost of pizza algebraically as

$$c(t) = 12(1.04)^t,$$

where $c(t)$ represents the cost and t represents the number of years from now.

Definition

A quantity y that increases by the same percent each year is said to increase exponentially. Such a quantity can be expressed algebraically as a function of time t by the rule

$$y = ab^t,$$

where a is its starting value (at $t = 0$), the base b is the fixed growth factor, and t represents the number of years that have elapsed. The function defined by $y = ab^t$ is called an **exponential function**. Note that the input variable t occurs as the exponent of the growth factor.

---

4. Under a constant annual inflation rate of 4%, what will a pizza cost after 20 years?

5. a. Complete the following table. Results from Problems 2 and 4 are already entered.

t	0	10	20	30	40	50
$c(t)$	12	17.76	26.29			

b. Plot the data on an appropriately scaled and labeled axis.

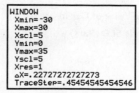

c. As the input t increases, how does the corresponding cost change?

d. Using the graph in part b, determine how many years it will take for the price of a $12 pizza to double.

6. a. A graph of the function $c(t) = 12(1.04)^t$ is shown here in a window for t between -30 and 30 and $c(t)$ between 0 and 30. Use the same window to graph this function on a graphing calculator.

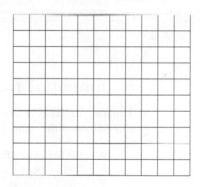

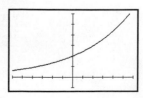

b. Use the trace or table feature of the graphing calculator to examine the coordinates of some points on the graph. Does your plot of the numerical data in Problem 5 agree with the graph shown in part a?

c. Is the entire graph in part a really relevant to the cost of pizza problem? Explain.

d. Resize your window to include only the first quadrant from $x = 0$ to $x = 30$ and $y = 0$ to $y = 30$. Regraph the function.

e. Use the calculator to determine how many years it will take for the price of a pizza to double? Explain how you determined your answer.

Exponential Decay

You just purchased a new automobile for $21,000. Much to your dismay, you learned that you should expect the value of your car to depreciate by 15% per year!

7. What is the decay factor (see page 69, Activity 1.9) for the yearly depreciation rate 15%?

8. Use the decay factor from Problem 7 to determine the car's values for the years given in the table. Record your results in the table.

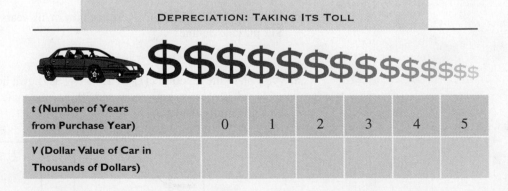

DEPRECIATION: TAKING ITS TOLL

t (Number of Years from Purchase Year)	0	1	2	3	4	5
V (Dollar Value of Car in Thousands of Dollars)						

In Problem 8, you could calculate the retail value of the car in any given year by multiplying the previous year's value by 0.85, the depreciation decay factor. Thus, to obtain the retail value after 5 years, you could multiply the original value by 0.85 a total of 5 times, as follows:

$$21(0.85)(0.85)(0.85)(0.85)(0.85) \approx 9.3 \text{ thousand, or } \$9300,$$

or equivalently in exponential form,

$$21(0.85)^5 \approx 9.3 \text{ thousand, or } \$9300.$$

Therefore, you can model the value of the car, V, algebraically by the formula

$$V(t) = 21(0.85)^t,$$

where t represents the number of years you own the car and V represents the car's value in thousands of dollars.

9. Determine the value of the car in 6 years.

10. Graph the depreciation formula for the car's value, $V(t) = 21(0.85)^t$, as a function of the time from the year of purchase. Use the following grid or a graphing calculator. Extend your graph to include 10 years from the date of purchase.

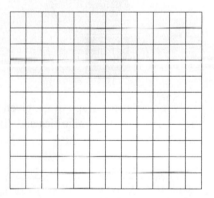

11. How long will it take the value of the car to decrease to half its original value? Explain how you determined your answer.

SUMMARY: ACTIVITY 4.8

An **exponential function** is a function in which the input variable appears as the exponent of the growth (or decay) factor, as in

$$y = ab^t,$$

where a is the starting output value (at $t = 0$), the base b is the fixed growth or decay factor, and t represents the time that has elapsed.

EXERCISES: ACTIVITY 4.8

1. a. Complete the following table in which $f(x) = \left(\dfrac{1}{2}\right)^x$ and $g(x) = 2^x$.

x	−2	−1	0	1	2
f(x)					
g(x)					

b. Use the points in part a to sketch a graph of the given functions f and g on the same coordinate axes. Compare the graphs. List the similarities and differences.

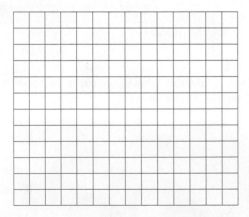

c. Use technology to graph the functions $f(x) = \left(\frac{1}{2}\right)^x$ and $g(x) = 2^x$ given in part a in the window Xmin $= -2$, Xmax $= 2$, Ymin $= -2$, and Ymax $= 5$. Compare the graph with the graphs you obtained in part b.

2. Suppose the inflation rate is 5% per year and remains the same for the next 6 years.

   a. Determine the growth factor for a 5% inflation rate.

   b. A pair of athletic shoes costs $110. If the inflation rate remains constant at 5% per year, write an algebraic rule to determine the cost, $c(t)$, of the shoes after t years.

   c. Complete the following table for the cost of a pair of athletic shoes that costs $110 now. Round to the nearest cent.

t, Years from Now	0	1	2	3	4	5	6
$c(t)$, Cost of Athletic Shoes ($)							

   d. What would be the cost of the athletic shoes in 10 years if the inflation rate stayed at 5%?

3. An exponential function may be increasing or decreasing. Determine which is the case for each of the following functions. Explain how you determined each answer.

   a. $f(x) = 5^x$ b. $g(x) = \left(\frac{1}{2}\right)^x$ c. $h(t) = 1.5^t$ d. $k(p) = 0.2^P$

4. a. Evaluate the functions in the following table for the input values, x.

Input x	0	1	2	3	4	5
$g(x) = 3x$						
$f(x) = 3^x$						

b. Compare the rate of increase of the functions $f(x) = 3^x$ and $g(x) = 3x$ from $x = 0$ to $x = 5$ by calculating the average rate of change for each function from $x = 0$ to $x = 5$. Determine which function grows faster on average in the given interval.

5. a. For investment purposes, you recently bought a house for $150,000. You expect the price of the house to increase $18,000 a year. How much will your investment be worth in 1 year? in 2 years?

b. Suppose the $18,000 increase in your house continues for many years. What type of function characterizes this growth? If P represents the housing price (in $1000s) as a function of t (in elapsed years), write this function rule.

c. You decided to buy a second investment house, again for $150,000, but in another area. Here, the prices have been rising at a rate of 12% a year. If this rate of increase continues, how much will your investment be worth in 1 year? In 2 years?

d. Suppose the 12% increase in housing prices continues for many years. What type of function characterizes this growth? If P represents the housing price (in $1000s) as a function of t (in elapsed years), write this function rule.

e. Which investment will give you the better return?

6. As a radiology specialist, you use the radioactive substance iodine-131 to diagnose conditions of the thyroid gland. Iodine-131 decays at the rate of 8.3% per day. Your hospital currently has a 20-gram supply.

a. What is the decay factor for the decay of the iodine?

b. Write an exponential decay formula for N, the number of grams of iodine-131 remaining, in terms of t, the number of days from the current supply of 20 grams.

c. Use the function rule from part b to determine the number of grams remaining for the days listed in the following table. Record your results in the table to the nearest hundredth.

t, Number of Days Starting from a 20-Gram Supply of Iodine-131	0	4	8	12	16	20	24
N, Number of Grams of Iodine-131 Remaining from a 20-Gram Supply	20.00						

d. Determine the number of grams of iodine-131 remaining from a 20-gram supply after 2 months (60 days).

e. Graph the decay formula for iodine-131, $N = 20(0.917)^t$, as a function of the time t (days). Use appropriate scales and labels on the following grid or an appropriate window on a graphing calculator.

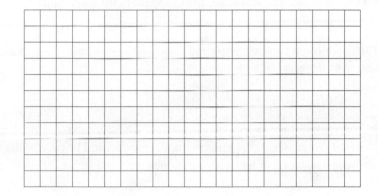

f. Use the graph or table to determine how long will it take iodine-131 to decrease to half its original value. Explain how you determined your answer.

Activity 4.9

A Thunderstorm

Objectives

1. Recognize the equivalent forms of the direct variation statement.

2. Determine the constant of proportionality in a direct variation problem.

3. Solve direct variation problems.

One of nature's more spectacular events is a thunderstorm. The sky lights up, delighting your eyes, and seconds later your ears are bombarded with the boom of thunder. Because light travels faster than sound, during a thunderstorm, you see the lightning before you hear the thunder. The formula

$$d = 1080t$$

describes the distance, d in feet, you are from the lightning flash if it takes t seconds for you to hear the thunder. This will approximate the distance to the storm's center.

1. Complete the following table for the model $d = 1080t$.

t, in Seconds	1	2	3	4
d, in Feet				

2. What does the ordered pair (3, 3240) from the above table mean in a practical sense?

3. As the value of t increases, what happens to the value of d?

The relationship between time t and distance d in this situation is an example of **direct variation**. As t increases, d also increases.

Definition

Two quantities are said to **vary directly** if whenever one quantity increases, the other quantity increases by the same multiplicative factor. The ratio of the two quantities is always constant. For instance, when one quantity doubles, so does the other.

4. Graph the distance, d, as a function of time, t, using the values in Problem 1.

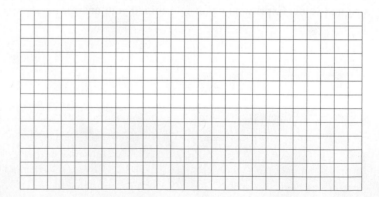

5. If the time it takes for you to hear the thunder after you see the lightning decreases, what is happening to the distance between you and the center of the storm?

If x represents the input variable and y represents the output variable, then the following statements are equivalent.

a. y varies directly as x.

b. y is directly proportional to x.

c. $y = kx$ for some constant k.

d. The ratio $\dfrac{y}{x}$ always has the same constant value, k.

The number represented by k is called the **constant of proportionality** or **constant of variation**.

6. The amount of garbage, A, varies directly with the population, P. The population of Grand Prairie, Texas, is 0.13 million and creates 2.6 million pounds of garbage each week. Determine the amount of garbage produced by Houston with a population of 2 million.

   a. Write an equation relating A and P and the constant of variation k.

   b. Use the Grand Prairie data to determine the value of k.

   c. Rewrite the equation in part a using the value of k from part b.

   d. Use the equation in part c and the population of Houston to determine the weekly amount of garbage produced.

Procedure

To solve direct variation problems, the procedure of Problem 6 is summarized, with x representing the input variable and y representing the output variable.

   i. Write an equation of the form $y = kx$ relating the variables and the constant of proportionality, k.

   ii. Determine a value for k, using a known relationship between x and y.

   iii. Rewrite the equation in part i, using the value of k found in part ii.

   iv. Substitute a known value for x or y, and determine the unknown value.

---

7. A worker's gross wages, w, vary directly as the number of hours the worker works, h. The following table shows the relationship between the wages and the hours worked.

Hours Worked, h	15	20	25	30	35
Wages, w	$172.50	$230.00	$287.50	$345.00	$402.50

a. Graph the gross wages, w, as a function of hours worked, h, using the values in the preceding table

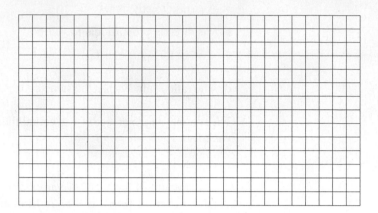

b. The relationship is defined by $w = kh$. Pick one ordered pair from the table, and use it to determine the value of k.

c. What does k represent in this situation?

d. Use the formula to determine the gross wages for a worker who works 40 hours.

Direct variation can involve higher powers of x, such as x^2, x^3, or in general, x^n.

Example 1   *Let s vary directly as the square of t. If s = 64 when t = 2, determine the direct variation equation.*

SOLUTION

Because s varies directly as the square of t, you have

$$s = kt^2,$$

where k is the constant of variation. Substituting 64 for s and 2 for t, you have

$$64 = k(2)^2 \quad \text{or} \quad 64 = 4k \quad \text{or} \quad k = 16.$$

Therefore, the direct variation equation is

$$s = 16t^2.$$

8. The approximate power P generated by a certain wind turbine varies directly as the square of the wind speed w. The turbine generates 750 watts of power in a 25-mile-per-hour wind. Determine the power it generates in a 45-mile-per-hour wind.

SUMMARY: ACTIVITY 4.9

1. Two variables are said to **vary directly** if whenever one variable increases, the other variable increases by the same multiplicative factor. The ratio of the two variables is always constant.

2. The following statements are equivalent. The number represented by k is called the **constant of proportionality or constant of variation**.

   a. y varies directly as x.

   b. y is directly proportional to x.

   c. $y = kx$ for some constant x.

   d. The ratio $\dfrac{y}{x}$ always has the same constant value, k.

3. The equation $y = kx^n$, where $k \neq 0$ and n is a positive integer, defines a **direct variation function** in which y varies directly as x^n. The constant, k, is called the **constant of proportionality** or **constant of variation**.

EXERCISES: ACTIVITY 4.9

1. The amount of sales tax, s, on any item varies directly as the list price of an item, p. The sales tax in your area is 8%.

   a. Write a formula that relates the amount of sales tax, s, to the list price, p, in this situation.

   b. Use the formula in part a to complete the following table:

LIST PRICE, p ($)	SALES TAX, s ($)
10	
20	
30	
50	
100	

c. Graph the sales tax, s, as a function of list price, p, using the values in the preceding table.

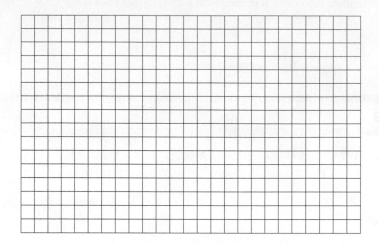

d. Determine the list price of an item for which the sales tax is $3.60.

2. The amount you tip a server, t, varies directly as the amount of the check, c.

a. Assuming that you tip your server 15% of the amount of the check, write a formula that relates the amount of the tip, t, to the amount of the check, c.

b. Use the formula in part a to complete the following table:

Amount of Check, c ($)	15	25	35	50	100
Amount of Tip, t ($)					

c. Graph the amount of the tip, t, as a function of the amount of the check, c, using the values in the preceding table.

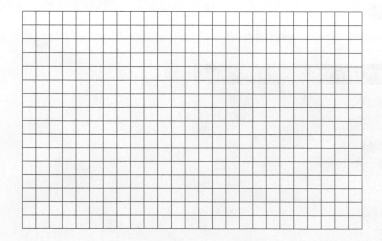

d. A company has a policy that no tip can exceed $12 and still be reimbursed by the company. What is the maximum cost of a meal for which all of the tip will be reimbursed if the tip is 15% of the check?

3. The distance, s, that an object falls from rest varies directly as the square of the time, t, of the fall. A ball dropped from the top of a building falls 144 feet in 3 seconds.

a. Write a formula that relates the distance the ball has fallen, s, in feet to the time, t, in seconds since it has been released.

b. Use the formula in part a to complete the following table.

TIME SINCE THE BALL WAS RELEASED, t (sec)	DISTANCE FALLEN, s (ft)
0.5	
1	
1.5	
2	
2.5	

c. Graph the distance fallen, s, as a function of the time since the ball was released, t, using the values in the preceding table.

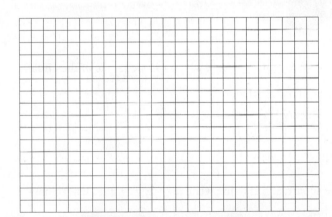

d. Estimate the height of the building if it takes approximately 2.3 seconds for the ball to hit the ground.

4. y varies directly as x, $y = 25$ when $x = 5$. Determine y when $x = 13$.

5. y varies directly as x, $y = 7$ when $x = 21$. Determine x when $y = 5$.

6. Given that y varies directly as x, consider the following table.

x	2	4	7		12
y	4	8		20	

    a. Determine a formula that relates x and y.

    b. Use the formula in part a to complete the preceding table.

7. For each table, determine the pattern and complete the table. Then write a direct variation equation for each table.

    a. y varies directly as x.

x	$\frac{1}{4}$	1	4	8
y		8		

    b. y varies directly as x^3.

x	$\frac{1}{2}$	1	3	6
y		1		

8. The area, A, of a circle is given by the function $A = \pi r^2$, where r is the radius of the circle.

    a. Does the area vary directly as the radius? Explain.

    b. What is the constant of variation k?

9. Assume that y varies directly as the square of x and that when $x = 2$, $y = 12$. Determine y when $x = 8$.

10. The distance, d, that you drive at a constant speed varies directly as the time, t, that you drive. If you can drive 150 miles in 3 hours, how far can you drive in 6 hours?

11. The number of meters, d, that a skydiver falls before her parachute opens varies directly as the square of the time, t, that she is in the air. A skydiver falls 20 meters in 2 seconds. How far will she fall in 2.5 seconds?

Activity 4.10

Diving Under Pressure, or Don't Hold Your Breath

Objectives

1. Recognize functions of the form $y = \dfrac{k}{x}, x \neq 0,$ as nonlinear.

2. Recognize equations of the form $xy = k$ as inverse variation.

3. Graph an inverse variation relationship from symbolic rules.

4. Solve equations of the form $\dfrac{a}{x} = b, x \neq 0.$

5. Identify functions of the form $y = \dfrac{k}{x^2}, x \neq 0,$ as an inverse variation relationship.

Do you know why you shouldn't hold your breath when scuba diving? Safe diving depends on a fundamental law of physics that states that the volume of a given mass of gas (air in your lungs) will decrease as the pressure increases (when the temperature remains constant). This law is known as Boyle's Law, named after its discoverer, the seventeenth-century scientist Sir Robert Boyle.

In this activity, you will discover the answer to the opening question.

1. You have a balloon filled with air that has a volume of 10 liters at sea level. The balloon is under a pressure of 1 atmosphere (atm) at sea level, 1 atmosphere is equal to 14.7 pounds of pressure per square inch.

   a. For every 33-foot increase in depth, the pressure will increase by 1 atmosphere. Therefore, if you take the balloon underwater to a depth of 33 feet below sea level, it will be under 2 atmospheres of pressure. At a depth of 66 feet, the balloon will be under 3 atmospheres of pressure. Complete column 2 (pressure) in the following table.

   b. At sea level, the balloon is under 1 atmosphere of pressure. At a depth of 33 feet, the pressure is 2 atmospheres. Because the pressure is now twice as much as it was at sea level, Boyle's Law states that the volume of the balloon decreases to one-half of its original volume, or 5 liters. Taking the balloon down to 66 feet, the pressure compresses the balloon to one-third of its original volume, or 3.33 liters; at 99 feet down, the volume is one-fourth the original, or 2.5 liters, etc. Complete the third column (volume) in the following table.

DEPTH OF WATER (ft.) BELOW SEA LEVEL	PRESSURE (atm)	VOLUME OF AIR IN BALLOON (L)	PRODUCT OF PRESSURE AND VOLUME
Sea level	1	10	$1 \cdot 10 = 10$
33	2	5	
66	3	$3.3\overline{3}$	
99		2.5	
132			
297			

   c. Note in column 4 of the table that the product of the volume and pressure at sea level is 10. Compute the product of the volume and pressure for the other depths in the table, and record them in the last column.

   d. What is the result of your computations in part c?

2. Let p represent the pressure of a gas and v the volume at a given (constant) temperature. Use what you observed in Problem 1 to write an equation for Boyle's Law for this experiment.

Boyle's Law is an example of **inverse variation** between two variables. In the case of Boyle's Law, when one variable changes by a factor n, the other changes by the factor $\dfrac{1}{n}$ so that their product is always the same positive constant. In general, an inverse variation is represented by an equation of the form $xy = k$, where x and y represent the variables and k represents the constant.

Sometimes when you are investigating an inverse variation, you may want to consider one variable as a function of the other variable. In that case, you can solve the equation $xy = k$ for y as a function of x. By dividing each side of the equation by x, you obtain $y = \dfrac{k}{x}.$

3. Write an equation for Boyle's Law where the volume, v, is expressed as a function of the pressure, p.

4. a. Graph the volume, v, as a function of pressure, p, using the values in the preceding table.

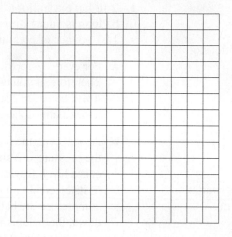

   b. If you decrease the pressure (by rising to the surface of the water), what will happen to the volume? (Use the graph in part a to help answer this question.)

5. a. Suppose you are 99 feet below sea level, where the pressure is 4 atmospheres and where the volume is one-fourth of the sea-level volume. You use a scuba tank to fill the balloon back up to a volume of 10 liters. What is the product of the pressure and volume in this case?

   b. Now suppose you take the balloon up to 66 feet, where the pressure is 3 atmospheres. What is the volume now?

   c. You continue to take the balloon up to 33 feet, where the pressure is 2 atmospheres. What is the volume?

   d. Finally, you are at sea level. What is the volume of the balloon?

   e. Suppose the balloon can expand to only 30 liters. What will happen before you reach the surface?

6. Explain why the previous problem shows that when scuba diving, you should not hold your breath if you use a scuba tank to fill your lungs.

7. a. Use the graph in Problem 4 to determine the volume of the balloon when the pressure is 12 atmospheres.

b. From the graph, what do you estimate the pressure to be when the volume is 11 liters?

c. Use the equation $v = \dfrac{10}{p}$ to determine the volume when the pressure is 12 atmospheres. How does your answer compare with the result in part a?

d. Use the equation $v = \dfrac{10}{p}$ to determine the pressure when the volume is 11 liters. How does your answer compare with the result in part b?

8. Solve each of the following.

a. $\dfrac{20}{x} = 10$

b. $\dfrac{150}{x} = 3$

Functions Defined by $y = \dfrac{k}{x^2}$, Where k Is a Nonzero Constant

9. The loudness (or intensity) of any sound is a function of the listener's distance from the source of the sound. In general, the relationship between the intensity I and the distance d can be modeled by an equation of the form

$$I = \frac{k}{d^2},$$

where I is measured in microwatts per square meter, d is measured in meters, and k is a constant determined by the source of the sound and nature of the surroundings.

a. The intensity, I, of a typical iPod at maximum setting can be given by the formula $I = \dfrac{64}{d^2}$. Complete the following table.

d(m)	0.1	0.5	1	2	5	10	20	30
$I(\mu W/m^2)$								

b. What is the practical domain of the function?

c. Sketch a graph that shows the relationship between intensity of sound and distance from the source of the sound. Use the table in part a to help you determine an appropriate scale.

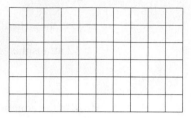

d. As you move closer to the person speaking, what happens to the intensity of the sound?

e. As you move away from the person speaking, what happens to the intensity of the sound?

The function defined by $I = \dfrac{64}{d^2}$ is another type of inverse variation function.

10. For the function defined by $I = \dfrac{64}{d^2}$, answer the following questions.

a. What is the constant of proportionality?

b. If d is doubled, what is the effect on I?

11. In Problem 9, the relationship between the intensity, I, of an iPod and the distance, d, from the individual was given by $I = \dfrac{64}{d^2}$, where I is measured in microwatts per square meter, d is measured in meters, and 64 is the constant of proportionality. The constant of proportionality depends on the source of the sound and the surroundings. If the source of the sound changes, the value of the constant of proportionality will also change.

a. The intensity of the sound made by a heavy truck 20 meters away is 1000 microwatts per square meter. Determine the constant of proportionality.

b. Write a formula for the intensity, I, of the sound made by a truck when it is d meters away.

c. Use the formula from part b to determine the intensity of the sound made by the truck when it is 100 meters away.

SUMMARY: ACTIVITY 4.10

1. For a function defined by $y = \dfrac{k}{x}$, where $k \neq 0$, as the input x increases, the output y decreases.

2. The function rule $y = \dfrac{k}{x}$, $k \neq 0$, can also be expressed in the form $xy = k$. This form shows explicitly that as one variable changes by a factor n, the other changes by a factor $\dfrac{1}{n}$ so that the product of the two variables is always the constant value, k. The relationship between the variables defined by this function is known as **inverse variation**.

3. The input value $x = 0$ is not in the domain of the function defined by $y = \dfrac{k}{x}$.

EXERCISES: ACTIVITY 4.10

1. According to the blueprint, the floor area of the stage in the new auditorium at your college must be rectangular and equal to 1200 square feet. The width of the stage is key to all of the theater productions. Therefore, in this situation, the stage's depth is a function of its width.

a. Let d represent the depth (in feet) and w represent the width (in feet). Write an equation that expresses d as a function of w.

b. Complete the following table using the equation from part a.

w (ft.)	30	35	40	50	60
d (ft.)					

c. What happens to the depth as the width increases?

d. What happens if the width is 100 feet? Is this realistic? Explain.

e. Can the width be zero? Explain.

f. What do you think is the practical domain for this function?

g. What type of a function do you have in this situation?

Exercise numbers appearing in color are answered in the Selected Answers appendix.

2. A bicyclist, runner, jogger, and walker each completed a journey of 30 miles. Their results are recorded in the table below.

	CYCLIST	RUNNER	JOGGER	WALKER
v, Average Speed (mph)	30	10	7.5	5
t, Time Taken (hr.)	1	3	4	6
d = *vt*				

a. Compute the product *vt* for each person, and record the products in the last row of the table.

b. Write the time to complete the journey as a function of average speed.

c. On the following grid, graph the function you wrote in part b for a journey of 30 miles.

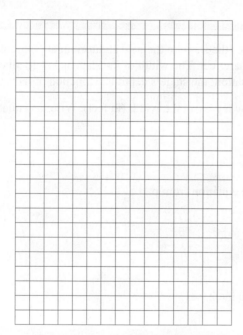

d. From the graph, determine how long the cyclist would take if she decreased her average speed to 20 miles per hour. Verify your answer from the graph, using the equation you wrote in part b.

e. From the graph, determine how long it would take the runner to complete her run if she sped up 12 miles per hour. Verify your answer from the graph, using your equation from part b.

3. The weight, w, of a body varies inversely as the square of its distance, d, measured in miles, from the center of Earth. The equation $wd^2 = 3.2 \times 10^9$ represents this relationship for a 200-pound person.

   a. Rewrite the equation to express the weight of a body, w, as a function of the distance, d, from the center of Earth.

   b. How much would a 200-pound man weigh if he were 1000 miles above the surface of Earth? The radius of Earth is approximately 4000 miles. Explain how you solve this problem.

4. a. Solve $xy = 120$ for y. Is y a function of x? What is its domain?

   b. Solve $xy = 120$ for x. Is x a function of y? What is its domain?

5. Solve the following equations.

   a. $30 = \dfrac{120}{x}$ **b.** $2 = \dfrac{250}{x}$ **c.** $-8 = \dfrac{44}{x}$

6. The amount of current, I, in a circuit varies inversely as the resistance, R. A circuit containing a resistance of 10 ohms has a current of 12 amperes. Determine the current in a circuit containing a resistance of 15 ohms.

7. The intensity, I, of light varies inversely as the square of the distance, d, between the source of light and the object being illuminated. A light meter reads 0.25 unit at a distance of 2 meters from a light source. What will the meter read at a distance of 3 meters from the source?

8. You are investigating the relationship between the volume, V, and pressure, P, of a gas. In a laboratory, you conduct the following experiment: While holding the temperature of a gas constant, you vary the pressure and measure the corresponding volume. The data that you collect appears in the following table.

P (psi)	20	30	40	50	60	70	80
V (ft.³)	82	54	41	32	27	23	20

a. Sketch a graph of the data.

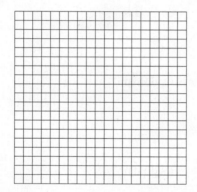

b. One possible model for the data is that V varies inversely as the square of P. Does the data fit the model $V = \dfrac{k}{P^2}$? Explain.

c. Another possible model for the data is that V varies inversely as P. Does $V = \dfrac{k}{P}$ model the data? Explain.

d. Predict the volume of the gas if the pressure is 65 pounds per square inch.

Activity 4.11

Hang Time

Objectives

1. Recognize functions of the form $y = a\sqrt{x}$ as nonlinear.

2. Evaluate and solve equations that involve square roots.

3. Graph square root functions from numerical data.

4. Graph square root functions from symbolic rules.

The time, t (output, in seconds), that a basketball player spends in the air during a jump is called hang time and depends on the height, s (input, in feet), of the jump according to the formula

$$t = \frac{1}{2}\sqrt{s}.$$

To determine the hang time of a jump 1 foot above the court floor, you perform a sequence of two operations:

Start with s (1 ft.) $\rightarrow$ take square root $\rightarrow$ multiply by $\frac{1}{2}$ $\rightarrow$ to obtain $t\left(\frac{1}{2}\text{ sec. in the air}\right)$.

A function such as $f(s) = \frac{1}{2}\sqrt{s}$ is known as a square root function because determining the output involves taking the square root of an expression involving the input.

1. A good college athlete can usually jump about 2 feet above the court floor. Determine his hang time.

2. Michael Jordan is known to have jumped as high as 39 inches above the court floor. What was his hang time on that jump?

Example 1 *How high must a jump be for someone to have a hang time of 0.5 second?*

SOLUTION

You can determine the height from the hang time formula. You can do this by "reversing direction," or replacing each operation with its inverse operation in a sequence of algebraic steps. Note that the inverse of the square root operation is squaring.

Start with 0.5 sec. $\rightarrow$ multiply by 2 $\rightarrow$ square $\rightarrow$ to obtain 1 ft.

You can also solve the problem in a formal algebraic way by applying these inverse operations to each side of the equation.

Substitute 0.5 second for t in the formula $t = \frac{1}{2}\sqrt{s}$ to obtain the equation $0.5 = \frac{1}{2}\sqrt{s}$, and proceed as follows:

$2(0.5) = 2 \cdot \frac{1}{2}\sqrt{s}$	**Multiply by 2.**
$1 = \sqrt{s}$	**Simplify.**
$(1)^2 = (\sqrt{s})^2$	**Square both sides.**
$1 \text{ ft.} = s$	**Simplify.**

3. Use the hang time formula and the data regarding Michael Jordan's hang time on a 39-inch jump to determine whether it is reasonable for a professional basketball player to have a hang time of 3 seconds. Explain your answer.

4. a. What would be a reasonable domain and range for the hang time function having equation $t = \frac{1}{2}\sqrt{s}$?

b. Complete the following table for the given jump heights.

HEIGHT OF JUMP, s (ft.)	HANG TIME, t (sec.)
0	
1	
2	
3	
4	

c. Plot the points (t, s) using the table you completed in part b, and connect them with a smooth curve.

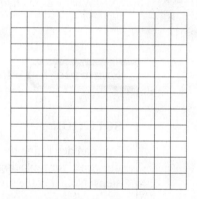

5. a. Is the hang time function $t = \frac{1}{2}\sqrt{s}$ a linear function? Use the graph in problem 4c to explain.

b. Solve the hang time formula $t = \frac{1}{2}\sqrt{s}$ for s in terms of t.

c. Use the formula you obtained in part b to explain why the relationship between s and t is not linear.

6. Solve each of the following equations. Check your result in the original equation.

a. $\sqrt{x} = 9$

b. $\sqrt{t} + 3 = 10$

c. $4\sqrt{x} = 12$

d. $\sqrt{t+1} = 5$

e. $\sqrt{5t} = 10$

f. $\sqrt{2t} - 3 = 0$

7. Your friend solved the equation $2\sqrt{x} + 10 = 0$ for x and obtained $x = 25$ as the result.

a. Use the result, $x = 25$, to evaluate the left side of the original equation, $2\sqrt{x} + 10 = 0$. Does the value of the left side equal the value of the right side? What does that indicate about the result, $x = 25$, that your friend obtained?

b. You and your friend want to determine why $x = 25$ is not a solution. One way is to review the steps used to solve the equation. Where is the error?

c. Another way to investigate why $x = 25$ is not a solution to the equation $2\sqrt{x} + 10 = 0$ is to consider the graphs of $y_1 = 2\sqrt{x} + 10$ and $y_2 = 0$. Do the graphs intersect? What does that indicate about the solution(s) to the equation $2\sqrt{x} + 10 = 0$?

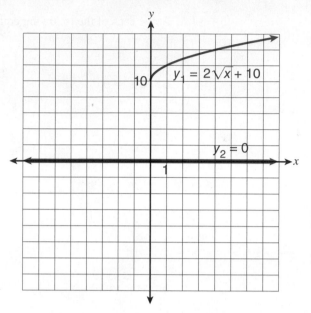

SUMMARY: ACTIVITY 4.11

1. The function defined by $y = a\sqrt{x}$, where $a \neq 0$ and x is nonnegative, is called a **square root function** because determining the output involves the square root of an expression containing the input.

2. For $a > 0$, the graph of a square root function has the following general shape.

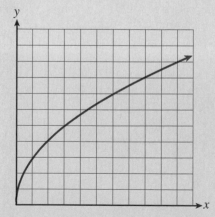

3. For non-negative numbers, squaring and taking square roots are inverse operations.

4. To evaluate an output of a square root function $y = a\sqrt{x}$ for a specified input, do the following.

   Start with $x \rightarrow$ take square root $\rightarrow$ multiply by $a \rightarrow$ to obtain y.

For example, if $a = 3$ and $x = 4$, then $y = 3\sqrt{4} \rightarrow y = 3 \cdot 2 \rightarrow y = 6$.

5. To solve a square root equation $y = a\sqrt{x}$ for a specified output, do the following:

Start with $y \rightarrow$ divide by $a \rightarrow$ square $\rightarrow$ to obtain x.

For example, if $a = 3$ and $y = 6$, then $6 = 3\sqrt{x} \rightarrow 2 = \sqrt{x} \rightarrow 2^2 = x \rightarrow x = 4$.

EXERCISES: ACTIVITY 4.11

1. You are probably aware that the stopping distance for a car, once the brakes are firmly applied, depends on its speed and on the tire and road conditions. The following formula expresses this relationship.

$$S = \sqrt{30fd},$$

where S is the speed of the car (in miles per hour); f is the drag factor, which takes into account the condition of tires, pavement, etc.; and d is the stopping distance (in feet), which can be determined by measuring the skid marks.

Assume that the drag factor, f, has the value 0.83.

a. At what speed is your car traveling if it leaves skid marks of 100 feet?

b. Your friend stops short at an intersection when she notices a police car nearby. She is pulled over and claims that she was doing 25 miles per hour in a 30-mile-per-hour zone. The skid marks measure 60 feet. Should she be issued a speeding ticket?

c. Solve the formula $S = \sqrt{30fd}$ for d.

d. Determine the stopping distances (length of the skids) for the speeds given in the following table. Round your answers to the nearest whole foot.

STOPPING DISTANCE, (ft.)	SPEED, S (mph)
	50
	55
	60
	65
	70
	75

Exercise numbers appearing in color are answered in the Selected Answers appendix.

e. Use the data in the preceding table to graph speed as a function of stopping distance.

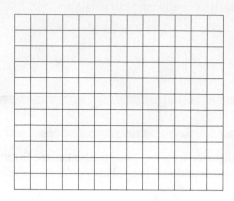

2. Solve each of the following equations (if possible) using an algebraic approach. Check your answers.

a. $3x^2 = 27$ **b.** $5x^2 + 1 = 16$

c. $t^2 + 9 = 0$ **d.** $3r^2 + 10 = 2r^2 + 14$

e. $\sqrt{x} - 4 = 3$ **f.** $5\sqrt{r} = 10$

g. $\sqrt{x + 4} - 3 = 0$ **h.** $4 = \sqrt{2w}$

3. a. Complete the following table. Round your answers to the nearest tenth.

x	0	1	2	3	4	5
$f(x) = 3\sqrt{x}$						
$g(x) = \sqrt{3x}$						

b. Sketch a graph of each function on the same coordinate axes.

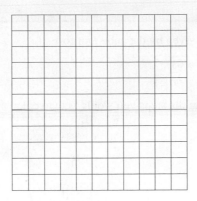

c. Both of the functions are increasing. Which function is increasing more rapidly?

4. In your science class, you are learning about pendulums. The period of a pendulum is the time it takes the pendulum to make one complete swing back and forth. The relationship between the period and the pendulum length is described by the formula $T = \dfrac{\pi}{4}\sqrt{\dfrac{l}{6}}$, where l is the length of the pendulum in inches and T is the period in seconds.

a. The length of the pendulum inside your grandfather clock is 24 inches. What is its period?

b. A metal sculpture hanging from the ceiling of an art museum contains a large pendulum. In science class, you are asked to determine the length of the pendulum without measuring it. You use a stopwatch and determine that its period is approximately 3 seconds. What is the length of the pendulum in feet?

Cluster 3   What Have I Learned?

1. The diameter, d, of a sphere having volume V is given by the cube root formula

$$d = \sqrt[3]{\frac{6V}{\pi}}.$$

   a. Use what you learned in this cluster to explain the steps you would take to calculate a value for d, given a value for V.

   b. Use what you learned to explain the steps you would take to solve the formula for V in terms of d.

2. The starting salary at your new job is \$22,000 per year. You are offered two options for salary increases:

     Plan 1: an annual increase of \$1000 per year

     Plan 2: an annual percentage increase of 4% of your salary

Your salary is a function of the number of years of employment at your job.

   a. Write an equation to determine the salary S after x years on the job using each plan described above.

   b. Complete the following table using the equations from part a.

x	0	1	3	5	10	15
S, Plan 1						
S, Plan 2						

   c. Which plan would you choose? Explain.

3. The equation $xy = 4$ describes the relationship between the width, x, and length, y, of a rectangle with a fixed area of 4 square centimeters. Answer the following questions based on what you learned in this cluster.

Exercise numbers appearing in color are answered in the Selected Answers appendix.

a. Write the length as a function of the width for this rectangle, and graph the function for $0 < x \leq 10$ and $0 < y \leq 10$.

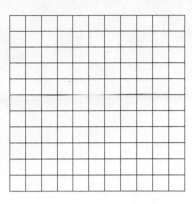

b. Describe the graph in a few sentences, noting its domain and range and whether it is an increasing or decreasing function.

c. Answer the following questions regarding the relationship between the length and width of the rectangle.

   i. For what value will the width and length be the same? What shape will the rectangle take for this value?

ii. When will the rectangle be twice as long as its width? When will it be half as long?

d. Determine several other possible dimensions for the rectangle. Use a ruler to draw those rectangles and the rectangles from parts i and ii above. Is there any one shape whose proportions are most pleasing to your eye? Explain your answer.

4. a. Is every function defined by an equation of the form $y = kx$, $k \neq 0$, a linear function? If yes, identify the slope and vertical intercept.

b. Does every function defined by an equation of the form $y = kx$, $k \neq 0$, represent a direct variation? If yes, what is the constant of proportionality?

c. Do all linear functions represent a direct variation? Explain.

Cluster 3   How Can I Practice?

1. Solve each of the following formulas for the specified variable.

a. $V = \pi r^2 h$, solve for r

b. $a^2 + b^2 = c^2$, solve for b

2. The formula $V = \sqrt{\dfrac{1000p}{3}}$ shows the relationship between the velocity of the wind, V (in miles per hour), and wind pressure, p (in pounds per square foot).

a. The pressure gauge on a bridge indicates a wind pressure, p, of 10 pounds per square foot. What is the velocity, V, of the wind?

b. If the velocity of the wind is 50 miles per hour, what is the wind pressure?

3. Your union contract provides that your salary will increase 3% in each of the next 5 years. Your current salary is $32,000.

a. Represent your salary symbolically as a function of years, and state the practical domain.

b. Use the formula from part a to determine your salary in each of the next 5 years.

YEARS, y	SALARY, s(y)
0	$32,000
1	
2	
3	
4	
5	

c. What is the percent increase in your salary over the first 2 years of the contract?

d. What is the total percent increase in your salary over the life of the contract?

e. If the next 5-year contract also includes a 3% raise each year, how much will you be earning in 10 years?

4. Your online company has determined that the price-demand function, $N(p) = \dfrac{5700}{p}$, models quite well the relationship between the price, p (in dollars), of the product that you make and the number, $N(p)$, of units that you will sell.

a. How many units does your company expect to sell if the price is $1.50?

b. What price should you charge if you want to sell 6000 units?

c. Use technology to verify your answers to parts a and b graphically. Use the window: Xmin = 0, Xmax = 4, Xscl = 0.25, Xmin = 0, Xmax = 7000, and Xscl = 200.

d. What is the domain of this price-demand function, based on its algebraic definition?

e. Determine a reasonable practical domain for the price-demand function?

5. a. Solve $xy = 25$ for y. Is y a function of x? What is its domain?

   b. Solve $xy = 25$ for x. Is x a function of y? What is its domain?

Solve the following equations.

6. $24 = \dfrac{36}{x}$

7. $2500 = \dfrac{50}{x}$

8. $18 = \dfrac{1530}{x + 2}$

9. $3\sqrt{x} = 3$

10. $\sqrt{x} - 9 = 0$

11. $\sqrt{x + 1} = 7$

12. a. Suppose you are taking a trip of 145 miles. Assume that you drive the entire distance at a constant speed. Express your time to take this trip as a function of your speed.

   b. What is the practical domain of this function?

   c. Using the equation for the function, determine the domain.

13. When you take medicine, your body metabolizes and eliminates the medication until there is none left in your body. The half-life of a medication is the time is takes your body to eliminate one-half of the amount present. For many people, the half-life of the medicine Prozac is 1 day.

   a. What fraction of a dose is left in your body after 1 day?

   b. What fraction of the dose is left in your body after 2 days? (This is one-half of the result from part a.)

c. Complete the following table. Let t represent the number of days after a dose of Prozac is taken, and let Q represent the fraction of the dose of Prozac remaining in your body.

t (days)	Q (mg)
0	1
1	
2	
3	
4	

d. The values of the fraction of the dosage, Q, can be written as powers of $\frac{1}{2}$. For example, $1 = \left(\frac{1}{2}\right)^0$, $\frac{1}{2} = \left(\frac{1}{2}\right)^1$, etc. Complete the following table by writing each value of Q in the preceding table as a power of $\frac{1}{2}$.

t (days)	Q (mg)
0	$\left(\frac{1}{2}\right)^0$
1	$\left(\frac{1}{2}\right)^1$
2	
3	
4	

e. Use the result of part d to write an equation for Q in terms of t. Is this function linear? Explain.

f. Sketch a graph of the data in part a and the function in part e on an appropriately scaled and labeled coordinate axis.

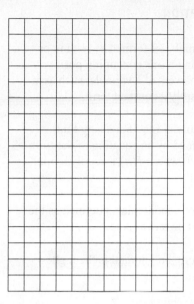

14. The length, L (in feet), of skid distance left by a car varies directly as the square of the initial velocity, v (in miles per hour), of the car.

a. Write a general equation for L as a function of v. Let k represent the constant of variation.

b. Suppose a car traveling 40 miles per hour leaves a skid distance of 60 feet. Use this information to determine the value of k.

c. Use the function to determine the length of the skid distance left by the car traveling 60 miles per hour.

15. The number of computers infected by a virus t days after it first appears often increases exponentially. Recently, a computer worm spread from about 2.4 million computers on June 8 to approximately 3.2 million computers on June 9.

a. Determine the growth factor.

b. Write an exponential equation that can be used to predict the number N of computers infected t days after June 8.

c. Predict the number of computers infected by the virus after 3 days.

The bracketed numbers following each concept indicate the activity in which the concept is discussed.

CONCEPT/SKILL	DESCRIPTION	EXAMPLE
Polynomial [4.1]	Polynomials are formed by adding or subtracting terms containing positive integer powers of a variable, such as x, and possibly a constant term.	$25x^2 - 10x + 1$
Polynomial function [4.1]	A polynomial function is any function defined by an equation of the form $y = f(x)$, where $f(x)$ is a polynomial expression.	$f(x) = 3x^3 - 2x - 7$
Degree of a monomial [4.1]	Degree of a monomial is defined as the exponent on its variable.	Degree of $11x^5$ is 5.
Degree of a polynomial [4.1]	Degree of a polynomial is the highest degree among all its terms.	Degree of $6x^3 - 7x^2 - 5x + 4$ is 3.
Classifying polynomials by the number of terms they contain [4.1]	A monomial has one term, a binomial has two terms, and a trinomial has three terms.	monomial: $6x$ binomial: $2x^3 - 54$ trinomial: $25x^2 - 10x + 1$
Simplifying expressions containing polynomials [4.1]	1. If necessary, apply the distributive property to remove parentheses. 2. Combine like terms.	$5x^2 - 3x(x + 2) + 2x$ $= 5x^2 - 3x^2 - 6x + 2x$ $= 2x^2 - 4x$
Adding or subtracting polynomials [4.1]	Apply the same process used to simplify polynomials.	
Properties of Exponents [4.2]	Property 1: $b^m \cdot b^n = b^{m+n}$ Property 2: $\dfrac{b^m}{b^n} = b^{m-n}, b \neq 0$ Property 3: $(a \cdot b)^n = a^n b^n$ and $\left(\dfrac{a}{b}\right)^n = \dfrac{a^n}{b^n}$ Property 4: $(b^m)^n = b^{m \cdot n}$ Definition: $b^0 = 1, b \neq 0$ Definition: $b^{-n} = \dfrac{1}{b^n}, b \neq 0$	1. $5^7 \cdot 5^6 = 5^{13}$ 2. $\dfrac{7^8}{7^5} = 7^3$ 3. $(2 \cdot 3)^4 = 2^4 \cdot 3^4$ and $\left(\dfrac{3}{4}\right)^2 = \dfrac{3^2}{4^2}$ 4. $(3^2)^4 = 3^8$ 5. $10^0 = 1$ 6. $2^{-3} = \dfrac{1}{2^3}$

CONCEPT/SKILL	DESCRIPTION	EXAMPLE
Multiplying a series of monomial factors [4.2]	1. If present, remove parentheses using Property of Exponents. 2. Multiply the numerical coefficients. 3. Apply Property 1 of Exponents to variable factors that have the same base.	$3x^4(x^3)^2 \cdot 5x^3$ $= 3x^4 x^6 \cdot 5x^3$ $= 15x^{13}$
Writing an algebraic expression in expanded form [4.2]	Use the distributive property to expand, applying the Properties of Exponents.	$-2x^2(x^3 - 5x + 3)$ $= -2x^5 + 10x^3 - 6x^2$
Factoring completely [4.2]	Determine the GCF, and then use the properties of exponents to write the polynomial in factored form.	$24x^5 - 16x^3 + 32x^2$ $= 8x^2(3x^3 - 2x + 4)$
Multiplying a monomial times a polynomial [4.2]	Multiply each term of the polynomial by the monomial.	$2x(x^2 - 5x + 3)$ $= 2x \cdot x^2 - 2x \cdot 5x + 2x \cdot 3$ $= 2x^3 - 10x^2 + 6x$
Multiplying two binomials (also known as the FOIL Method) [4.3]	Multiply every term in the second binomial by each term in the first. The FOIL Method (**F**irst + **O**uter + **I**nner + **L**ast) is a multiplication process in which each of the letters F, O, I, L refers to the position of the factors in the two binomials.	$(12 + x)(9 + x)$ $(12 + x)(9 + x)$ $= 108 + 12x + 9x + x^2$ $= 108 + 21x + x^2$
Multiplying two polynomials [4.3]	Multiply every term in the second polynomial by each term in the first.	$(x - 3)(x^2 - 2x + 1)$ $= x(x^2 - 2x + 1) - 3(x^2 - 2x + 1)$ $= x^3 - 2x^2 + x - 3x^2 + 6x - 3$ $= x^3 - 5x^2 + 7x - 3$
Special Products [4.3]	Difference of Squares: $(A + B)(A - B) = A^2 - B^2$ Perfect-square trinomial: $(x + a)^2 = x^2 + 2ax + a^2$ $(x - a)^2 = x^2 - 2ax + a^2$	$(2x + 3)(2x - 3) = 4x^2 - 9$ $(x + 4)^2 = x^2 + 8x + 16$ $(x - 3)^2 = x^2 - 6x + 9$
Parabola [4.4]	A parabola is the U-shaped graph described by a function of the form $y = ax^2 + bx + c, a \neq 0$.	

CONCEPT/SKILL	DESCRIPTION	EXAMPLE
Solving an equation of the form $ax^2 = c$, $ac > 0$ [4.4]	Solve an equation of this form algebraically by dividing both sides of the equation by a and then taking the positive and negative square roots of both sides.	$2x^2 = 8$ $$\frac{2x^2}{2} = \frac{8}{2}$$ $x^2 = 4$ $x = \pm \sqrt{4} = \pm 2.$
Solving an equation of the form $(x \pm a)^2 = c$ [4.4]	Solve an equation of this form algebraically by taking the square root of both sides of the equation and then solving each of the resulting linear equations.	$(x + 3)^2 = 16$ $x + 3 = \pm 4$ $x = -3 + 4 = 1$ $x = -3 - 4 = -7$
Zero-product property [4.5]	If the product of two factors is zero, then at least one of the factors must also be zero.	$(x + 6)(x - 1) = 0$ $x + 6 = 0$ or $x - 1 = 0$ $x = -6$ or $x = 1$
Solving an equation of the form $ax^2 + bx = 0$, $a \neq 0$ [4.5]	Solve an equation of this form algebraically by factoring the left side and then using the zero-product property.	$x^2 - 3x = 0$ $x(x - 3) = 0$ $x = 0$ or $x - 3 = 0$ $x = 0$ or $x = 3$
x-intercepts of the graph of $y = ax^2 + bx$, $a \neq 0$ [4.5]	The x-intercepts of the graph of $y = ax^2 + bx$ are precisely the solutions of the equation $ax^2 + bx = 0$.	
Quadratic function [4.6]	A quadratic function is a polynomial of degree 2 that can be written in the general form $y = ax^2 + bx + c$, where a, b, and c are arbitrary constants and $a \neq 0$.	$y = 5x^2 - 6x + 1$
Quadratic equation [4.6]	A quadratic equation is any equation that can be written in the general form $ax^2 + bx + c = 0$, $a \neq 0$.	$5x^2 - 6x + 1 = 0$
Quadratic equation in standard form [4.6]	$ax^2 + bx + c = 0$, $a \neq 0$	The equation $5x^2 - 6x + 1 = 0$ is written in standard form; $5x^2 - 6x = -1$ is not.

CONCEPT/SKILL	DESCRIPTION	EXAMPLE
Solving quadratic equations by factoring [4.6]	1. Write the equation in standard form (i.e., with one side equal to zero). 2. Divide each term in the equation by the greatest common numerical factor. 3. Factor the nonzero side. 4. Apply the zero-product property to obtain two linear equations. 5. Solve each linear equation. 6. Check your solutions in the original equation.	$5x^2 - 10x = 15$ $5x^2 - 10x - 15 = 0$ $x^2 - 2x - 3 = 0$ $(x - 3)(x + 1) = 0$ $x - 3 = 0$ or $x + 1 = 0$ $x = 3$ or $x = -1$
Quadratic formula [4.7]	The solutions of the quadratic equation $ax^2 + bx + c = 0$ are $$x = \frac{-b \pm \sqrt{b^2 - 4ac}}{2a}, a \neq 0.$$	$5x^2 - 6x + 1 = 0$ $a = 5, b = -6, c = 1$ $$x = \frac{-b \pm \sqrt{b^2 - 4ac}}{2a}$$ $$x = \frac{-(-6) \pm \sqrt{(-6)^2 - 4 \cdot 5 \cdot 1}}{2 \cdot 5}$$ $$= \frac{6 \pm \sqrt{16}}{10}$$ $x = \frac{1}{5}$ or $x = 1$
Exponential function [4.8]	An exponential function is a function in which the input variable appears as the exponent of the growth (or decay) factor, as in $$y = ab^t,$$ where a is the starting output value (at $t = 0$), b is the fixed growth or decay factor, and t represents the time that has elapsed.	$y = 16(2.1)^x$
Direct variation [4.9]	Two variables are said to vary directly if whenever one variable increases, the other variable increases by the same multiplicative factor. The ratio of the two variables is always constant.	$y = 12x$ When $x = 3$, $y = 12(3) = 36$. If x doubles to 6, y also doubles, becoming 72.

CONCEPT/SKILL	DESCRIPTION	EXAMPLE
The constant of proportionality or constant of variation, k [4.9]	The following statements are equivalent: a. y varies directly as x b. y is directly proportional to x c. $y = kx$ for some constant k	In the direct variation function $y = 2x$, 2 is the constant of variation.
Inverse variation function [4.10]	An inverse variation function is defined by the equation $xy = k \left(\text{or } y = \dfrac{k}{x} \right)$. When $k > 0$ and the input x changes by a constant factor p, the output y changes by the factor $\dfrac{1}{p}$. The input value $x = 0$ is not in the domain of this function.	$y = \dfrac{360}{x}$ When $x = 4$, $y = \dfrac{360}{4} = 90$. If x doubles to 8, y is halved and becomes 45.
Square root function [4.11]	A square root function is a function of the form $y = a\sqrt{x}$, where $a \neq 0$. The domain of the function is $x \geq 0$.	$t = \dfrac{1}{2}\sqrt{s}$
Graph of a square root function [4.11]	The graph of $y = a\sqrt{x}$, where $a > 0$, increases and is not linear.	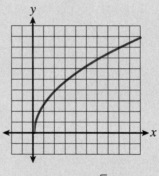 $y = a\sqrt{x}, a > 0$

In Exercises 1–11, perform the indicated operations and simplify as much as possible.

1. $-3(2x - 5) + 7(3x - 8)$

2. $-4(5x^2 - 7x + 8) + 2(3x^2 - 4x - 2)$

3. Add $5x^3 - 12x^2 - 7x + 4$ and $-3x^3 + 3x^2 + 6x - 9$.

4. Subtract $3x^3 + 5x^2 - 12x - 17$ from $2x^3 - 8x^2 + 4x - 19$.

5. $x^4(3x^3 + 2x^2 - 5x - 20)$

6. $-2x(3x^3 - 4x^2 + 8x - 5) + 3x(5x^3 - 8x^2 - 7x + 2)$

7. $(2x^3)^4$ **8.** $(5xy^3)^3$ **9.** $(-8a^2b^3)(-2a^3b^2)$

10. $\dfrac{y^{11}}{y^3}$ **11.** $\dfrac{45t^8}{9t^5}$

In Exercises 12 and 13, factor the given expression completely.

12. $27x^5 - 9x^3 - 12x^2$ **13.** $8x^2 + 4x$

14. A car travels at the average speed of $5a - 13$ miles per hour for 2 hours. Write a polynomial expression in expanded form for the distance traveled.

15. Write each of the following products in expanded form.

   a. $(x - 4)(x - 6)$ **b.** $(x + 3)(x - 8)$

c. $(x + 3)(x - 3)$ **d.** $(x - 6)^2$

e. $(2x - 3)(x - 5)$ **f.** $(3x + 4)(2x - 7)$

g. $(2x + 5)(2x - 5)$ **h.** $(x + 4)(x + 4)$

i. $(x + 3)(2x^2 + x - 3)$ **j.** $(x + 5)(3x^2 - 2x - 4)$

16. Your circular patio has radius 11 feet. To enlarge the patio, you increase the radius by y feet. Write a formula for the area of the new patio in terms of y in both factored and expanded form.

17. You have a square garden. Let x represent one side of your garden. If you increase one side 7 feet and decrease the other side 2 feet, write a formula for the area of the new garden in terms of x in both factored and expanded form.

18. Solve each of the following equations.

    a. $75x^2 = 300$ **b.** $9x^2 = 81$ **c.** $4x^2 = -16$

    d. $(x + 2)^2 = 64$ **e.** $(x - 3)^2 = 25$

19. Solve each of the following equations.

    a. $x(x + 8) = 0$ **b.** $4x^2 - 12x = 0$ **c.** $5x^2 - 25x = 0$

    d. $3x(x - 5) = 0$ **e.** $(4x - 1)(x - 6) = 0$

20. You are sitting on the front lawn and toss a ball straight up into the air. The height, s (in feet), of the ball above the ground t seconds later is described by the function $s = -16t^2 + 48t$.

a. Determine the height of the ball after 0.5 second.

b. After how many seconds does the ball reach a height of 32 feet?

c. When does the ball reach the ground?

21. Solve the following equations by factoring and using the zero-product property.

a. $x^2 + 4x - 12 = 0$ **b.** $x^2 = 5x + 14$

c. $x^2 - 5x = 24$ **d.** $x^2 - 10x + 21 = 0$

e. $x^2 + 2x - 24 = 0$ **f.** $y^2 + 3y = 40$

22. Solve the following equations using the quadratic formula.

    a. $3x^2 - 2x - 6 = 0$ **b.** $5x^2 + 7x = 12$

    c. $-2x^2 + 4x = -9$ **d.** $2x^2 - 9x = -12$

23. Solve the following equations.

    a. $\dfrac{90}{x} = 15$ **b.** $\dfrac{56}{x} = -8$ **c.** $3\sqrt{x} = 21$

    d. $\sqrt{x} - 25 = 0$ **e.** $\sqrt{x + 3} = 6$

24. Due to inflation, your college's tuition is predicted to increase by 6% each year.

    a. If tuition is $300 per credit now, determine how much it will be in 5 years and in 10 years.

    b. Calculate the average rate of change in tuition over the next 5 years.

    c. Calculate the average rate of change in tuition over the next 10 years.

    d. If the inflation stays at 6%, approximately when will the tuition double? Explain how you determined your answer.

25. Aircraft design engineers use the following formula to determine the proper landing speed, V (in feet per second)

$$V = \sqrt{\frac{841L}{cs}},$$

where L is the gross weight of the aircraft in pounds, c is the coefficient of lift, and s is the wing surface area in square feet.

Determine L if $V = 115$, $c = 2.81$, and $s = 200$.

26. A propane gas bill varies directly as the amount of gas used. The bill for 56 gallons of propane was $184.80. What is the bill for 70 gallons of propane?

27. The time it takes to drive a certain distance varies inversely as the rate of travel. If it takes 6 hours at 48 miles per hour to drive the distance, how long will it take at 72 miles per hour?

28. The sound intensity of a speaker varies inversely as the square of the distance from the speaker. The intensity is 20 microwatts per square meter when you are 10 feet from the speaker. Determine the intensity when you are 5 feet from the speakers.

29. An object is dropped from the roof of a five-story building whose floors are vertically 12 feet apart. The function that relates the distance fallen, s (in feet), to the time, t (in seconds), that it falls is $s = 16t^2$.

To answer the questions in parts a–c, do the following.

- Determine the distance that the object falls.

- Substitute the distance in the equation $s = 16t^2$.

- Solve the resulting equation.

a. How long will it take the object to reach the floor level of the fifth floor?

b. How long will it take the object to reach the floor level of the third floor?

c. How long will it take the object to reach ground level?

30. In recent years, there has been a great deal of discussion about global warming and the possibility of developing alternative energy sources that are not carbon-based. One of the fastest-growing sources has been wind energy.

The following table shows the rapid increase in installed capacity of wind power in the United States from 2001 to 2008. The wind power capacity is measured in megawatts, MW. Note that the input variable (year) increases in steps of 1 unit (year).

Year	2001	2002	2003	2004	2005	2006	2007	2008
Installed Capacity, MW	4261	4685	6374	6740	9149	11,575	16,596	25,176

a. Is this a linear function? Explain.

b. The wind data in the first table can be modeled by an exponential function defined by $N(t) = 3716(1.282)^t$, where $N(t)$ represents the generating capacity of windmills in the United States (in MW) and t represents the number of years since 2001. Note that $t = 0$ corresponds to 2001, $t = 1$ to 2002, etc. Complete the following table.

t	CALCULATION FOR THE GENERATING CAPACITY OF WINDMILLS IN THE U.S.	EXPONENTIAL FORM	GENERATING CAPACITY IN MW
0	3716	$3716(1.282)^0$	3716
1	3716(1.282)	$3716(1.282)^1$	
2	3716(1.282)(1.282)	$3716(1.282)^2$	
3	3716(1.282)(1.282)(1.282)	$3716(1.282)^3$	

c. How well do the data determined by the exponential model fit the actual data in part a?

d. Use the model to estimate the generating capacity of windmills in the year 2012.

31. You just obtained a special line of credit at the local electronics store. You immediately purchase a stereo system for $415. Your credit limit is $500. Assume that you make no payments and purchase nothing more and that there are no other fees. The monthly interest rate is 1.18%.

a. What is your initial credit balance?

b. What is the growth factor of your credit balance?

c. Write an exponential function to determine how much you will owe (represented by y) after x months with no more purchases or payments.

d. How much will you owe after 10 months?

e. When you reach your credit limit of $500, the company will expect a payment. How long do you have before you must start paying back the money? Use technology to approximate the solution.

Appendix A

Fractions

Proper and Improper Fractions

A fraction in the form $\frac{a}{b}$ is called **proper** if a and b are counting numbers and a is less than b ($a < b$). If a is greater than or equal to b ($a \geq b$), then $\frac{a}{b}$ is called **improper**.

Note: a can be zero but b cannot because you may not divide by zero.

Example 1: $\frac{2}{3}$ is proper because $2 < 3$.

Example 2: $\frac{7}{5}$ is improper because $7 > 5$.

Example 3: $\frac{0}{4} = 0$

Example 4: $\frac{8}{0}$ has no numerical value and, therefore, is said to be **undefined**.

Reducing a Fraction

A fraction, $\frac{a}{b}$, is in **lowest terms** if a and b have no common divisor other than 1.

Example 1: $\frac{3}{4}$ is in lowest terms. The only common divisor is 1.

Example 2: $\frac{9}{15}$ is not in lowest terms. 3 is a common divisor of 9 and 15.

To **reduce a fraction** $\frac{a}{b}$ **to lowest terms**, divide both a and b by a common divisor until the numerator and denominator no longer have any common divisors.

Example 1: Reduce $\frac{4}{8}$ to lowest terms.

Solution: $\frac{4}{8} = \frac{4 \div 4}{8 \div 4} = \frac{1}{2}$

Example 2: Reduce $\frac{54}{42}$ to lowest terms.

Solution: $\frac{54}{42} = \frac{54 \div 2}{42 \div 2} = \frac{27}{21} = \frac{27 \div 3}{21 \div 3} = \frac{9}{7}$

Mixed Numbers

> A **mixed number** is the sum of a whole number plus a proper fraction.

Example 1: $3\frac{2}{5} = 3 + \frac{2}{5}$

Changing an Improper Fraction, $\frac{a}{b}$, to a Mixed Number

1. Divide the numerator, a, by the denominator, b.

2. The quotient becomes the whole-number part of the mixed number.

3. The remainder becomes the numerator and b remains the denominator of the fractional part of the mixed number.

Example 1: Change $\frac{7}{5}$ to a mixed number.

Solution: $7 \div 5 = 1$ with a remainder of 2.

$$\text{So } \frac{7}{5} = 1 + \frac{2}{5} = 1\frac{2}{5}.$$

Example 2: Change $\frac{22}{6}$ to a mixed number.

Solution: First, reduce $\frac{22}{6}$ to lowest terms.

$$\frac{22}{6} = \frac{22 \div 2}{6 \div 2} = \frac{11}{3}$$

$11 \div 3 = 3$ with a remainder of 2.

$$\text{So } \frac{11}{3} = 3 + \frac{2}{3} = 3\frac{2}{3}.$$

Changing a Mixed Number to an Improper Fraction

1. To obtain the numerator of the improper fraction, multiply the denominator by the whole number and add the original numerator to this product.

2. Place this sum over the original denominator.

Example 1: Change $4\frac{5}{6}$ to an improper fraction.

Solution: $6 \cdot 4 + 5 = 29$

$$\text{So } 4\frac{5}{6} = \frac{29}{6}.$$

Exercises

1. Reduce these fractions if possible.

   a. $\dfrac{12}{16}$ **b.** $\dfrac{33}{11}$ **c.** $\dfrac{21}{49}$ **d.** $\dfrac{13}{31}$

2. Change each improper fraction to a mixed number. Reduce if possible.

   a. $\dfrac{20}{16}$ b. $\dfrac{34}{8}$ c. $\dfrac{48}{12}$ 0   d. $\dfrac{33}{5}$

3. Change each mixed number to an improper fraction. Reduce if possible.

   a. $7\dfrac{2}{5}$ b. $8\dfrac{6}{10}$ c. $9\dfrac{6}{7}$ d. $11\dfrac{3}{4}$

Finding a Common Denominator of Two Fractions

A **common denominator** is a number that is divisible by each of the original denominators.

The **least common denominator** is the smallest possible common denominator.

A common denominator can be obtained by multiplying the original denominators. Note that this product may not necessarily be the *least* common denominator.

Example 1: Determine a common denominator for $\dfrac{5}{6}$ and $\dfrac{7}{9}$.

Solution: $6 \cdot 9 = 54$, so 54 is a common denominator.

However, 18 is the least common denominator.

Equivalent Fractions

Equivalent fractions are fractions with the same numerical value.

To obtain an equivalent fraction, multiply or divide both numerator and denominator by the same nonzero number.

Note: Adding the same number to or subtracting the same number from the original numerator and denominator does *not* yield an equivalent fraction.

Example 1: Determine a fraction that is equivalent to $\dfrac{3}{5}$ and has denominator 20.

Solution: $\dfrac{3}{5} = \dfrac{3 \cdot 4}{5 \cdot 4} = \dfrac{12}{20}$ **Note:** $\dfrac{3 + 15}{5 + 15} = \dfrac{18}{20} = \dfrac{9}{10} \neq \dfrac{3}{5}$

Comparing Fractions: Determining Which Is Greater, Is $\dfrac{a}{b} < \dfrac{c}{d}$ or is $\dfrac{a}{b} > \dfrac{c}{d}$?

1. Obtain a common denominator by multiplying b and d.

2. Write $\dfrac{a}{b}$ and $\dfrac{c}{d}$ as equivalent fractions, each with denominator $b \cdot d$.

3. Compare numerators. The fraction with the greater (or lesser) numerator is the greater (or lesser) fraction.

Example 1: Determine whether $\dfrac{3}{5}$ is less than or greater than $\dfrac{7}{12}$.

Solution: A common denominator is $5 \cdot 12 = 60$.

$$\frac{3}{5} = \frac{3 \cdot 12}{5 \cdot 12} = \frac{36}{60}, \quad \frac{7}{12} = \frac{7 \cdot 5}{12 \cdot 5} = \frac{35}{60}$$

Because $36 > 35, \dfrac{3}{5} > \dfrac{7}{12}$.

Exercises

Compare the two fractions, and indicate which one is larger.

1. $\dfrac{4}{7}$ and $\dfrac{5}{8}$ **2.** $\dfrac{11}{13}$ and $\dfrac{22}{39}$ **3.** $\dfrac{5}{12}$ and $\dfrac{7}{16}$ **4.** $\dfrac{3}{5}$ and $\dfrac{14}{20}$

Addition and Subtraction of Fractions with the Same Denominators

To add (or subtract) $\dfrac{a}{c}$ and $\dfrac{b}{c}$:

1. Add (or subtract) the numerators, a and b.

2. Place the sum (or difference) over the common denominator, c.

3. Reduce to lowest terms.

Example 1: Add $\dfrac{5}{16} + \dfrac{7}{16}$.

Solution: $\dfrac{5}{16} + \dfrac{7}{16} = \dfrac{5 + 7}{16} = \dfrac{12}{16} \qquad \dfrac{12}{16} = \dfrac{12 \div 4}{16 \div 4} = \dfrac{3}{4}$

Example 2: Subtract $\dfrac{19}{24} - \dfrac{7}{24}$.

Solution: $\dfrac{19}{24} - \dfrac{7}{24} = \dfrac{19 - 7}{24} = \dfrac{12}{24} \qquad \dfrac{12}{24} = \dfrac{12 \div 12}{24 \div 12} = \dfrac{1}{2}$

Addition and Subtraction of Fractions with Different Denominators

To add (or subtract) $\dfrac{a}{b}$ and $\dfrac{c}{d}$, where $b \neq d$:

1. Obtain a common denominator by multiplying b and d.

2. Write $\dfrac{a}{b}$ and $\dfrac{c}{d}$ as equivalent fractions with denominator $b \cdot d$.

3. Add (or subtract) the numerators of the equivalent fractions, and place the sum (or difference) over the common denominator, $b \cdot d$.

4. Reduce to lowest terms.

Example 1: Add $\dfrac{3}{5} + \dfrac{2}{7}$.

Solution: A common denominator is $5 \cdot 7 = 35$.

$$\frac{3}{5} = \frac{3 \cdot 7}{5 \cdot 7} = \frac{21}{35}, \frac{2}{7} = \frac{2 \cdot 5}{7 \cdot 5} = \frac{10}{35},$$

$$\frac{3}{5} + \frac{2}{7} = \frac{21}{35} + \frac{10}{35} = \frac{31}{35}$$

$\dfrac{31}{35}$ is already in lowest terms.

Example 2: Subtract $\dfrac{5}{12} - \dfrac{2}{9}$.

Solution: Using a common denominator, $12 \cdot 9 = 108$:

$$\frac{5}{12} = \frac{5 \cdot 9}{12 \cdot 9} = \frac{45}{108}$$

$$\frac{2}{9} = \frac{2 \cdot 12}{9 \cdot 12} = \frac{24}{108}$$

$$\frac{5}{12} - \frac{2}{9} = \frac{45}{108} - \frac{24}{108} = \frac{21}{108}$$

$$\frac{21}{108} = \frac{21 \div 3}{108 \div 3} = \frac{7}{36}$$

Using the least common denominator, 36:

$$\frac{5}{12} = \frac{5 \cdot 3}{12 \cdot 3} = \frac{15}{36}$$

$$\frac{2}{9} = \frac{2 \cdot 4}{9 \cdot 4} = \frac{8}{36}$$

$$\frac{5}{12} - \frac{2}{9} = \frac{15}{36} - \frac{8}{36} = \frac{7}{36}$$

Addition and Subtraction of Mixed Numbers

To add (or subtract) mixed numbers:

1. Add (or subtract) the whole-number parts.

2. Add (or subtract) the fractional parts. In subtraction, this may require borrowing.

3. Add the resulting whole-number and fractional parts to form the mixed-number sum (or difference).

Example 1: Add $3\dfrac{2}{3} + 5\dfrac{3}{4}$.

Solution: Whole-number sum: $3 + 5 = 8$

Fractional sum: $\dfrac{2}{3} + \dfrac{3}{4} = \dfrac{2 \cdot 4}{3 \cdot 4} + \dfrac{3 \cdot 3}{4 \cdot 3} = \dfrac{8}{12} + \dfrac{9}{12} = \dfrac{17}{12} = 1\dfrac{5}{12}$

Final result: $8 + 1\dfrac{5}{12} = 8 + 1 + \dfrac{5}{12} = 9\dfrac{5}{12}$

Example 2: Subtract $4\dfrac{1}{3} - 1\dfrac{7}{8}$.

Solution: Because $\dfrac{1}{3}$ is smaller than $\dfrac{7}{8}$, you must borrow as follows:

$$4\frac{1}{3} = 3 + 1 + \frac{1}{3} = 3 + 1\frac{1}{3} = 3\frac{4}{3}$$

The original subtraction now becomes $3\frac{4}{3} - 1\frac{7}{8}$.

Subtracting the whole-number parts: $3 - 1 = 2$.

Subtracting the fractional parts:

$$\frac{4}{3} - \frac{7}{8} = \frac{4 \cdot 8}{3 \cdot 8} - \frac{7 \cdot 3}{8 \cdot 3} = \frac{32}{24} - \frac{21}{24} = \frac{11}{24}$$

Final result is $2 + \frac{11}{24} = 2\frac{11}{24}$.

Alternatively, to add (or subtract) mixed numbers:

1. Convert each mixed number to an improper fraction.

2. Add (or subtract) the fractions.

3. Rewrite the result as a mixed number.

Example 3: Subtract $4\frac{1}{3} - 1\frac{7}{8}$.

Solution: $4\frac{1}{3} = \frac{13}{3}$ and $1\frac{7}{8} = \frac{15}{8}$

$$\frac{13}{3} - \frac{15}{8} = \frac{13 \cdot 8}{3 \cdot 8} - \frac{15 \cdot 3}{8 \cdot 3} = \frac{104}{24} - \frac{45}{24} = \frac{59}{24} = 2\frac{11}{24}$$

Exercises

Add or subtract as indicated.

1. $1\frac{3}{4} + 3\frac{1}{8}$ 2. $8\frac{2}{3} - 7\frac{1}{4}$ 3. $6\frac{5}{6} + 3\frac{11}{18}$ 4. $2\frac{1}{3} - \frac{4}{5}$

5. $4\frac{3}{5} + 2\frac{3}{8}$ 6. $12\frac{1}{3} + 8\frac{7}{10}$ 7. $14\frac{3}{4} - 5\frac{7}{8}$ 8. $6\frac{5}{8} + 9\frac{7}{12}$

Multiplying Fractions

To multiply fractions, $\dfrac{a}{b} \cdot \dfrac{c}{d}$:

1. Multiply the numerators, $a \cdot c$, to form the numerator of the product fraction.

2. Multiply the denominators, $b \cdot d$, to form the denominator of the product fraction.

3. Write the product fraction by placing $a \cdot c$ over $b \cdot d$.

4. Reduce to lowest terms.

Example 1: Multiply $\dfrac{3}{4} \cdot \dfrac{8}{15}$.

Solution: $\dfrac{3}{4} \cdot \dfrac{8}{15} = \dfrac{3 \cdot 8}{4 \cdot 15} = \dfrac{24}{60}, \dfrac{24}{60} = \dfrac{24 \div 12}{60 \div 12} = \dfrac{2}{5}$

Note: It is often simpler and more efficient to cancel any common factors of the numerators and denominators *before* multiplying.

Dividing by a Fraction

Dividing *by* a fraction, $\dfrac{c}{d}$, is equivalent to multiplying by its reciprocal, $\dfrac{d}{c}$.

To divide fraction $\dfrac{a}{b}$ by $\dfrac{c}{d}$, written $\dfrac{a}{b} \div \dfrac{c}{d}$:

1. Rewrite the division as an equivalent multiplication, $\dfrac{a}{b} \cdot \dfrac{d}{c}$.

2. Proceed by multiplying as described above.

Example 1: Divide $\dfrac{2}{3}$ by $\dfrac{1}{2}$.

Solution: $\dfrac{2}{3} \div \dfrac{1}{2} = \dfrac{2}{3} \cdot \dfrac{2}{1} = \dfrac{2 \cdot 2}{3 \cdot 1} = \dfrac{4}{3}$

Note: $\dfrac{4}{3}$ is already in lowest terms.

Exercises

Multiply or divide as indicated.

1. $\dfrac{11}{14} \cdot \dfrac{4}{5}$ **2.** $\dfrac{3}{7} \div \dfrac{3}{5}$ **3.** $\dfrac{8}{15} \cdot \dfrac{3}{4}$ **4.** $6 \div \dfrac{2}{5}$

Multiplying or Dividing Mixed Numbers

To multiply (or divide) mixed numbers:

1. Change the mixed numbers to improper fractions.

2. Multiply (or divide) the fractions as described earlier.

3. If the result is an improper fraction, change to a mixed number.

Example 1: Divide $8\dfrac{2}{5}$ by 3.

Solution: $8\dfrac{2}{5} = \dfrac{42}{5}, 3 = \dfrac{3}{1}$

$\dfrac{42}{5} \div \dfrac{3}{1} = \dfrac{42}{5} \cdot \dfrac{1}{3} = \dfrac{42}{15}$

$\dfrac{42}{15} = \dfrac{42 \div 3}{15 \div 3} = \dfrac{14}{5}$

$\dfrac{14}{5} = 2\dfrac{4}{5}$

Exercises

Multiply or divide as indicated.

1. $3\dfrac{1}{2} \cdot 4\dfrac{3}{4}$ **2.** $10\dfrac{3}{5} \div 4\dfrac{2}{3}$ **3.** $5\dfrac{4}{5} \cdot 6$ **4.** $4\dfrac{3}{10} \div \dfrac{2}{5}$

Fractions and Percents

To convert a fraction or mixed number to a percent:

1. If present, convert the mixed number to an improper fraction.

2. Multiply the fraction by 100, and attach the % symbol.

3. If the result is an improper fraction, change it to a mixed number.

Example 1: Write $\dfrac{1}{12}$ as a percent.

Solution: $\dfrac{1}{12} \cdot 100\% = \dfrac{100}{12}\% = 8\dfrac{1}{3}\%$

Example 2: Write $2\dfrac{1}{4}$ as a percent.

Solution: $2\dfrac{1}{4} = \dfrac{9}{4} \cdot 100\% = 225\%$

Exercises

Convert each fraction to a percent.

1. $\dfrac{3}{4}$ **2.** $2\dfrac{3}{5}$ **3.** $\dfrac{2}{3}$ **4.** $\dfrac{5}{9}$

To convert a percent to a fraction:

1. Remove the % symbol, and multiply the number by $\dfrac{1}{100}$.

Example 1: Write $8\dfrac{1}{3}\%$ as a fraction.

Solution: $8\dfrac{1}{3}\% = 8\dfrac{1}{3} \cdot \dfrac{1}{100} = \dfrac{25}{3} \cdot \dfrac{1}{100} = \dfrac{1}{12}$

Exercises

Convert each percent to a fraction.

1. 75%   **2.** 60%   **3.** $66\dfrac{2}{3}\%$ **4.** $55\dfrac{5}{9}\%$

Decimals

Reading and Writing Decimal Numbers

Decimal numbers are written numerically according to a place value system. The following table lists the place values of digits to the *left* of the decimal point.

hundred million	ten million	million,	hundred thousand	ten thousand	thousand,	hundred	ten	one	decimal point

The following table lists the place values of digits to the *right* of the decimal point.

decimal point	tenths	hundredths	thousandths	ten-thousandths	hundred-thousandths	millionths

To read or write a decimal number in words:

1. Use the first place value table to read the digits to the left of the decimal point, in groups of three from the decimal point.

2. Insert the word *and*.

3. Read the digits to the right of the decimal point as though they were not preceded by a decimal point, and then attach the place value of its rightmost digit.

Example 1: Read and write the number 37,568.0218 in words.

				3	7	5	6	8	.
hundred million	ten million	million,	hundred thousand	ten thousand	thousand,	hundred	ten	one	decimal point

and

.	0	2	1	8		
decimal point	tenths	hundredths	thousandths	ten-thousandths	hundred-thousandths	millionths

Therefore, 37,568.0218 is read thirty-seven thousand, five hundred sixty-eight and two hundred eighteen ten-thousandths.

Example 2: Write the number seven hundred eighty-two million, ninety-three thousand, five hundred ninety-four and two thousand four hundred three millionths numerically in standard form.

7	8	2,	0	9	3,	5	9	4	.
hundred million	ten million	million,	hundred thousand	ten thousand	thousand,	hundred	ten	one	*decimal point*

and

.	0	0	2	4	0	3
decimal point	tenths	hundredths	thousandths	ten-thousandths	hundred-thousandths	millionths

That is, 782,093,594.002403

Exercises

  1. Write the number 9467.00624 in words.

  2. Write the number 35,454,666.007 in words.

  3. Write the number numerically in standard form: four million, sixty-four and seventy-two ten-thousandths.

  4. Write the number numerically in standard form: seven and forty-three thousand fifty-two millionths.

Rounding a Number to a Specified Place Value

  1. Locate the digit with the specified place value (target digit).

  2. If the digit directly to its right is less than 5, keep the target digit. If the digit is 5 or greater, increase the target digit by 1.

  3. If the target digit is to the right of the decimal point, delete all digits to its right.

  4. If the target digit is to the left of the decimal point, replace any digits between the target digit and the decimal point with zeros as placeholders. Delete the decimal point and all digits that follow it.

Example 1: Round 35,178.2649 to the nearest hundredth.

The digit in the hundredths place is 6. The digit to its right is 4. Therefore, keep the 6 and delete the digits to its right. The rounded value is 35,178.26.

Example 2: Round 35,178.2649 to the nearest tenth.

The digit in the tenths place is 2. The digit to its right is 6. Therefore, increase the 2 to 3 and delete the digits to its right. The rounded value is 35,178.3.

Example 3: Round 35,178.2649 to the nearest ten thousand.

The digit in the ten thousands place is 3. The digit to its right is 5. Therefore, increase the 3 to 4 and insert four zeros to its right as placeholders. The rounded value is 40,000, where the decimal point is not written.

Exercises

1. Round 7456.975 to the nearest hundredth.

2. Round 55,568.2 to the nearest hundred.

3. Round 34.6378 to the nearest tenth.

Converting a Fraction to a Decimal

To convert a fraction to a decimal, divide the numerator by the denominator.

Example 1: Convert $\frac{4}{5}$ to a decimal.

Solution: On a calculator:

Key in ④ ÷ ⑤ =
to obtain 0.8.

Using long division:
$$\begin{array}{r} 0.8 \\ 5\overline{)4.0} \\ \underline{4.0} \\ 0 \end{array}$$

Example 2: Convert $\frac{1}{3}$ to a decimal.

Solution: On a calculator:

Key in ① ÷ ③ =
to obtain 0.3333333.

Using long division:
$$\begin{array}{r} 0.333 \\ 3\overline{)1.000} \\ \underline{-9} \\ 10 \\ \underline{-9} \\ 1 \end{array}$$

Because a calculator's display is limited to a specified number of digits, it will cut off the decimal's trailing right digits.

Because this long division process will continue indefinitely, the quotient is a repeating decimal, 0.33333 . . . and is instead denoted by $0.\overline{3}$. The bar is placed above the repeating digit or above a repeating sequence of digits.

Exercises

Convert the given fractions into decimals. Use a repeating bar, if necessary.

1. $\frac{3}{5}$ **2.** $\frac{2}{3}$ **3.** $\frac{7}{8}$ **4.** $\frac{1}{7}$ **5.** $\frac{4}{9}$

Converting a Terminating Decimal to a Fraction

1. Read the decimal.

2. The place value of the rightmost nonzero digit becomes the denominator of the fraction.

3. The original numeral, with the decimal point removed, becomes the numerator. Drop all leading zeros.

Example 1: Convert 0.025 to a fraction.

Solution: The rightmost digit, 5, is in the thousandths place.

So as a fraction, $0.025 = \dfrac{25}{1000}$, which reduces to $\dfrac{1}{40}$.

Example 2: Convert 0.0034 to a fraction.

Solution: The rightmost digit, 4, is in the ten-thousandths place.

So as a fraction, $0.0034 = \dfrac{34}{10,000}$, which reduces to $\dfrac{17}{5000}$.

Exercises

Convert the given decimals to fractions.

1. 0.4   **2.** 0.125   **3.** 0.64   **4.** 0.05

Converting a Decimal to a Percent

Multiply the decimal by 100 (that is, move the decimal point 2 places to the right, inserting placeholding zeros when necessary), and attach the % symbol.

Example 1: 0.78 written as a percent is 78%.

Example 2: 3 written as a percent is 300%.

Example 3: 0.045 written as a percent is 4.5%.

Exercises

Write the following decimals as equivalent percents.

1. 0.35   **2.** 0.076   **3.** 0.0089   **4.** 6.0

Converting a Percent to a Decimal

Divide the percent by 100 (that is, move the decimal point 2 places to the left, inserting place-holding zeros when necessary), and drop the % symbol.

Example 1: 5% written as a decimal is 0.05.

Example 2: 625% written as a decimal is 6.25.

Example 3: 0.0005% written as a decimal is 0.000005.

Exercises

Write the following percents as decimals.

1. 45%   **2.** 0.0987%   **3.** 3.45%   **4.** 2000%

Comparing Decimals

1. Write the decimals one below the next, lining up their respective decimal points.

2. Read the decimals from left to right, comparing corresponding place values. The decimal with the first and largest nonzero digit is the greatest number.

Example 1: Order from greatest to least: 0.097, 0.48, 0.0356.

Solution: Align by decimal point: 0.097

0.48

0.0356

Because 4 (in the second decimal) is the first nonzero digit, 0.48 is the greatest number. Next, because 9 is greater than 3, 0.097 is next greatest. Finally, 0.0356 is the least number.

Example 2: Order from greatest to least: 0.043, 0.0043, 0.43, 0.00043.

Solution: Align by decimal point: 0.043

0.0043

0.43

0.00043

Because 4 (in the third decimal) is the first nonzero digit, 0.43 is the greatest number. Similarly, 0.043 is next, followed by 0.0043. Finally, 0.00043 is the least number.

Exercises

Place each group of decimals in order, from greatest to least.

1. 0.058   0.0099   0.105   0.02999

2. 0.75   1.23   1.2323   0.9   0.999

3. 13.56   13.568   13.5068   13.56666

Adding and Subtracting Decimals

1. Write the decimals one below the next, lining up their respective decimal points. If the decimals have a different number of digits to the right of the decimal point, place trailing zeros to the right in the shorter decimals.

2. Place the decimal point in the answer, lined up with the decimal points in the problem.

3. Add or subtract the numbers as usual.

Example 1: Add: 23.5 + 37.098 + 432.17.

Solution: 23.500
          37.098
      + 432.170
      ‾‾‾‾‾‾‾‾‾‾
         492.768

Example 2: Subtract 72.082 from 103.07.

Solution:

$$
\begin{array}{r}
103.070 \\
-\ 72.082 \\
\hline
30.988
\end{array}
$$

Exercises

1. Calculate: $543.785 + 43.12 + 3200.0043$.

2. Calculate: $679.05 - 54.9973$.

Multiplying Decimals

1. Multiply the numbers as usual, ignoring the decimal points.

2. Sum the number of decimal places in each number to determine the number of decimal places in the product.

Example 1: Multiply 32.89 by 0.021.

Solution:

$$
\begin{array}{r}
32.89 \\
\times\ 0.021 \\
\hline
3289 \\
6578 \\
\hline
0.69069
\end{array}
$$

$32.89 \longrightarrow$ 2 decimal places
$\times\ 0.021 \longrightarrow$ 3 decimal places
$0.69069 \longrightarrow$ 5 decimal places

Example 2: Multiply 64.05 by 7.3.

Solution:

$$
\begin{array}{r}
64.05 \\
\times\ \ \ \ 7.3 \\
\hline
19215 \\
44835 \\
\hline
467.565
\end{array}
$$

$64.05 \longrightarrow$ 2 decimal places
$\times\ \ \ 7.3 \longrightarrow$ 1 decimal places
$467.565 \longrightarrow$ 3 decimal places

Exercises

Multiply the following decimals.

1. 12.53×8.2 2. 115.3×0.003 3. 14.62×0.75

Dividing Decimals

1. Write the division in long division format.

2. Move the decimal point the same number of places to the right in both divisor and dividend so that the divisor becomes a whole number. Insert zeros, if necessary.

3. Place the decimal point in the quotient directly above the decimal point in the dividend, and divide as usual.

Example 1: Divide 92.4 by 0.25.

(*dividend*) (*divisor*)

Solution: $0.25\overline{)92.4}$ becomes $25\overline{)9240}$:

$$
\begin{array}{r}
369.6 \\
25\overline{)9240.0} \\
-75 \\
\hline
174 \\
-150 \\
\hline
240 \\
-225 \\
\hline
150 \\
-150 \\
\hline
0
\end{array}
$$

So $92.4 \div 0.25 = 369.6$.

Example 2: $\underset{(\textit{dividend})}{0.00052} \div \underset{(\textit{divisor})}{0.004}$

Solution: $0.004\overline{)0.00052}$ becomes $4\overline{)0.52}$:

$$
\begin{array}{r}
0.13 \\
4\overline{)0.52} \\
-4 \\
\hline
12 \\
-12 \\
\hline
0
\end{array}
$$

So $0.00052 \div 0.004 = 0.13$.

Exercises

1. Divide 12.05 by 2.5.

2. Divide 18.9973 by 78.

3. Divide 14.05 by 0.0002.

4. Calculate $150 \div 0.03$.

5. Calculate $0.00442 \div 0.017$.

6. Calculate $69.115 \div 0.0023$.

Skills Checks

Skills Check 1

1. Subtract 187 from 406.

2. Multiply 68 by 79.

3. Write 16.0709 in words.

4. Write three thousand four hundred two and twenty-nine thousandths in standard form.

5. Round 567.0468 to the nearest hundredth.

6. Round 2.59945 to the nearest thousandth.

7. Add $48.2 + 36 + 2.97 + 0.743$.

8. Subtract 0.48 from 29.3.

9. Multiply 2.003 by 0.36.

10. Divide 28.71 by 0.3.

11. Change 0.12 to a percent.

12. Change 3 to a percent.

13. The sales tax rate in some states is 6.5%. Write this percent as a decimal.

Answers to all Skills Check exercises are included in the Selected Answers appendix.

14. Write 360% in decimal form.

15. Write $\dfrac{2}{15}$ as a decimal.

16. Write $\dfrac{3}{7}$ as a percent. Round the percent to the nearest tenth.

17. The membership in the Nautilus Club increased by 12.5%. Write this percent as a fraction.

18. Write the following numbers in order from smallest to largest:

3.027   3.27   3.0027   3.0207

19. Find the average of 43, 25, 37, and 58.

20. In a recent survey, 14 of the 27 people questioned preferred Coke to Pepsi. What percent of the people preferred Coke?

21. If the sales tax is 7%, what is the cost, including tax, of a new baseball cap that is priced at $8.10?

22. If you spend $63 a week for food and you earn $800 per month, what percent of your monthly income is spent on food?

23. The enrollment at a local college has increased 5.5%. Last year's enrollment was 9500 students. How many students are expected this year?

24. You decide to decrease the number of calories in your diet from 2400 to 1500. Determine the percent decrease.

25. The total area of the world's land masses is about 57,000,000 square miles. If the area of 1 acre is about 0.00156 square mile, determine the number of acres of land in the world.

26. You and a friend have been waiting for the price of a winter coat to come down sufficiently from the manufacturer's suggested retail price (MSRP) of $500 so you each can afford to buy one. The retailer always discounts the MSRP by 10%. Between Thanksgiving and New Year's, the price is further reduced by 40%, and in January, it is reduced again by 50%.

    a. Your friend remarks that it looks like you can get the coat for free in January. Is that true? Explain.

    b. What does the coat cost before Thanksgiving, in December, and in January?

    c. How would the final price differ if the discounts had been taken in the reverse order (50% off, 40% off, and 10% off)? How would the intermediate prices be affected?

27. A local education official proudly declares that although math scores had fallen by nearly 55% since 2004, by 2014 they had rebounded over 65%. This sounds like great news for your district, doesn't it? Determine how the 2014 math scores actually compare with the 2004 scores.

28. A telecommunications company noted that its January sales were up 10.8% over December's sales. Sales took a dip of 4.7% in February and then increased 12.4% in March. If December's sales were approximately $6 million, what was the sales figure for March?

29. Each year, Social Security payments are adjusted for cost of living. Your grandmother relies on Social Security for a significant part of her retirement income. Suppose her Social Security income was $1250 per month. If cost-of-living increases over the next 3-year period were 2.1%, 2.7% and 4.1%, what was the amount of your grandmother's monthly Social Security checks at the end of the three years?

Skills Check 2

1. Which number is greater, 1.0001 or 1.001?

2. Evaluate: $32.09 \cdot 0.0006$

3. Evaluate 27^0.

4. Evaluate $|-9|$.

5. Determine the length of the hypotenuse of a right triangle if each of the legs has a length of 6 meters. Use the formula $c = \sqrt{a^2 + a^2}$, where c represents the length of the hypotenuse and a represents the length of the leg.

6. Evaluate $|9|$.

7. Write 203,000,000 in scientific notation.

8. How many quarts are there in 8.3 liters?

9. Write 2.76×10^{-6} in standard form.

10. How many ounces are there in a box of crackers that weighs 283 grams?

11. Determine the area of the front of a U.S. $1 bill.

12. Determine the circumference of a quarter. Use the formula $C = \pi d$, where C represents the circumference and d represents the diameter of the circle.

13. Determine the area of a triangle with a base of 14 inches and a height of 18 inches. Use the formula $A = \dfrac{1}{2} bh$, where A represents area, b represents the base, and h represents the height.

14. Twenty-five percent of the students in your history class scored between 70 and 80 on the last exam. In a class of 32 students, how many students does this include?

15. You have decided to accept word processing jobs to earn some extra money. Your first job is to type a 420-page thesis. If you can type 12 pages per hour, how long will it take you to finish the job?

16. How many pieces of string 1.6 yards long can be cut from a length of string 9 yards long?

17. Reduce $\dfrac{16}{36}$ to lowest terms.

18. Change $\dfrac{16}{7}$ to a mixed number.

19. Change $4\dfrac{6}{7}$ to an improper fraction.

20. Add: $\dfrac{2}{9} + \dfrac{3}{5}$

21. Subtract: $4\dfrac{2}{5} - 2\dfrac{1}{10}$

22. Evaluate: $\dfrac{4}{21} \cdot \dfrac{14}{9}$

23. Calculate: $1\dfrac{3}{4} \div \dfrac{7}{12}$

24. You mix $2\dfrac{1}{3}$ cups of flour, $\dfrac{3}{4}$ cup of sugar, $\dfrac{3}{4}$ cup of mashed bananas, $\dfrac{1}{2}$ cup of walnuts, and $\dfrac{1}{2}$ cup of milk to make banana bread. How many cups of mixture do you have?

25. Your stock starts the day at $30\dfrac{1}{2}$ points and goes down $\dfrac{7}{8}$ of a point during the day. What is its value at the end of the day?

26. You just opened a container of orange juice that has 64 fluid ounces. You have three people in your household who drink one 8-ounce serving of orange juice per day. In how many days will you need a new container?

27. You want to serve quarter-pound hamburgers at your barbecue. There will be seven adults and three children at the party, and you estimate that each adult will eat two hamburgers and each child will eat one. How much hamburger meat should you buy?

28. Your friend tells you that her height is 1.67 meters and her weight is 58 kilograms. Convert her height and weight to feet and pounds, respectively.

29. Convert the following measurements to the indicated units:

    a. 10 miles = _____ feet

    b. 3 quarts = _____ ounces

    c. 5 pints = _____ ounces

    d. 6 gallons = _____ ounces

    e. 3 pounds = _____ ounces

30. The following table lists the minimum distance of each listed planet from Earth, in millions of miles. Convert each distance into scientific notation.

PLANET	DISTANCE (IN MILLIONS OF MILES)	DISTANCE IN MILES (IN SCIENTIFIC NOTATION)
Mercury	50	
Venus	25	
Mars	35	
Jupiter	368	
Saturn	745	
Uranus	1606	
Neptune	2674	

31. A recipe you obtained on the Internet indicates that you need to melt 650 grams of chocolate in $\frac{1}{5}$ liter of milk to prepare icing for a cake. How many pounds of chocolate and how many ounces of milk do you need?

32. $\left(3\dfrac{2}{3} + 4\dfrac{1}{2}\right) \div 2$

33. $18 \div 6 \cdot 3$

34. $18 \div (6 \cdot 3)$

35. $\left(3\dfrac{3}{4} - 2\dfrac{1}{3}\right)^2 + 7\dfrac{1}{2}$

36. $\left(5\dfrac{1}{2}\right)^2 - 8\dfrac{1}{3} \cdot 2$

37. $6^2 \div 3 \cdot 2 + 6 \div (-3 \cdot 2)^2$

38. $6^2 \div 3 \cdot (-2) + 6 \div 3 \cdot 2^2$

39. $-4 \cdot 9 + (-9)(-8)$

40. $-6 \div 3 - 3 + 5^0 - 14 \cdot (-2)$

Algebraic Extensions

Properties of Exponents

The basic properties of exponents are summarized as follows:

> If a and b are both positive real numbers and n and m are any real numbers, then
>
> **1.** $a^n a^m = a^{n+m}$ **2.** $\dfrac{a^n}{a^m} = a^{n-m}$ **3.** $(a^n)^m = a^{nm}$
>
> **4.** $a^{-n} = \dfrac{1}{a^n}$ **5.** $(ab)^n = a^n b^n$ **6.** $a^0 = 1$

Note: The properties of exponents are covered in detail in Activity 4.2.

Property 1: $a^n a^m = a^{n+m}$ If you are multiplying two powers of the same base, add the exponents.

Example 1: $x^4 \cdot x^7 = x^{4+7} = x^{11}$

Note: The exponents were added, and the base did not change.

Property 2: $\dfrac{a^n}{a^m} = a^{n-m}$ If you are dividing two powers of the same base, subtract the exponents.

Example 2: $\dfrac{6^6}{6^4} = 6^{6-4} = 6^2 = 36$

Note: The exponents were subtracted, and the base did not change.

Property 3: $(a^n)^m = a^{nm}$ If a power is raised to a power, multiply the exponents.

Example 3: $(y^3)^4 = y^{12}$

Note: The exponents were multiplied. The base did not change.

Property 4: $a^{-n} = \dfrac{1}{a^n}$ Sometimes presented as a definition, Property 4 states that any base raised to a negative power is equivalent to the reciprocal of the base raised to the positive power. Note that the negative exponent does not have any effect on the sign of the base. This property could also be viewed as a result of the second property of exponents as follows:

Consider $\dfrac{x^3}{x^5}$. Using Property 2, $x^{3-5} = x^{-2}$. If you view this expression algebraically, you have three factors of x in the numerator and five in the denominator. If you divide out the three common factors, you are left with $\dfrac{1}{x^2}$. Therefore, if Property 2 is true, then $x^{-2} = \dfrac{1}{x^2}$.

Example 4: Write each of the following without negative exponents.

$$\textbf{a. } 3^{-2} \qquad\qquad\qquad \textbf{b. } \frac{2}{x^{-3}}$$

Solution: $\quad \textbf{a. } 3^{-2} = \frac{1}{3^2} = \frac{1}{9} \qquad\qquad \textbf{b. } \frac{2}{x^{-3}} = \frac{2}{\frac{1}{x^3}} = 2 \div \frac{1}{x^3} = 2 \cdot x^3 = 2x^3$

Property 5: $(ab)^n = a^n b^n$ If a product is raised to a power, each factor is raised to that power.

Example 5: $(2x^2 y^3)^3 = 2^3 \cdot (x^2)^3 \cdot (y^3)^3 = 8x^6 y^9$

Note: Because the base contained three factors, each of those was raised to the third power. The common mistake in an expansion such as this is not to raise the coefficient to the power.

Property 6: $a^0 = 1, a \neq 0$. Often presented as a definition, Property 6 states that any nonzero base raised to the zero power is 1. This property or definition is a result of Property 2 of exponents as follows:

Consider $\dfrac{x^5}{x^5}$. Using Property 2, $x^{5-5} = x^0$. However, you know that any fraction in which the numerator and the denominator are equal is equivalent to 1. Therefore, $x^0 = 1$.

Example 6: $\left(\dfrac{2x^3}{3yz^5}\right)^0 = 1$

Given a nonzero base, if the exponent is zero, the value is 1.

> A factor can be moved from a numerator to a denominator or from a denominator to a numerator by changing the *sign of the exponent*.

Example 7: Simplify and express your result with positive exponents only.

$$\frac{x^3 y^{-4}}{2x^{-3} y^{-2} z}$$

Solution: Move the x^{-3} and the y^{-2} factors from the denominator to the numerator, making sure to change the sign of the exponents. This results in

$$\frac{x^3 y^{-4} x^3 y^2}{2z}.$$

Simplify the numerator using Property 1 and then use Property 4:

$$\frac{x^{3+3} y^{-4+2}}{2z} = \frac{x^6 y^{-2}}{2z} = \frac{x^6}{2y^2 z}$$

Exercises

Simplify and express your results with positive exponents only. Assume that all variables represent only nonzero values.

1. 5^{-3}

2. $\dfrac{1}{x^{-5}}$

3. $\dfrac{3x}{y^{-2}}$

4. $\dfrac{10x^2 y^5}{2x^{-3}}$

5. $\dfrac{5^{-1}z}{x^{-1}z^{-2}}$

6. $5x^0$

7. $(a + b)^0$

8. $-3(x^0 - 4y^0)$

9. $x^6 \cdot x^{-3}$ **10.** $\dfrac{4^{-2}}{4^{-3}}$ **11.** $(4x^2y^3) \cdot (3x^{-3}y^{-2})$

12. $\dfrac{24x^{-2}y^3}{6x^3y^{-1}}$ **13.** $\dfrac{(14x^{-2}y^{-3}) \cdot (5x^3y^{-2})}{6x^2y^{-3}z^{-3}}$

14. $\left(\dfrac{2x^{-2}y^{-3}}{z^2}\right) \cdot \left(\dfrac{x^5y^3}{z^{-3}}\right)$ **15.** $\dfrac{(16x^4y^{-3}z^{-2})(3x^{-3}y^4)}{15x^{-3}y^{-3}z^2}$

Addition Method for Solving a System of Two Linear Equations

The basic strategy for the addition method is to reduce a system of two linear equations to a single linear equation by eliminating a variable.

For example, consider the x-coefficients of the linear system

$$2x + 3y = 1$$
$$4x - y = 9.$$

The coefficients are 2 and 4. The LCM of 2 and 4 is 4. Use the multiplication principle to multiply each side of the first equation by -2. The resulting system is as follows:

$$\begin{matrix} -2(2x + 3y = 1) \\ 4x - y = 9 \end{matrix} \quad \text{or, equivalently,} \quad \begin{matrix} -4x - 6y = -2 \\ 4x - y = 9 \end{matrix}$$

Multiplying by -2 produces x-coefficients that are additive inverses, or opposites. Now add the two equations together to eliminate the variable x.

$$-7y = 7$$

Solving for y, $y = -1$ is the y-value of our solution. To find the value of x, substitute -1 for y and solve for x in any equation that involves x and y. For an alternative to determining the value of x, consider the original system and the coefficients of y, 3 and -1 in the original system. The LCM is 3. Because the signs are already opposites, multiply the second equation by 3.

$$\begin{matrix} 2x + 3y = 1 \\ 3(4x - y = 9) \end{matrix} \quad \text{or, equivalently,} \quad \begin{matrix} 2x + 3y = 1 \\ 12x - 3y = 27 \end{matrix}$$

Adding the two equations will eliminate the y-variable.

$$14x = 28$$

Solving for x, $x = 2$. Therefore, the solution is $(2, -1)$. This should be checked to make certain that it satisfies both equations.

Depending on the coefficients of the system, you may need to change both equations when using the addition method. For example, the coefficients of x in the linear system

$$5x - 2y = 11$$
$$3x + 5y = -12$$

are 5 and 3. The LCM is 15. Multiply the first equation by 3 and the second by -5 as follows:

$$3(5x - 2y = 11)$$
$$-5(3x + 5y = -12)$$

or, equivalently,

$$15x - 6y = 33$$
$$-15x - 25y = 60$$

Add the two equations to eliminate the x terms from the system.

$$-31y = 93$$

Solving for y, $y = -3$. Substituting this value for y in the first equation of the original system yields

$$5x - 2(-3) = 11$$
$$5x + 6 = 11$$
$$5x = 5$$
$$x = 1$$

Therefore, $(1, -3)$ is the solution of the system.

Exercises

Solve the following systems using the addition method. If the system has no solution or both equations represent the same line, state this as your answer.

1. $x - y = 3$
$x + y = -7$

2. $x + 4y = 10$
$x + 2y = 4$

3. $-5x - y = 4$
$-5x + 2y = 7$

4. $4x + y = 7$
$2x + 3y = 6$

5. $3x - y = 1$
$6x - 2y = 5$

6. $4x - 2y = 0$
$3x + 3y = 5$

7. $x - y = 9$
$-4x - 4y = -36$

8. $-2x + y = 6$
$4x + y = 1$

9. $\dfrac{3}{2}x + \dfrac{2}{5}y = \dfrac{9}{10}$

$\dfrac{1}{2}x + \dfrac{6}{5}y = \dfrac{3}{10}$

10. $0.3x - 0.8y = 1.6$

$0.1x + 0.4y = 1.2$

Factoring Trinomials with Leading Coefficient $\neq$ 1

With patience, it is possible to factor many trinomials by trial and error, using the FOIL method in reverse.

Factoring Trinomials by Trial and Error

1. Factor out the greatest common factor.

2. Try combinations of factors for the first and last terms in the two binomials.

3. Check the outer and inner products to match the middle term of the original trinomial.

4. If the check fails, repeat Steps 2 and 3.

Example 1: Factor $6x^2 - 7x - 3$.

Solution:

Step 1. There is no common factor, so go to Step 2.

Step 2. You could factor the first term as $6x(x)$ or as $2x(3x)$. The last term has factors of 3 and 1, disregarding signs. Suppose you try $(2x + 1)(3x - 3)$.

Step 3. The outer product is $-6x$. The inner product is $3x$. The sum is $-3x$. The check fails.

Step 4. Suppose you try $(2x - 3)(3x + 1)$. The outer product is $2x$. The inner product is $-9x$. The sum is $-7x$. It checks. Therefore, $6x^2 - 7x - 3 = (2x - 3)(3x + 1)$.

Exercises

Factor each of the following completely.

1. $x^2 + 7x + 12$ **2.** $6x^2 - 13x + 6$ **3.** $3x^2 + 7x - 6$

4. $6x^2 + 21x + 18$ **5.** $9x^2 - 6x + 1$ **6.** $2x^2 + 6x - 20$

7. $15x^2 + 2x - 1$ **8.** $4x^3 + 10x^2 + 4x$

Solving Equations by Factoring

Many quadratic and higher-order polynomial equations can be solved by factoring using the zero-product property.

Solving an Equation by Factoring

1. Use the addition principle to move all terms to one side so that the other side of the equation is zero.

2. Simplify and factor the nonzero side.

3. Use the zero-product property to set each factor equal to zero, and then solve the resulting equations.

4. Check your solutions in the original equation.

Example 1: Solve the equation $3x^2 - 2 = -x$.

Solution: Adding x to both sides, you obtain $3x^2 + x - 2 = 0$. Because there are no like terms, factor the trinomial.

$$(3x - 2)(x + 1) = 0$$

Use the zero-product property:

$$
\begin{aligned}
3x - 2 &= 0 && \text{or} && x + 1 = 0 \\
3x &= 2 && \text{or} && x = -1 \\
x &= \frac{2}{3}
\end{aligned}
$$

The two solutions are $x = \dfrac{2}{3}$ and $x = -1$. The check is left to the reader.

Example 2: Solve the equation $3x^3 - 8x^2 = 3x$.

Solution: Subtracting $3x$ from both sides, you obtain $3x^3 - 8x^2 - 3x = 0$. Because there are no like terms, factor the trinomial.

$$x(3x^2 - 8x - 3) = 0$$
$$x(3x + 1)(x - 3) = 0$$

Use the zero-product property:

$$
\begin{aligned}
x &= 0 && \text{or} && 3x + 1 = 0 && \text{or} && x - 3 = 0 \\
& && && 3x = -1 && \text{or} && x = 3 \\
& && && x = -\frac{1}{3}
\end{aligned}
$$

The three solutions are $x = 0$, $x = -\dfrac{1}{3}$, and $x = 3$. The check is left to the reader.

Exercises

Solve each of the following equations by factoring if possible.

1. $x^2 - x - 63 = 0$

2. $3x^2 - 9x - 30 = 0$

3. $-7x + 6x^2 = 10$

4. $3y^2 = 2 - y$

5. $-28x^2 + 15x - 2 = 0$

6. $4x^2 - 25 = 0$

7. $(x + 4)^2 - 16 = 0$

8. $(x + 1)^2 - 3x = 7$

9. $2(x + 2)(x - 2) = (x - 2)(x + 3) - 2$

10. $18x^3 = 15x^2 + 12x$

Getting Started with the TI-84 Plus Family of Calculators

The calculator screens in this appendix are from the TI-84 Plus C calculator. The procedures given in this appendix apply to the TI-84 Plus with the latest operating system as well.

ON-OFF

To turn on the TI-84 Plus C, press the (ON) key. To turn off the TI-84 Plus C, press (2nd) and (ON) in sequence.

In general, to access any of the white commands, simply press the black or gray key. To access the blue commands, press (2nd) and then the black or gray key below the desired command. Similarly, to access any of the green commands or symbols, press (ALPHA) followed by the appropriate black or gray key.

MODE

The (MODE) key controls many calculator settings. The activated settings are highlighted. For most of your work in this course, the settings in the left-hand column should be highlighted.

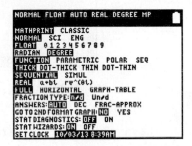

To change a setting, move the cursor to the desired setting and press (ENTER).

The Home Screen

The home screen is used for calculations.

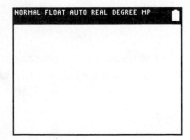

Note: The status at the top gives selected mode settings as well as the battery level.

You may return to the home screen at any time using the QUIT command. This command is accessed by pressing (2nd) (MODE). All calculations in the home screen are subject to the order of operations.

Enter all expressions as you would write them. Always observe the order of operations. Once you have typed the expression, press (ENTER) to obtain the simplified result. Before you hit (ENTER), you may edit your expression by using the arrow keys, the delete command (DEL), and the insert command (2nd) (DEL).

Three keys of special note are the reciprocal key (X⁻¹), the caret key (^) and the negative key ((−)).

The reciprocal command (X⁻¹) reciprocates the number in the home screen.

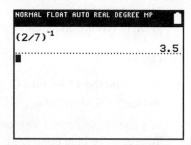

If a decimal result represents a rational number, it can be displayed as a fraction in lowest terms as follows. Press (ALPHA) (Y=) (4) (ENTER).

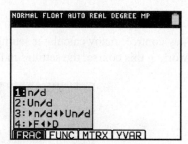
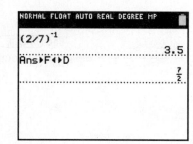

Pressing (ALPHA) (Y=) accesses the F1 menu which has functions used for working with fractions.

The carat key (^) is used to raise numbers to powers.

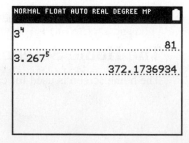

The square of a number can be evaluated two ways. Using the ^ key as above or using the (x²) key. To use the (x²) key first enter the value to be squared then press the (x²) key and (ENTER).

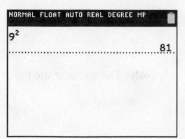

The key colors vary from calculator to calculator. The negative key is different from the minus key. To enter a negative number, use the $\boxed{(-)}$ key on the bottom row, not the blue $\boxed{-}$ key on the right side of the calculator.

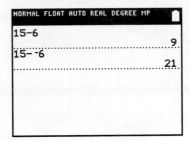

A table of keys and their functions follows.

KEY	FUNCTION DESCRIPTION
ON	Turns calculator on (or off)
Clear	Clears text screen
ENTER	Executes a command
(−)	Calculates the additive inverse
MODE	Displays current operating settings
DEL	Deletes the character at the cursor
^	Symbol used for exponentiation
ANS	Storage location of the last calculation
ENTRY	Retrieves a previously Executed expression

ANS and ENTRY

The last two commands in the above table can be real time-savers. The result of your last calculation is always stored in a memory location known as ANS. It is accessed by pressing $\boxed{2\text{nd}}$ $\boxed{(-)}$, or it is automatically accessed by pressing any operation button.

Suppose you want to evaluate $12.5\sqrt{1 + 0.5 \cdot (0.55)^2}$. It could be evaluated in one expression and checked with a series of calculations using ANS.

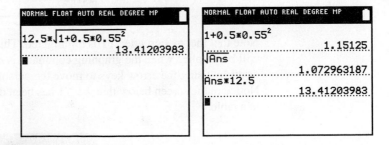

After you have keyed in an expression and pressed $\boxed{\text{ENTER}}$, you cannot move the cursor back up to edit or recalculate this expression. This is where the ENTRY $\boxed{2\text{nd}}$ $\boxed{\text{ENTER}}$ command is used. The ENTRY command retrieves the previous expression and places the cursor at the end of the expression. You can use the left and right arrow keys to move the cursor to any location in the expression that you want to modify.

Suppose you want to evaluate the compound interest expression $P\left(1 + \dfrac{r}{n}\right)^{nt}$, where P is the principal, r is the interest rate, n is the number of compounding periods annually, and t is the number of years when $P = \$1000$, $r = 6.5\%$, $n = 1$, and $t = 2, 5$, and 15 years. Using the ENTRY command, this expression would be entered once and edited twice.

```
NORMAL FLOAT AUTO REAL DEGREE MP

1000(1+.065)²
                        1134.225
1000(1+.065)⁵
                     1370.086663
1000(1+.065)¹⁵
                     2571.841007

```

Note that many "last expressions" are stored in the ENTRY memory location. You can repeat the ENTRY command as many times as you want to retrieve a previously entered expression.

An alternative to ANS and ENTRY is simply to use the up arrow button to select previous answers or entries and then press (ENTER) to insert the selected expression at the cursor.

If you repeat the proceeding example using the up arrow button to retrieve the previous entries, the resulting screen will look exactly the same.

Functions and Graphing with the TI-84 Plus C

Y= menu

Functions of the form $y = f(x)$ can be entered in the TI-84 Plus C using the Y= menu. To access the Y= menu, press the (Y=) key. Type the expression $f(x)$ after Y1 using the (X,T,θ,n) key for the variable x, and press (ENTER).

For example, enter the function $f(x) = 3x^5 - 4x + 1$. Remember to press the right arrow key after you input the exponents.

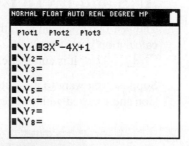

Note that the $=$ sign after Y1 is highlighted. This indicates that the function Y1 is active and will be graphed when the graphing command is executed. The highlighting may be turned on or off by using the arrow keys to move the cursor to the $=$ symbol and then pressing (ENTER). Notice in the screen below that the Y1 has been deactivated and will not be graphed or appear in a table.

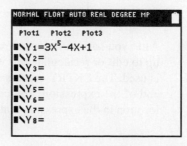

Once the function has been entered in the Y= menu, function values may be evaluated in the Home screen.

For example, given $f(x) = 3x^5 - 4x + 1$, evaluate $f(4)$. With the cursor on the home screen press the (ALPHA) (TRACE) buttons to access the F4 menu. Then press (1) to select Y1 and place it at the cursor. Then enter (( ) (4) ( )) (ENTER) to display the result.

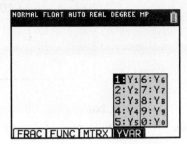
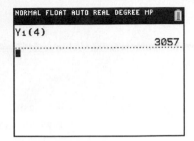

Tables of Values

If you are interested in viewing several function values for the same function, you may want to construct a table.

Before constructing the table, make sure the function appears in the "Y−" menu with its "−" highlighted. You may also want to deactivate or clear any functions that you do not need to see in your table. Next, you need to check the settings in the Table Setup menu. To do this, use the TBLSET command (2nd) (WINDOW).

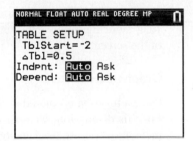

As shown in the screen above, the default setting for the table highlights the Auto options for both the independent (x) and the dependent (y) variables. Choosing this option will display ordered pairs of the function with equally spaced x-values. TblStart is the first x-value to be displayed and here is assigned the value -2. ΔTbl represents the equal spacing between consecutive x-values and here is assigned the value 0.5. The TABLE command (2nd) (GRAPH) brings up the table displayed in the screen below.

Use the up and down arrows to view additional ordered pairs of the function.

If the input values of interest are not evenly spaced, you may want to choose the Ask mode for the independent variable from the Table Setup menu.

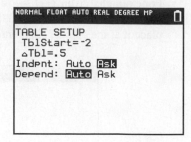

The resulting table is blank, but you can fill it by choosing any values for x that you like and pressing (ENTER) after each value.

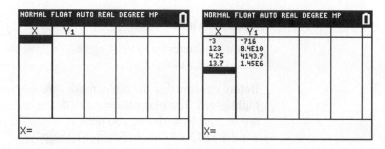

Note that the number of digits shown in the output is limited by the table width. But if you want more digits, move the cursor to the desired output, more digits then appear at the bottom of the screen.

Graphing a Function

Once a function is entered in the "Y=" menu and activated, it can be displayed and analyzed. For this discussion, we will use the function $f(x) = -x^2 + 10x + 12$. Enter this as Y1, making sure to use the negation key ((−)) and not the subtraction key ((−)).

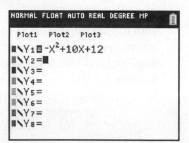

The Viewing Window

The viewing window is the portion of the rectangular coordinate system that is displayed when you graph a function.

Xmin defines the left edge of the window.

Xmax defines the right edge of the window.

Xscl defines the distance between horizontal tick marks.

Ymin defines the bottom edge of the window.

Ymax defines the top edge of the window.

Yscl defines the distance between vertical tick marks.

In the standard viewing window, Xmin $= -10$, Xmax $= 10$, Xscl $= 1$, Ymin $= -10$, Ymax $= 10$, and Yscl $= 1$.

To select the Standard Viewing Window, press $\boxed{\text{ZOOM}}$ $\boxed{6}$.

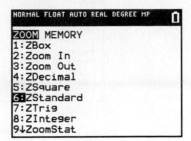

You will view the following.

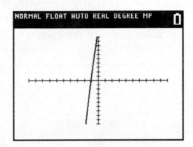

Is this an accurate and/or complete picture of your function, or is the window giving you a misleading impression? You may want to use your table function to view the output values from -10 to 10.

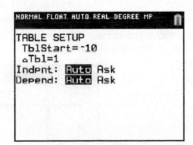

The table indicates that the minimum output value on the interval from $x = -10$ to $x = 10$ is -188, occurring at $x = -10$, and that the maximum output value is 37, occurring at $x = 5$. Press $\boxed{\text{WINDOW}}$ and reset the settings to approximately the following.

$$\text{Xmin} = -10, \text{Xmax} = 10, \text{Xscl} = 1,$$

$$\text{Ymin} = -190, \text{Ymax} = 40, \text{Yscl} = 10$$

Press ⌈ GRAPH ⌋ to view the graph with these new settings.

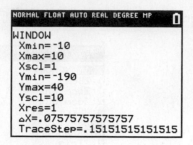 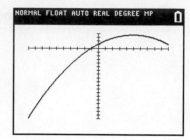

The new graph gives us a more complete picture of the behavior of the function on the interval $[-10, 10]$.

The coordinates of specific points on the curve can be viewed by activating the trace feature. While in the graph window, press ⌈ TRACE ⌋. The function equation will be displayed at the top of the screen, a flashing cursor will appear on the curve at the middle of the screen, and the coordinates of the cursor will be displayed at the bottom of the screen.

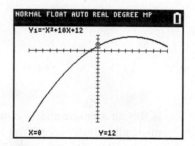

The left arrow key will move the cursor toward smaller input values. The right arrow key will move the cursor toward larger input values. If the cursor reaches the edge of the window and you continue to move the cursor, the window will adjust automatically.

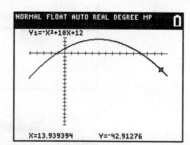

Zoom Menu

The Zoom menu offers several options for changing the window very quickly.

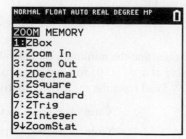 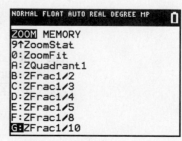

The features of many of the commands are summarized in the following table.

ZOOM COMMAND	DESCRIPTION
1:ZBox	Draws a box to define the viewing window
2:Zoom In	Magnifies the graph near the cursor
3:Zoom Out	Increases the viewing window around the cursor
4:ZDecimal	Sets a window so that Xscl and Yscl are 0.1
5:ZSquare	Sets equal-size pixels on the *x*- and *y*-axes
6:ZStandard	Sets the window to standard settings
7:ZTrig	Sets built-in trig window variables
8:ZInteger	Sets integer values on the *x*- and *y*-axes
9:Zoom Stat	Sets window based on the current values in the stat lists
0:ZoomFit	Replots graph to include the max and min output values for the current Xmin and Xmax

Solving Equations Graphically Using the TI-84 Plus C

The Intersection Method

This method is based on the fact that solutions to the equation $f(x) = g(x)$ are input values of x that produce the same output for the functions f and g. Graphically, these are the x-coordinates of the intersection points of $y = f(x)$ and $y = g(x)$.

The following procedure illustrates how to use the intersection method to solve $x^3 + 3 = 3x$ graphically.

Step 1. Enter the left-hand side of the equation as Y1 and the right-hand side as Y2 in the "Y=" editor. Select the standard viewing window.

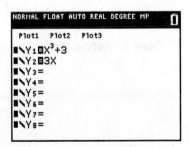

Step 2. Examine the graphs to determine the number of intersection points.

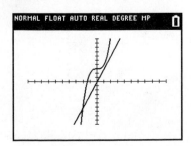

You may need a couple of windows to be certain of the number of intersection points.

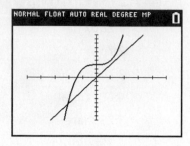

Step 3. Access the Calculate menu by pressing <u>2nd</u> <u>TRACE</u>, then choose option 5: **intersect**.

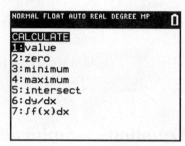
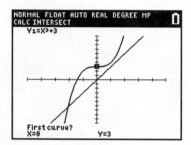

Step 4. Move the cursor close to an intersection point for the first curve and press <u>ENTER</u>.

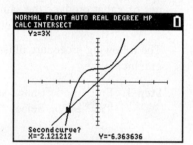

Step 5. Repeat Step 4 for the second curve.

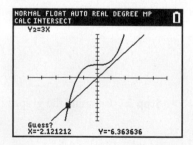

Step 6. To use the cursor's current location as your guess, press (ENTER) in response to the question on the screen that asks Guess? If you want to move to a better guess, do so before you press (ENTER). You can also enter a Guess using the calculator keypad and pressing (ENTER).

The coordinates of the intersection point appear below the word *Intersection*.

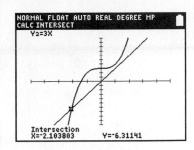

The x-coordinate is a solution of the equation.
If there are other intersection points, repeat the process as necessary.

Using the TI-84 Plus C to Determine the Linear Regression Equation for a Set of Paired Data Values

Example 1:

INPUT	OUTPUT
2	2
3	5
4	3
5	7
6	9

Enter the data into the calculator as follows:

1. Press the (STAT) button, and choose edit.

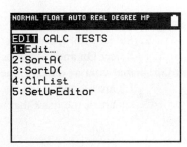

2. The TI-84 Plus C has six built-in lists: L1, L2, . . . , L6. If there is data in L1, clear the list as follows:

a. Use the arrows to place the cursor on L1 at the top of the list. Press (CLEAR) followed by (ENTER) followed by the down arrow.

b. Follow the same procedure to clear L2, if necessary.

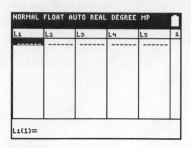

c. Enter the input values into L1 and the corresponding output values into L2.

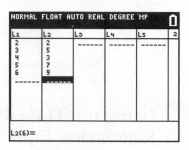

To see a scatterplot of the data, proceed as follows:

1. STAT PLOT is the second function of the (Y=) button. You must press the (2nd) button before pressing the (Y=) button to access the Stat Plot menu.

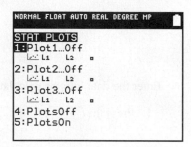

2. Select Plot 1, and make sure the other Plots are OFF. The screen below will appear. Select On and then choose the scatterplot option (first icon) on the Type line. Confirm that your *x*- and *y*-values are stored, respectively, in L1 and L2. The symbols L1 and L2 are second functions of the (1) and (2) keys, respectively. Finally, select the small square as the mark that will be used to plot each point.

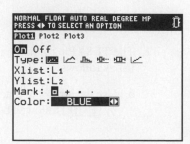

3. Press $\boxed{Y=}$ and clear or turn off any functions entered.

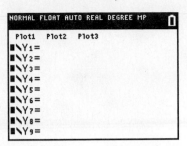

4. To display the scatterplot, have the calculator determine an appropriate window by pressing $\boxed{\text{ZOOM}}$ and then $\boxed{9}$ (ZoomStat).

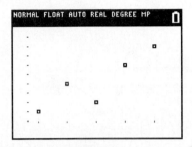

The following instructions will calculate the linear regression equation and store it in Y1.

1. Press $\boxed{\text{STAT}}$ and right-arrow to CALC.

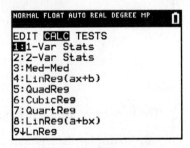

2. Choose 4 LinReg(ax+b). To tell the calculator where the data is, press $\boxed{\text{2nd}}$ and $\boxed{1}$ (for L1), then $\boxed{,}$, and then $\boxed{\text{2nd}}$ and $\boxed{2}$ (for L2) because the Xlist and the Ylist were stored in L1 and L2, respectively. The display looks like this.

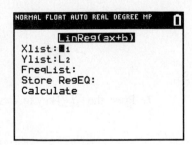

If you press $\boxed{\text{ENTER}}$ until Calculate is highlighted, the calculator will calculate the linear regression equation from the data in L1 and L2. However, the TI-84 Plus C will automatically paste the equation into Y1 (or Y2, Y3, . . .) as follows:

3. Arrow down to Store RegEQ: Press the (ALPHA) (TRACE) buttons to access the F4 menu.

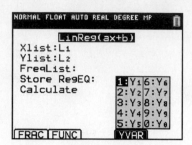

4. Then press (1) to select Y1 (or (2) for Y2, etc.) and place it at the cursor.

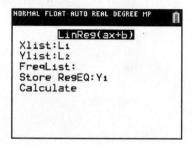

5. Move the cursor to Calculate and press (ENTER).

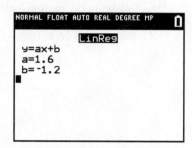

The linear regression equation for this data is $y = 1.6x - 1.2$.

6. To display the regression line, press (GRAPH).

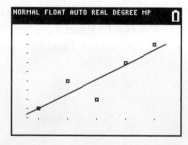

7. Press the (Y=) key to view the equation.

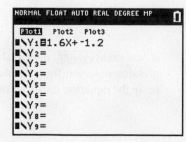

Selected Answers

Chapter 1

Activity 1.2 Exercises: 1. a. Nine justices; **3. a.** 2, **b.** 2, **c.** 5; **4. a.** 2, **b.** 5; **5. b.** Each number is generated by multiplying the preceding number by 2 and then adding 1.

6.

COLUMN 2	COLUMN 3	COLUMN 4
5	3	8
8	5	13
13	8	21
21	13	34

Each number is generated by adding the two numbers that precede it.

Activity 1.3. Exercises: 1. a. 175, **c.** 55; **2. a.** 1960, **c.** 60; **3. b.** $83 \cdot 5 = 415$; **4. a.** $15 + 12 = 27$, **c.** $26 + 14 - 3 = 40 - 3 = 37$; **6. b.** $\frac{20}{4} = 5$, **d.** $64 \div 4 \cdot 2 = 16 \cdot 2 = 32$, **f.** 156, **h.** $100 - (0) \div 3 = 100 - 0 = 100$; **7. b.** 32, **c.** 17; **8. b.** 49, **d.** 5, **f.** 25, **g.** 36; **9. a.** 3, **c.** $4\sqrt{100} = 40$; **10. a.** 2.55, **c.** 7.35, **d.** 18.52; **11. h.** 5.551405×10^{14}; **12. b.** 45,320,000; **13.** 5.859×10^{12} miles; **14.** $7484; **15.** 2.94×10^9 times

What Have I Learned? 1. b. Each of the 14 students kicks the ball to the 13 other students exactly once for a total of $14 \cdot 13 = 182$ kicks. **3. d.** $3 \cdot 10 + 4 \cdot 5 - 46 = 30 + 20 - 46 = \$4$; **6. b.** 36

How Can I Practice? 1. b. Each number is generated by multiplying the preceding number by 2. **2. b.** 13 **d.** 4 **f.** 146, **h.** 3, **j.** 5, **l.** 98, **r.** 8; **3. b.** 144, **d.** 64; **6.** 2×10^{25} molecules; **7. b.** associative property of multiplication; **8. b.** false, left side = 11, right side = 19

Activity 1.4. Exercises: 1. 434 classmates; **2.** $1250 each per semester; **4.** $7\frac{5}{12}$ hours; **6.** $4\frac{19}{24}$ cups; **8. b.** $7\frac{5}{6}$, **d.** $5\frac{8}{35}$, **f.** $1\frac{13}{50}$; **10. b.** 273,000,000 internet users

Activity 1.5. Exercises: 2. My GPA is 2.589 if I earn an F in economics. **4. b.** 60 points. **5.** The GPA is 2.978.

What Have I Learned? 2. b. Each part is $\frac{1}{10}$ unit because $\frac{1}{10} + \frac{1}{10} = \frac{1}{5}$.

How Can I Practice? 2. $4\frac{9}{10}$ in. **4.** $1\frac{3}{4}$ ft. of the floor will show on each side. **6. b.** $1\frac{1}{8}$, **d.** $1\frac{19}{30}$, **f.** $\frac{1}{2}$, **h.** $\frac{5}{12}$, **j.** $\frac{11}{20}$, **l.** $\frac{7}{12}$, **n.** $1\frac{1}{2}$, **p.** 16, **r.** $8\frac{1}{18}$, **t.** $1\frac{23}{26}$, **v.** $5\frac{1}{4}$; **8. a.** 20 hr., **b.** 15.4 hr.

Activity 1.6. Exercises:

2. b. 1. and E, $\frac{12}{27} = \frac{20}{45} = 0.44\overline{4} - 44.\overline{4}\%$

2. and C, $\frac{28}{36} = \frac{21}{27} = 0.77\overline{7} = 77.\overline{7}\%$

3. and D, $\frac{45}{75} = \frac{42}{70} = 0.6 = 60\%$

4. and A, $\frac{64}{80} = \frac{60}{75} = 0.8 = 80\%$

5. and B, $\frac{35}{56} = \frac{25}{40} = 0.625 = 62.5\%$

4. b. 0.313; **6.** The relative number of women at the university is 51.2%; $\frac{2304}{4500} = 0.512 = 51.2\%$. The relative number of women is greater at the community college. **8.** 75.3%; **9. b.** 47.2%, **d.** 69.3%, **f.** 355 students, **h.** 32.7%, **j.** 11.3%

Activity 1.7. Exercises: 1. c. 159.6 million; **4.** 313 million; **5.** $13.00; **7.** $680; **9. c.** 8.003%; **10. b.** 92, 71, 10, 21, 34; **12.** 1000 phone calls; **14.** 69 students; **15.** $34

Activity 1.8. Exercises: 3. c. 21.8%; **4.** First company: 33.3%; Second company: 10%; **6. a.** 400%, **b.** 80%; **8.** 200%

Activity 1.9. Exercises: 2. a. 1.08875, **b.** $21,496.28; **4. a.** $\frac{12,144}{11,952} \approx 1.016$, **b.** 1.016, **c.** 1.6%; **6. c.** $92,970; **8.** $3055.50; **13.** 0.964, 3.6%; **14. b.** $0.011 = 1.1\%$; **16.** $90.97; **18.** 162.5 mg

Activity 1.10.   Exercises: 4. $47,239.92;   **6.** $2433;
9. $29,925, which is less.

Activity 1.11.   Exercises: 1. 6.2 mi.;   **2.** approximately
29.81 mi./day;   **4. b.** 73.4 EUR, **d.** 10 USD, **f.** 61.0 GBP,
73.4 EUR, 10,441.30 JPY;   **5. b.** 45 gallons per month;
8. approximately 5.5 mi., 8.8481 km, 8848 m
10. approximately 4.8 quarts or 9.5 pints;
13. 4.8 grams or 0.168 ounces;   **15. a.** 3 tablets, **d.** 125 ml

What Have I Learned? 3. a. 18.3%, **b.** 17.5%, **c.** 14.81%,
d. 9.2%;   **6. a.** $199 \cdot 0.90 \approx 179$ lb.; my relative
will weigh 179 lb. after he loses 10% of his body
weight. **b.** $179 \cdot 0.90 \approx 161$ lb.; $161 \cdot 0.90 \approx 145$ lb.;
he must lose 10% of his body weight 3 times to reach 145 lb.

How Can I Practice? 1. a. 0.25, **d.** 0.035, **e.** 2.50;
2.

COLUMN 4
$7 \div 100 = 0.07 = 7\%$
$0.7 \div 5.0 = 0.14 = 14\%$
$0.7 \div 5.0 = 0.14 = 14\%$
$1 \div 12.5 = 0.08 = 8\%$

4. 4850 students;   **6.** 89.2%;   **8. a.** $168, **b.** 44%;
10. a. $826.92   **12.** actual increase: 702,294, relative
increase: $\approx 35\%$

Skills Check 1: Appendix C   1. 219;   **2.** 5372;
3. Sixteen and seven hundred nine ten-thousandths;
4. 3402.029;   **5.** 567.05;   **6.** 2.599;   **7.** 87.913;
8. 28.82;   **9.** 0.72108;   **10.** 95.7;   **11.** 12%;
12. 300%;   **13.** 0.065;   **14.** 3.6;   **15.** $0.1\overline{3}$;
16. $\frac{3}{7} \approx 0.4286 \approx 42.86\% \approx 42.9\%$;   **17.** $\frac{1}{8}$;
18. 3.0027, 3.0207, 3.027, 3.27;   **19.** 40.75;   **20.** 52%;
21. $8.67;   **22.** 32%;   **23.** $9500 \, (1.055) \approx 10,023$
students;   **24.** 37.5%;   **25.** 36.54 billion acres
26. a. No, the percent reductions are based on the price at
the time of the specific reduction, so it is not correct to sum
up the percent reductions. **b.** $450 before Thanksgiving,
$270 in December, and $135 in January. **c.** The final price
in January would be the same as before, $135. The interme-
diate prices would be different; before Thanksgiving, the
price would be $250; in December, the cost would be $150.
27. The initial decay factor was 0.45. The subsequent
growth factor was 1.65. Therefore, the 2014 math scores
were $(0.45)(1.65) = 0.7425$, or approximately 74% of
the 2004 scores. This is not good news.   **28.** 7,121,151.5;
a little over $7 million;   **29.** $1364.45; approximately
$1364.

Activity 1.12   Exercises: 4. 43-yd. gain;   **7.** $-11°C$;
9. 35°C;   **15.** Dropped 9 cents;   **17. a.** 0, **b.** 50

Activity 1.13   Exercises: 4. The sign of a product will be
positive if there is an even number of factors with negative
signs; the sign of a product will be negative if there is an
odd number of factors with negative signs.   **6.** -65.58;
My balance decreased by $65.58;   **8.** -823.028, a loss of
$823.03 per month;   **10. a.** $100,000, **b.** $50,000;
11. d. $-\$50$.   **12.** I suffered a $0.20 loss per share, for a
total loss of $36.

Activity 1.14   Exercises: 2. $39\frac{7}{12}$ ft.

6. c. $-1 \cdot 6 \cdot 6 = -36$, **d.** $\dfrac{1}{(-6)^2} = \dfrac{1}{36}$;

10. a. 8.8×10^{-4} lb. = 0.00088lb., **b.** (Answers will
vary.) I would use grams because the number is easier to
say, and it may sound larger and impress my friend.

What Have I Learned? 4. a. The result is positive
because there are two negatives being multiplied, resulting
in a positive value. **c.** The sign of the product is negative
because there is an odd number of negative factors.
5. c. A negative number raised to an even power produces
a positive result.   **6. b.** A negative number raised to an
odd power produces a negative result.   **8. b.** Any number
other than zero raised to the zero power equals 1.
9. b. 0.03 cm = 3.0×10^{-2} cm

How Can I Practice? 37. -6; I still need to lose 6 lb.
39. a. $-7°C$; The temperature drops 7°C. **b.** $-19°C$;
The evening temperature is expected to be $-19°C$. **d.** a 9°C
drop, **e.** The temperature rises 8°C;   **41. b.** $-3°F$;
42. a. -9, **c.** -5, **e.** 17, **g.** -9;   **43. b.** 0.03937 =
3.937×10^{-2} in.

Skills Check 2: Appendix C   1. 1.001;   **2.** 0.019254;
3. 1;   **4.** 9;   **5.** $c = \sqrt{72} \approx 8.49$ m; approximately
8.5 m;   **6.** 9;   **7.** 2.03×10^8;   **8.** approximately 8.8 qt. in
8.3 L;   **9.** 0.00000276;   **10.** approximately 9.96 oz.;
11. 16.1 sq. in.;   **12.** I measured the diameter to be
$d = \dfrac{15}{16}$ in. Then $C = \pi \left(\dfrac{15}{16} \right) \approx 2.95$ in. The
circumference is approximately 2.95 in.   **13.** 126 sq. in.;
14. 8;   **15.** 35 hr.;   **16.** $\dfrac{9}{1.6} = 5.625$; You can cut
5 pieces that are 1.6 yd. long.   **17.** $\dfrac{4}{9}$;   **18.** $2\dfrac{2}{7}$;   **19.** $\dfrac{34}{7}$;
20. $\dfrac{37}{45}$;   **21.** $1\dfrac{7}{10}$;   **22.** $\dfrac{8}{27}$;   **23.** 3;   **24.** $4\dfrac{5}{6}$ cups;
I have almost 5 cups of mixture.   **25.** $29\dfrac{5}{8}$ points at the end
of the day;   **26.** $2.\overline{6}$ days;   **27.** I should buy at least $4\dfrac{1}{4}$ lb.
of hamburger;   **28.** 5.48 ft., 128.2 lb.;   **29. a.** 52,800,
b. 96, **c.** 80, **d.** 768, **e.** 48;

30.

COLUMN 3
5×10^7
2.5×10^7
3.5×10^7
3.68×10^8
7.45×10^8
1.606×10^9
2.674×10^9

31. about 1.5 lb. of chocolate and 7 oz. of milk;

32. $4\frac{1}{12}$; **33.** 9; **34.** 1; **35.** $9\frac{73}{144}$; **36.** $13\frac{7}{12}$;

37. $24\frac{1}{6}$; **38.** -16; **39.** 36; **40.** 24

Gateway Review 1. two and two hundred two ten-thousandths; **2.** 14.003; **3.** hundred thousandths; **4.** 10.524; **5.** 12.16; **6.** 0.30015; **7.** 0.007; **8.** 2.05;

9. 450%; **10.** 0.003; **11.** $7\frac{3}{10} = \frac{73}{10}$; **12.** 60%;

13. $1\frac{1}{8}$, 1.1, 1.01, 1.001; **14.** 27; **15.** 1; **16.** $\frac{1}{4^2} = \frac{1}{16}$;

17. -64; **18.** 16; **19.** -25 **20.** 12; **21.** 5.43×10^{-5};

22. 37,000; **23.** $\frac{1}{3}\left(\frac{25}{30} + \frac{85}{100} + \frac{60}{70}\right) = .8468 \approx 85\%$;

24. $4\frac{1}{2}$; **25.** $\frac{23}{4}$; **26.** $\frac{3}{5}$; **27.** $\frac{4}{24} + \frac{15}{24} = \frac{19}{24}$;

28. $\frac{21}{4} - \frac{15}{4} = \frac{6}{4} = \frac{3}{2} = 1\frac{1}{2}$; **29.** $\frac{2}{9} \cdot \frac{\overset{3}{\cancel{27}}}{\underset{4}{\cancel{8}}} = \frac{3}{4}$;

30. $\frac{\overset{3}{\cancel{6}}}{\cancel{11}} \cdot \frac{\overset{2}{\cancel{22}}}{\underset{4}{\cancel{8}}} = \frac{6}{4} = \frac{3}{2} = 1\frac{1}{2}$; **31.** $\frac{\overset{3}{\cancel{21}}}{8} \cdot \frac{\overset{2}{\cancel{16}}}{7} = 6$;

32. 20% of 80 is 16; **33.** I did better on the second exam. **34.** There were 2500 complaints last year. **35.** 6 students; **36.** decreased by 30%; **37.** 162 students; **38.** 40 pages in 4 hours; **39.** 1.364 lb; **40.** $\approx 1.3\frac{\text{in.}}{\text{sec.}}$; **41.** -1;

42. -14; **43.** -13; **44.** 3; **45.** -25; **46.** -1;

47. 2; **48.** -11; **49.** 0; **50.** $\frac{1}{10}$; **51.** 19; **52.** 80;

53. 183; **54.** 20; **55.** 160 feet below the surface; **56.** a loss of $600; **57.** over drawn and am $63 in debt;

58. 4.4×10^{-27} pound; **59. a.** 15, **b.** 3, **c.** $\frac{8}{3}$, **d.** 1, **e.** 3

Chapter 2

Activity 2.1   Exercises: 2. a. $n + m$, **b.** $0.99n + 1.29m$; **4. a.** $100 - y$; **5. a.** $3s + 15$; **8. a.** number of pieces of fruit bought, **b.** dollars paid for A apples, **c.** dollars paid for A apples and P pears, **d.** dollars left out of $10 after paying for A apples and P pears; **10. a.** -17.5 **c.** 32.5, **e.** -12, **g.** 60; **12.** approximately 87.9

Activity 2.2   Exercises: 1. a. The input variable is the year. The output variable is the number of monthly text messages (in billions); **2. a.** 20, 10,5, 2.5, 1.25; **4. a.** No. Owning a negative number of DVDs doesn't make sense; **b.** Yes. It means the student doesn't own any DVDs; **6. a.** $B \approx 22$, **b.** $B = \frac{126{,}540}{h^2}$, **c.** 35.15, 30.90, 27.37, 21.91, 19.77

Activity 2.3   Exercises: 2. b. i. $y = x + 10$, ii. $y = 3x + 2$; **3. a.** $209 gross pay; **b.** Gross pay is calculated by multiplying the number of hours worked by 9.50, the pay per hour, **c.** $P = 9.50h$, **d.** 57.00, 114.00, 171.00, 190.00, 209.00, 266.00, 285.00; **4. b.** $5(x - 6)$, **e.** $-2x - 20$; **5. a.** $y = x - 10$, **b.** $y = 3(x - 4)$, **c.** $y = 9 + \frac{x}{6}$, **e.** $y = -4x - 15$; **6. a.** The input variable is the age of the driver. The output variable is the number of arrests per 100,000 drivers for this age group, **c.** 17-year-olds and 45-year-olds have 200 arrests per 100,000 drivers.

Activity 2.4   Exercises: 1. c. Commutative Property of Multiplication; **2. b.** The output values in columns 2 and 3 are opposites. $3 - x$ is the opposite (or negative) of $x - 3$; **3. a.** i. To obtain the output, square the input, ii. To obtain the output, square the input and then change the sign; iii. To obtain the output, change the sign of the input and then square that result.

What Have I Learned? 3. a. If the distance between tick marks represents 10 units, the input values will fit.

How Can I Practice? 1. c. No, negative numbers would represent distance above the surface. **2. c.** The D.O. content changes the most between 11°C and 16°C, when it drops 1.6 ppm. **4. a.** $2x - 20$, **b.** $\frac{1}{2}x + 6$; **5. a.** 0, **c.** 22; **6. a.** $4p + 6q$, **b.** $20 - x$; **8. b.** The unemployment rate was highest in 2011. The rate was about 14%, **c.** The unemployment rate was lowest in 2000. That rate was about 2.5%.

Activity 2.5   Exercises: 1. b. $T = 275n$, **c.** any counting number up to 11, **e.** $1650, **f.** 7 credit hours; **4. c.** $15 million; **5. a.** $y = 52.5$, **b.** $x \approx 41.14$; **7. a.** $y = -198.9$, **b.** $x = 23$; **9. a.** $y = 1.8$, **b.** $x = 8.2$; **11.** $1200

Activity 2.6   Exercises: 2. b. $p = 10n - 2100$, **d.** 360 students; **3. c.** $97.50, **d.** 160 min. **5. a.** approximately 181 cm; **7.** 70 in.; **12. a.** $x = -1$, **b.** $x = 4$, **d.** $x = -7$, **f.** $x = 0$, **h.** $x = 18$, **14.** $y = 21.75$, $x = -60$,

Activity 2.7   Exercises: 2. b. 70 mi., **c.** $3.39 \approx t$, approximately 3.4 hr.; **5. a.** 409.2 ft., **b.** $\dfrac{33(p - 15)}{15} = d$; **7.** $\dfrac{C}{2\pi} = r$; **9.** $\dfrac{P - 2l}{2} = w$; **11.** $\dfrac{A - P}{Pt} = r$; **13.** $m + vt^2 = g$

Activity 2.8   Exercises: 2. a. $x = 24$; **6.** 6750 seniors; **8.** about 53.1 pounds; **11.** school tax = \$3064; **14.** 12.5 in.

What Have I Learned? 1. In both cases, you need to isolate x by performing the inverse operations indicated by $4x - 5$ in reverse order; that is add 5, and then divide by 4.

How Can I Practice? 2. a. $t = 2.5r + 10$; **3. a.** $x = -3$, **c.** $x = 15.5$, **f.** $x = 4$, **h.** $x = 11.\overline{6}$; **4. a.** Multiply the number of hours of labor by \$68, and add \$148 for parts to obtain the total cost of the repair. **d.** \$420, **e.** No, 3.5 hr work plus parts will cost \$386. **f.** 4 hr., **g.** $x = \dfrac{y - 148}{68}$;

Solving for x allows you to answer questions such as part f more quickly. It would be especially useful if you wanted to determine the number of mechanic hours for several different amounts of money.

7. a. $B = 655.096 + 9.563(55) + 1.85(172) - 4.676(70)$ ≈ 1171.9 calories. He is not properly fed.
b. $61.52 \approx A$. He is about 62 years old.

8. a. $\dfrac{d}{t} = r$, **c.** $\dfrac{A - P}{Pt} = r$, **e.** $\dfrac{7}{4}(w - 3) = h$

Activity 2.9   Exercises: 2. $24x - 30$; **4.** $10 - 5x$;
6. $-3p + 17$; **8.** $-12x^2 + 9x - 21$; **10.** $\dfrac{5}{8}x + \dfrac{5}{9}$;
12. $15x^2 - 12x$; **16.** $A = P + Prt$; **17. c.** $lw + 5l$;
19. a. $y - 5$, **b.** $12(y - 5) = 12y - 60$;
20. a. $3(x + 5)$, **c.** $xy(3 - 7 + 1)$, **e.** $4(1 - 3x)$,
g. $st(4rt - 3rt + 10)$; **23. a.** $5a + 8ab - 3b$,
c. $104r - 13s^2 - 18s^3$, **e.** $2x^3 + 7y^2 + 4x^2$,
g. $3x^2y - xy^2$, **i.** $2x - 2x^2 - 5$; **25. a.** $24 - x$,
c. $5x - 75$, **e.** $9 + x$, **g.** $7x - 15$, **i.** $5x - 5$,
k. $-x^2 - 9x$; **26. c.** $398 - 26x$

Activity 2.10   Exercises: 1. c. Yes, the news is good. The value of my stock increased sixfold.
2. a. $3\left[\dfrac{-2n + 4}{2} - 5\right] + 6$, **b.** $-3n - 3$;
4. b. $8x - 5y + 13$, **e.** $13x - 43$; **7. b.** \$395

Activity 2.11   Exercises: 1. a. $C = 0.75x + 60$,
e. Company 2 is lower if I drive less than 120 miles.
3. d. $y = 16x + 120$, **e.** \$312, **g.** \$280, **i.** $x = \dfrac{y - 120}{16}$;

This form of the equation would be useful when I am setting a salary goal and need to determine the number of hours I must work. **4. c.** $0.85(12 - x)$, **f.** 7 roses and 5 carnations; **6.** $x = -1$; **8.** $x = -2$; **10.** $x = 81$;
12. $t = 3$; **14.** $x = -11$; **16.** $x = 2500$;

20. a. $x = \dfrac{y - b}{m}$, **c.** $h = \dfrac{A - 2\pi r^2}{2\pi r}$, **e.** $y = \dfrac{3x - 5}{2}$,
g. $P = \dfrac{A}{1 + rt}$

What Have I Learned? 1. a. By order of operations, $-x^2$ indicates to square x first, then negate. For example, $-3^2 = -(3)(3) = -9$. **b.** The negative sign can be interpreted as -1 and by the distributive property reverses the signs of the terms in parentheses.

How Can I Practice?
1.

x	13 + 2(15x − 3)	(10x − 10)	10x + 7
1	17	20	17
5	57	60	57
10	107	110	107

Expression a and expression c appear to be equivalent.
2. b. $3x^2 + 15x$, **d.** $-2.4x - 2.64$, **f.** $-6x^2 - 4xy + 8x$;
3. b. $2x(3y - 4z)$, **e.** $2x(x - 3)$; **4. a.** 4, **b.** 5, **c.** -1,
d. -3, **e.** 4 and x^2 (or 4, x, x); **6. b.** $x^2y^2 + 2xy^2$;
7. b. i. add 5 to x; ii. square result; iii. subtract 15;
8. a. $6 - x$, **c.** $3x^2 + 3x$, **e.** $-4x - 10$;
9. $V = 25(2x - 3)$; **13. c.** $D = (198x - 45)$ mi.;
14. c. $600 = 280 + 0.20(x - 1000)$, **d.** $x = 2600$.
I must sell \$2600 worth of furniture in order to have a gross salary of \$600. **15. a.** $S = 200 + 0.30x$,
b. $S = 350 + 0.15x$, **d.** $x = \dfrac{150}{0.15} = 1000$. I would have to sell about \$1000 worth of electronics per week to earn the same weekly salary from either option.
16. c. $x + (x + 10) + (x + 65) = 3x + 75$

Gateway Review 1. a. $x + 5$, **b.** $18 - x$, **c.** $2x$, **d.** $\dfrac{4}{x}$,
e. $3x + 17$, **f.** $12(8 + x)$, **g.** $11(14 - x)$, **h.** $\dfrac{x}{7} - 49$;
2. a. $x = 9$, **b.** $y = -24$, **c.** $x = 2.5$, **d.** $x = -12$,
e. $y = -2$, **f.** $x = 16$, **g.** $x = 144$, **h.** $y = \dfrac{5}{6}$,
i. $x = 72$; **3. a.** $A = \dfrac{63 + 68 + 72 + x}{4} = \dfrac{203 + x}{4}$,
b. He must score 61. **4. a.** $x = 5$, **b.** $y = 42$,
c. $x = 6$, **d.** $x = -7.75$, **e.** $y = -66$ **f.** $x = 139.5$,
g. $x = -96$, **h.** $y = 27$; **5. a.** The input variable is the number of miles driven. **b.** The output variable is the cost of rental for a day. **c.** $y = 25 + 0.15x$;
d.

Input, x (mi)	100	200	300	400	500
Output, y (\$)	40	55	70	85	100

e.

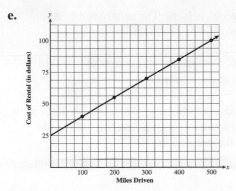

f. The cost to travel round trip from Buffalo to Syracuse is just under $75. **g.** $70.90, **h.** I can drive approximately 425 miles. **i.** approximately 433 mi., **j.** (Answers will vary.) **6.** My total annual home sales must be $714,286 in order to gross $30,000. **7. a.** $I = 2000(0.05)(I) = \$100$, **b.** $I = 3000(0.06)(2) = \$360$; **8. a.** $P = 2(2.8 + 3.4) = 12.4$,

b. $P = 2\left(7\frac{1}{3} + 8\frac{1}{4}\right) = 31\frac{1}{6}$;

9.

x	$(4x - 3)^2$	$4x^2 - 3$	$(4x)^2 - 3$
-1	49	1	13
0	9	-3	-3
3	81	33	141

None are equivalent. **10. a.** $3x + 3$, **b.** $-12x^2 + 12x - 18$, **c.** $4x^2 - 7x$, **d.** $-8 - 2x$, **e.** $12x - 25$, **f.** $17x - 8x^2$; **11. a.** $4(x - 3)$, **b.** $x(18z + 60 - y)$, **c.** $-4(3x + 5)$; **12. a.** $6x^2 - 6x + 3$, **b.** $x^2 - 3x + 7$; **13. a.** $10x - 35y$, **b.** $2a + b - 3$, **c.** $-10x + 20w - 10z$, **d.** $2c - 1$; **14. a.** $x = 8.75$, **b.** $x = 80$, **c.** $x = 20$, **d.** $x = 3$, **e.** $x = -13$, **f.** $x = 0$, **g.** $x = 4$, **h.** $x = 7$, **i.** $x = -3$, **15. a.** $280, **b.** $196, **c.** $0.70x$, **d.** $0.7(0.7x) = 0.49x$, **e.** $196. The price is the same **f.** The original price is $300. **16. a.** $500 - n$, **b.** $2.50n$, **c.** $4.00(500 - n)$, **d.** $2.50n + 4.00(500 - n) = 1550$, **e.** 300 student tickets and 200 adult tickets were sold. **17. a.** $C = 1200 + 25x$, **b.** $R = 60x$, **c.** $P = 35x - 1200$, **d.** 35 campers, **e.** 52 campers, **f.** $500; **18. a.** $P = \dfrac{l}{rt}$,

b. $t = \dfrac{f - v}{a}$, **c.** $y = \dfrac{2x - 7}{3}$; **19. a.** $C = 750 + 0.25x$, **b.** $875, **c.** A thousand booklets can be produced for $1000. **d.** $R = 0.75x$, **e.** To break even, 1500 booklets must be sold. **f.** To make a $500 profit, 2500 booklets must be sold. **20. a.** $22 - x$, **b.** $2.50x$, **c.** $(22 - x)(0.30)(15) = 4.50(22 - x)$, **d.** $2.50x + 4.50(22 - x) = 2.50x + 99 - 4.50x$
$= 99 - 2x$,
e. 7.5. I can drive 7 days. **f.** $77.

Chapter 3

Activity 3.1 **Exercises: 2. a.** Is a function because each input value is assigned a single output value, **b.** This data set does not represent a function because the input value 24 is assigned two output values, 53 and 29; **4. a. i.** -4, **ii.** $(2, -4)$, **b.** Yes, the out value of a function may be the same for several inputs; **6. b.** Yes, For any given height, there is only one given weight, **d.** $w(64) = 131$. The recommended weight for a $5'4''$ woman is 131 pounds, **e.** $h = 70$. If height is 70 inches, the corresponding weight is 149 pounds; **9. a.** A, B, C, D are in the first quadrant, **c.** I, J are in the third quadrant; **11. a.** quadrant III

Activity 3.2 **Exercises: 2.** $-0.05\,\dfrac{\text{yr. of age}}{\text{yr.}}$;

3. $0.024\,\dfrac{\text{yr. of age}}{\text{yr.}}$; **6. a.** yes, 1930–1940 and 1950–1960;

7. b. -4.93 gal./yr., **c.** 3.7 gal./yr.; **8. a.** approximately $-0.17°F/\text{hr.}$, **c.** $-1.5°F/\text{hr.}$

Activity 3.3 **Exercises: 2. b.** -4, **d.** 69; **3. b.** $x = 3$; **4.** Only the graphs of g and h contain the origin;

7. Only $\left(\dfrac{1}{2}, \dfrac{2}{5}\right)$ and $\left(-2, -\dfrac{2}{5}\right)$ lie on the graph;

8. e. i. $\dfrac{12x}{x + 1} = 5$, **ii.** $x = \dfrac{5}{7} \approx 0.714$, **iii.** Locate 5 on the vertical axis, move horizontally to the point on the graph $(1, 0.714)$

Activity 3.4 **Exercises: 2. d.** $v(a) = 26{,}700 - 5340a$, **f.** $5340, **g.** 2 years; **4. b.** $C = \dfrac{500a}{a + 12}$, **f.** 71 mg;

5. d. $h(4) = 12$ mi.,

h.

t	0	0.5	1	1.5	2	2.5	3
$h(t)$	0	1.5	3	4.5	6	7.5	9

What Have I Learned? 2. The statement $H(5) = 100$ means that for the function H, an input value of $x = 5$ corresponds to an output value of 100. **4.** $f(1)$ represents the output value when the input is 1. Because $f(1) = -3$, the point $(1, -3)$ is on the graph of f. **5.** After 10 minutes, the ice cube weighs 4 grams. **7.** If the rate of change of a function is negative, then the initial output value on any interval must be larger than the final value. We can conclude that the function is decreasing.

How Can I Practice? 1. b. Yes, for each number of credit hours (input), there is exactly one tuition cost (output). **e.** The most credit hours I can take for $700 is 5. **2. d.** $C(n) = 2n + 78$, **g.** The number of students who can sign up for the trip is restricted by the capacity of the bus. If the bus holds 45 students, then the practical domain is the set of integers from 0 to 45; **3. b.** quadrant 1, **c.** quadrant IV; **4. g.** 42.6, **h.** It is the numerical grade that corresponds to 213 points; **5. a.** This is a function, **d.** This is not a function, **f.** This is not a function. The input -3 has two different outputs

8. a. $F(x) = 2.50 + 2.20x$;   **10.** $g(-4) = 6$;
12. $h(-3) = 16$;   **15. c.** -5 lb./week,   **e.** The average rate of change indicates how quickly I am losing weight.

Activity 3.5   Exercises: 2. b. The average rate of change is not constant, so the data is not linear.   **4. a.** 50 ft. higher, **b.** 5.2% grade;   **5. b.** at least 240 in. long;

6. a. linear $\left(\text{constant average rate of change} = \dfrac{1}{2}\right)$,

c. not linear (average rate of change not constant);
7. d. Every week I hope to lose 2 pounds.   **8. b.** The slope of the line for this data is $-\dfrac{7}{10}$ beats per minute per year of age.   **9. a.** ii. $m = \dfrac{-10}{3}$;   **10. a.** 50 mph,   **b.** The line representing the distance traveled by car B is steeper, so car B is going faster. Car A a rate of 50 miles per hour; car B a rate of 58 miles per hour;   **11. b.** No, their slopes are all different. The number of units represented by each tick mark is different on each graph, producing different slopes.

Activity 3.6   Exercises: 3. a. The slope of the line is 2500, indicating that the value of my home increased at a constant rate of $2500 per year since 2009.   **b.** The v-intercept is (0, 125,000). In 2009, the value of my house was $125,000.   **c.** $V(8) = 2500(8) + 125,000 = 145,000$ and $2009 + 8 = 2017$. The market value of my house in 2017 will be $145,000;
7. slope is 3;
  y-intercept is $(0, -4)$;
  x-intercept is $\left(\dfrac{4}{3}, 0\right)$.

9. slope is 0;
  y-intercept is $(0, 8)$;
  There is no x-intercept.

11. slope is 2;
  y-intercept is $(0, -3)$;
  x-intercept is $\left(\dfrac{3}{2}, 0\right)$ or $(1.5, 0)$.

13. The lines all have the same slope but different y-intercepts. They are parallel.   **15.** $y = 12x + 3$;
17. b. The y-intercept is $(0, 5)$. The equation is $y = 3x + 5$.
19. a.

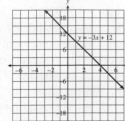

The x-intercept is $(4, 0)$; the y-intercept is $(0, 12)$.

Activity 3.7   Exercises: 1. b. The relative error is $\approx -0.2\%$;   **2. a.** 20.85, 21.28, 21.71,
c. $P(t) = 0.43t + 20.85$,   **f.** 27.3 million people;
3. c. $m \approx -19.6$, Detroit lost an average of 19,600 people in population each year from 2000 to 2012;
4. b. $y = -2.5x + 6$;   **5. a.** $P = 6.9 + 0.07792t$,
c. approximately 88 year's time, by 2098;   **6. d.** 450.2 ppm,
e. No, we are extrapolating too far into the future to be very confident of the prediction.

Activity 3.8   Exercises: 1. c. $t = 0.15i - 361$,
f. $13,647;
2. $y = 3x$;   **4.** $y = 7x + 16$;

6. $y = 2$;   **8.** $y = -\dfrac{2}{7}x + 2$;

11. $y = 3x + 9$;   **13.** $y = 2$;
16. a. $y = -3300x + 18,000$

What Have I Learned? 2. a. No, because the equation is not in the form $y = mx + b$,   **b.** $y = -2x + 1.5$,
c. $m = -2$ **4. a.** The graph is a straight line. **b.** It can be written in the form $y = mx + b$.   **c.** The rate of change is constant.

How Can I Practice? 1. a. y: 12, $m = 2$,   **c.** x: 6, y: 5, $m = -1$;   **2. a.** 7.5 ft;   **3. b.** Yes, the data represents a linear function because the average rate of change is constant, $28 per month.   **4. d.** For each second that passes during the first 5 seconds, speed increases 11 mph.
5. b. Yes; the three graphs represent the same linear function. They all have the same slope, 5, and the same vertical intercept $(0, 0)$.   **6. a.** The slope is undefined.   **c.** The slope is 0.5.   **7. b.** The vertical intercept is $(0, 10)$;
the horizontal intercept is $\left(\dfrac{20}{3}, 0\right)$.   **8. a.** $y = 2.5x - 5$,
b. $y = 7x + \dfrac{1}{2}$,   **c.** $y = -4$;   **11. a.** $y = 0x - 2$,
b. $y = 3x - 2$,   **c.** $y = x - 2$. The lines all have the same vertical intercept $(0, -2)$.

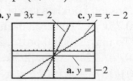

12. a. (Answers will vary.)

x	3	3	3	3
y	−4	0	2	5

c. The horizontal intercept is $(3, 0)$. There is no vertical intercept.   **13.** $m = 2$, y-intercept is $(0, 1)$, x-intercept is $\left(-\dfrac{1}{2}, 0\right)$;   **19.** $m = -2$, y-intercept is $(0, 2)$, x-intercept is $(1, 0)$;   **21.** $y = 9x - 4$;   **23.** $y = \dfrac{5}{3}x - 2$;

24. $y = 0$;   **29. a.** $y = 3x + 6$, **d.** $y = 4x - 27$;
30. a. $y = 2x + 6$, **e.** $x = 6$, a vertical line;
32. a. x-intercept is $(-2, 0)$

Activity 3.9   Exercises: 2. b. $P = -0.38t + 24.3$,
c. According to the model, approximately 16% of the total population 18 and older will smoke in 2020;
3. b. $y = 1124.09x + 25{,}928.25$ **c.** The slope is 1124.09. This means that on average, the per capita income in the United States increases by \$1124.09 each year;
4. c. $x \approx 31.76$. The maximum temperature is approximately 32°C.

Activity 3.11   Exercises: 1. a. $C = 19.99 + 0.79n$
b. $C = 29.99 + 0.59n$, **d.** $n = 50$ miles, **i.** $n = 50$ miles;
2. a. $V = -500x + 13{,}600$, **b.** $V = -800x + 16{,}000$,
f. $x = 8$, $y = 9600$;   **3. c.** $n = 9.5$ The total sales of both companies will be the same after 9.5 years (in 2024);
4. b. interest on credit union loan: $0.065c$, interest on Stafford loan: $0.0386s$, **c.** $0.065c + 0.0386s = 531.82$,
c. amount borrowed from the credit union: \$4500, amount borrowed from Stafford loan: \$6200.
6. $q = 2, p = 0$;

8. $x = 6, y = 1$;

Activity 3.12   Exercises: 1. a. $x = 0.5, y = 2.5$;
c. $x = 4, y = -1$;   **2. b.** $y = 5, x = 1$, **c.** $x = 0$, $y = -0.2$;   **4. a.** $8x + 5y = 106$, **b.** $x + 6y = 24$,
d. The cost of a centerpiece is \$12, and the cost of a glass is \$2.

Activity 3.14   Exercises: 1. $l + w + d \le 61$;
4. $x > -2$;   **6.** $x < 5$;   **8.** $x \ge 8$;   **10.** $x \ge -0.4$;
12. c. $30 + 1.00n < 60 + 0.75n$ **14. c.** $n > 18.6$

How Can I Practice? 2. $(-45, -40)$;
4. a. $C = 500 + 8n$, **b.** $R = 19.50n$,
c. 44 centerpieces;   **6. a.** $x = -1, y = 275$;
8. $x \approx -6.48, y \approx -15.78$;

10. d. After 80 boxcars are loaded, the management will realize a savings on the purchase of the forklift;
11. a. $-14.25t + 598.69 < 200$

Gateway Review 1. a. Yes, it is a function, **b.** No, two different outputs are paired with 2;   **2. a.** Yes. For each x-value, there is one y-value, **b.** The independent variable

is x, the number of hours worked, **c.** The dependent variable is y, the total cost, **d.** A negative number of hours worked does not make sense;   **3. a.** 400, 800, 1600, 2400,
b.

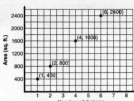

c. 2400, **d.** Four gallons of paint will cover 1600 square feet;   **4.** The slope is -2.5, the y-intercept is $(0, 4)$;
5. The equation is $y = -2x + 2$.   **6.** y-intercept is $(0, 8)$. An equation for the line is $y = \dfrac{-4}{3}x + 8$.   **7.** row 1: c, e, b;
row 2: f, a, d;   **8.** y-intercept is $(0, 5)$. Equation is $y = 3x + 5$.   **9.** The y-intercept is $(0, -2)$; the x-intercept is $(4, 0)$.

10.

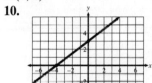

The vertical intercept is $(0, 3)$; the horizontal intercept is $(-4, 0)$.

11. $y = \dfrac{5}{3}x - 4$;   **12. a.** $(2, 2)$,

b.

c.

13. $a = -1; b - 6$;   **14.** $y = 2, x = 2$;
15. $q = -1.25p + 100$ **16.** $y = -0.56x + 108.8$;
17. a. $m = \dfrac{32 - 212}{0 - 100} = \dfrac{-180}{-100} = \dfrac{9}{5}$, **b.** $F = 1.8C + 32$,
c. 104° Fahrenheit corresponds with 40°C,
d. 170°F corresponds approximately to 77°C;
18. a. \$94.97, \$110.90, \$116.21, \$126.83,
b.

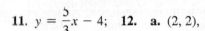

c. $C(k) = 0.1062k + 20.63$, **d.** The cost of using 875 kWh of electricity is \$113.56. **e.** I used approximately 1218 kWh of electricity.   **19.** $V(t) = -95t + 950$;
20. a. The regression equation is approximately $w(h) = 6.92h - 302$;

b. He would weigh about 252 lb. **c.** The player would be about 71.1 in. tall.   **21.** The solution is (2, 4).   **22.** The solution is (4, 3);   **23. a.** $x \geq 1$, **b.** $x \geq -6$;
24. Numerically:

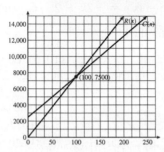

x	R(x)	C(x)
0	0	2500
50	3750	5000
100	7500	7500
150	11,250	10,000
200	15,000	12,500

Graphically:

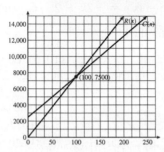

Algebraically: $x \geq 100$

Chapter 4

Activity 4.1   Exercises: 1. a. ii. Yes, it *is* a polynomial because the variable in each term contains a positive integer power of x or is a constant. **iv.** No, it is *not* a polynomial because the variable in the second term is inside a radical sign and in the denominator.   **4.** $7x - 1$;   **6.** $8x - 7y$;
8. $2x^2 + 6$;   **10.** $4x^3 + 5x^2 + 18x - 5$;   **12.** $2.8x + 7$;
14. 12.6;   **16.** 9.723;   **18.** -75;   **21. a.** $-x^2 - x + 2$
c. $-5x^3 + 6x^2 - x$;   **23. a.** $17,721$, **b.** $48,690$; $t = 56$
corresponds to the year 2016. If the polynomial function model continues to be valid past the year 2012, then the per capita income for U.S. residents in the year 2016 should be approximately $48,690.

Activity 4.2   Exercises: 2. a^4;   **4.** y^9;   **6.** $-12w^7$;
8. a^{15};   **10.** $-x^{50}$;   **12.** $-5.25x^9y^2$;   **14.** $\frac{2}{3}x^4$;   **16.** $\frac{1}{x^3}$;
18. $\frac{2}{5}z^{-2}$ or $\frac{2}{5z^2}$;   **20.** $11x^2$;   **22.** $\frac{81a^8}{16b^4}$;   **24.** $2x^2 + 6x$;
26. $2x^4 + 3x^3 - x^2$;   **28.** $10x^4 - 50x^3$;
30. $18t^6 - 6t^4 - 4.5t^2$;   **34. a.** $A = 5x(4x) = 20x^2$,
b. $V = 20x^2(x + 15) = 20x^3 + 300x^2$;
36. $A = (3xy^2)^2 = 9x^2y^4$;   **38.** $V = 9\pi h^3$;
40. $V = \pi r^3 + 4\pi r^2$;   **42. a.** $3x^2(6x^5 + 9x - 5)$,
c. $3x^3(2x^2 - 4x + 3)$

Activity 4.3   Exercises: 1. a. $x^2 + 8x + 7$,
e. $10 + 9c + 2c^2$, **i.** $12w^2 + 4w - 5$,
m. $2a^2 - 5ab + 2b^2$;
2. a. $3x^2 - 2xw + 2x + 2w - 5$,
c. $3x^3 - 11x^2 - 6x + 8$, **e.** $x^3 - 27$;
5. a. $V = 7(4)(8) = 224$ in.³,
b. $V = (7 + x)(4 + x)(8)$, **c.** $V = 224 + 88x + 8x^2$,

d. 432 in.³, 208 in.³, $\frac{208}{224} \approx 92.9\%$,
e. $V = (7 - x)(4 + x)(8)$,
f. $V = 224 + 24x - 8x^2$, **g.** $x = 2$;
$V = 224 + 24(2) - 8(2)^2 = 224 + 48 - 32 = 240$ cu. in.
Check: $V = (7 - 2)(4 + 2)(8) = 5(6)(8) = 240$ cu. in.

What Have I Learned? 2. The expression $-x^2$ instructs you to square the input first, and then change the sign. The expression $(-x)^2$ instructs you to change the sign of the input and then square the result.   **5.** It reverses the signs of the terms in parentheses: $-(x - y) = -1(x - y) = -x + y$.
7. Yes; in $3x^2$, only x is squared, but in $(3x)^2$, both the 3 and the x are squared: $(3x)^2 = (3x)(3x) = 9x^2$.

How Can I Practice? 2. e. $p^{20} \cdot p^6 = p^{26}$, **f.** $12x^{10}y^8$,
i. $3y^7$, **l.** a^{12}, **p.** $-3s^5t^{12}$;   **3. b.** $2x^2 + 14x$,
e. $6x^5 - 15x^3 - 3x$;   **4. b.** Start with x, then add 3, square the result, subtract 12;   **5. a.** $7x + 10$,
c. $4x^2 - 7x$;   **6. a.** $x^2 - x - 6$, **d.** $x^2 + 2xy - 8y^2$,
g. $x^3 + 2x^2 + 9$, **i.** $a^3 - 4a^2b + 4ab^2 - b^3$;
9. a. $A = (3x)(4x) = 12x^2$,
b. $A = (3x - 5)(4x) = 12x^2 - 20x$,
c. $A = (3x - 5)(4x + 5) = 12x^2 - 5x - 25$

Activity 4.4   Exercises: 1. e. $35 = \frac{8}{3}t^2$; approximately 3.62 sec;   **2. a.** $x = \pm 3$, **c.** $x^2 = -4$. There is no real number solution because the square of a real number cannot be negative. **f.** $x = \pm 8$;   **3. c.** $a = \pm 4$;
4. b. $x = -1$ or $x = 3$ **e.** $x = 7$ or $x = -3$
h. $x = 8$ or $x = -16$;   **5.** 13 in.;   **6.** 15 in.

Activity 4.5   Exercises: 1. b. The ball will hit the water 2.5 seconds after it is tossed in the air;   **3. a.** $x = 0$ or $x = 8$,
c. $x = 0$ or $x = 4$, **e.** $x = 0$ or $x = 5$;   **4. c.** The rocket is on the ground $t = 0$ seconds from launch and returns to the ground $t = 30$ seconds after launch.
f.

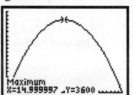

It will reach its maximum height of 3600 feet 15 seconds after launch.

5. a. $x = 0$ or $x = -10$, **c.** $y = 0$ or $y = -5$,
e. $t = 0$ or $t = 1.5$, **g.** $w = 0$ or $w = 4$

Activity 4.6   Exercises: 1. a. $x = -1$ or $x = -6$,
c. $y = -7$ or $y = -4$;   **2. a.** The rocket returns to the ground in 27 seconds. **c.** The rocket is 1760 feet above the ground 5 seconds after launch and again on the way down, 22 seconds after launch.   **5. a.** At $t = 0$, the apple is 96 feet above the ground. **b.** $-16t^2 + 16t + 96 = 0$,
c. $t = 3$ or $t = -2$. Discard the result $t = -2$.   **f.** The maximum point of the parabola is (0.5, 100). **g.** The coordinates (0.5, 100) indicate that the apple at its highest point is 100 feet above the ground half a second after I toss it.

Activity 4.7 **Exercises: 1.** $x = -5$ or $x = 3$;
3. $x = 1$ or $x = -\dfrac{1}{2}$; **5.** $x \approx 10.83$ or $x \approx -0.83$;

7. a. There are no solutions because $\sqrt{-16}$ is not a real number. Thus, there are no x-intercepts. **b.** The graph of $y = x^2 - 2x + 5$ does not cross the x-axis so there are no x-intercepts. **8. a.** According to the model, almost 81% of American adults will be overweight or obese in 2020.
b. $t \approx -74.9$ makes no sense because it is too far removed from the data points, but 54.5 corresponds to midyear 2014.

What Have I Learned? 1. b. $100 - x^2 = 40$;
2. a. Divide each term by 2 and then take square roots.
c. Factor the left side of the equation and then apply the zero-product property. **3. c.** $x = -5$ or $x = 3$;
7. There will always be *one* y-intercept. The x-coordinate of the y-intercept is always zero. Therefore, $y(0) = a(0)^2 + b(0) + c = c$. This shows that the y-intercept is always $(0, c)$.

How Can I Practice? 2. a. $x = 4$ or $x = 5$,
e. $x = 3$ or $x = -1$, **g.** $x = 5$ or $x = 2$,
i. $x = 0$ or $x = -\dfrac{3}{2}$;

4. a. $x = -6$ or $x = 1$, **d.** $x = -8$ or $x = 1$,
f. $x \approx 4.414$ or $x \approx 1.586$, **g.** $x \approx 3.414$ or $x \approx 0.586$;
7. a. 10 ft, **b.** 8 ft, **c.** -80. This result is possible only if there is a hole in the ground. By the time 3 seconds pass, the ball is already on the ground. **d.** 10.24, 10.2464, 10.2496, 10.25, 10.2496, 10.2464, 10.24. Both before and after 0.625 second, the height is less than 10.25 feet. Therefore, the ball reaches its maximum height in 0.625 second.

Activity 4.8 **Exercises: 2. a.** The growth factor is $1 + 0.05 = 1.05$; **c.** $110, $115.50, $121.28, $127.34, $133.71, $140.39, $147.41; **3. a.** The function is increasing because the base 5 is greater than 1 and is a growth factor. **b.** The function is decreasing because the base $\dfrac{1}{2}$ is less than 1 and is a decay factor. **c.** The base 1.5 is a growth factor (greater than 1), so the function is increasing. **d.** The base 0.2 is a decay factor (less than 1), so the function is decreasing. **6. a.** The decay factor is $1.00 - 0.083 = 0.917$. **b.** The formula is $N = 20(0.917)^t$. **c.** 14.14, 10.00, 7.07, 5.00, 3.54, 2.50, **f.** It will take about 8 days for iodine-131 to decay from 20 grams to 10 grams.

Activity 4.9 **Exercises: 1. a.** $s = 0.08p$,

b.

LIST PRICE, p ($)	SALES TAX s ($)
10	0.80
20	1.60
30	2.40
50	4.00
100	8.00

c.

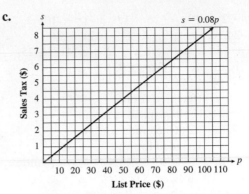

List Price ($)

4. $y = kx$, $25 = k(5)$, $k = 5$, $y = 5x$, $y = 5(13)$, $y = 65$;
8. a. No, area varies directly as the square of the radius.
b. $k = \pi$

Activity 4.10 **Exercises: 1. a.** $d = \dfrac{1200}{w}$, **b.** 40, 34.286, 30, 24, 20. **c.** As the width increases, the depth decreases.
d. The depth is 12 feet, not enough room for most theater sets; so it is not realistic, **e.** No, division by zero is undefined, **f.** (Answers will vary.) $30 \le w \le 60$, **g.** an inverse variation function. **3. b.** The man weighs 128 pounds when he is 1000 miles above the surface of Earth.
4. a. $y = \dfrac{120}{x}$. Yes, y is a function of x. The domain is all real numbers except $x = 0$. **5. a.** $x = 4$, **c.** $x = -5.5$,

Activity 4.11 **Exercises: 1. a.** 49.9 mph; **2. a.** $x = \pm 3$,
c. $t^2 = -9$. There is no real number solution. There is no real number whose square is a negative number.
e. $x = 49$, **g.** $x = 5$

What Have I Learned? 2. a. Plan 1: $S = 22,000 + 1000x$, Plan 2: $S = 22,000(1.04)^x$ **3. c. i.** $x = 2$, a square;
ii. The length is twice the width when the width is $\sqrt{2}$, or approximately 1.4 centimeters. The length is half the width when the width is $\sqrt{8}$, or approximately 2.8 centimeters.

How Can I Practice? 1. a. $r = \sqrt{\dfrac{V}{\pi h}}$, **b.** $b = \sqrt{c^2 - a^2}$;
3. a. $s(y) = 32,000(1.03)^y$, where s represents salary and y represents year. The practical domain covers the length of the contract, 0 to 5 years. **e.** $43,005; **4. b.** $0.95,
d. The domain is all real numbers except 0. **e.** The practical domain is all real numbers greater than 0 and less the $2 (or so). **6.** $x = 1.5$; **7.** $x = 0.02$; **8.** $x = \dfrac{1494}{18} = 83$;

13. a. One-half of the dose remains after 1 day.
d. $\left(\dfrac{1}{2}\right)^2, \left(\dfrac{1}{2}\right)^3, \left(\dfrac{1}{2}\right)^4$, **e.** $Q = \left(\dfrac{1}{2}\right)^t$; no, this function is exponential.

Gateway Review 1. $15x - 41$; **2.** $-14x^2 + 20x - 36$;
3. $2x^3 - 9x^2 - x - 5$; **4.** $-x^3 - 13x^2 + 16x - 2$;
5. $3x^7 + 2x^6 - 5x^5 - 20x^4$; **6.** $9x^4 - 16x^3 - 37x^2 + 16x$;
7. $16x^{12}$; **8.** $125x^3y^9$; **9.** $16a^5b^5$; **10.** $y^{11-3} = y^8$;
11. $5t^3$; **12.** $3x^2(9x^3 - 3x - 4)$; **13.** $4x(2x + 1)$;

14. $(5a - 13) \cdot 2 = 10a - 26$; **15. a.** $x^2 - 10x + 24$,
b. $x^2 - 5x - 24$, **c.** $x^2 - 9$, **d.** $x^2 - 12x + 36$,
e. $2x^2 - 13x + 15$, **f.** $6x^2 - 13x - 28$, **g.** $4x^2 - 25$,
h. $x^2 + 8x + 16$, **i.** $2x^3 + 7x^2 - 9$,
j. $3x^3 + 13x^2 - 14x - 20$;
16. $A = \pi(11 + y)^2$ (factored form),
$A = \pi y^2 + 22\pi y + 121\pi$ (expanded form);
17. $A = (x + 7)(x - 2)$ (factored form),
$A = x^2 + 5x - 14$ (expanded form);
18. a. $x = \pm 2$, **b.** $x = \pm 3$, **c.** $x^2 = -4$, no solution,
d. $x = 6$ or $x = -10$, **e.** $x = 8$ or $x = -2$,
19. a. $x = 0$ or $x = -8$, **b.** $x = 0$ or $x = 3$,

c. $x = 0$ or $x = 5$, **d.** $x = 0$ or $x = 5$, **e.** $x = \dfrac{1}{4}$ or $x = 6$;

20. a. $s = 20$ ft, **b.** The ball reaches a height of 32 feet
after 1 second and then again on its way down after 2 seconds.
c. The ball reaches the ground after 3 seconds.
21. a. $x = -6$ or $x = 2$, **b.** $x = 7$ or $x = -2$,

c. $x = 8$ or $x = -3$, **d.** $x = 3$ or $x = 7$,
e. $x = 4$ or $x = -6$, **f.** $y = -8$ or $y = 5$;
22. a. $x \approx 1.786$ or $x \approx -1.120$, **b.** $x = 1$ or $x = -2.4$,
c. $x \approx -1.345$ or $x \approx 3.345$, **d.** There is no real solution
because $\sqrt{-15}$ is not a real number.
23. a. $x = 6$, **b.** $x = -7$, **c.** $x = 49$,
d. $x = 625$, **e.** $x = 33$; **24. a.** In 5 years the tuition
will be \$401.47 per credit; in 10 years, \$537.25.
b. approximately \$20.29 per credit per year,
c. approximately \$23.73 per credit per year,
d. about 12 yrs; **25.** approximately 8840 lb;
26. \$231; **27.** 4 hours, **28.** The intensity is 80 microwatts
per square meter, **29. a.** approximately 0.87 sec.,
b. 1.5 sec., **c.** approximately 1.94 sec. **30. a.** not linear;
b. 4764, 6107, 7830; **c.** close, but not exact;
d. approximately 57,127 MW **31. a.** \$415, **b.** 1.0118,
c. $y = 415(1.0118)^x$, **d.** \$466.65, **e.** during the sixteenth
month.

Glossary

absolute value   The size or magnitude of a number. It is represented by a pair of vertical line segments enclosing the number and indicates the distance of the number from zero on the number line. The absolute value of equals 0. The absolute value of any nonzero number is always positive.

actual change   The difference between a new value and the original value of a quantity.

algebraic approach for solving an equation for a given variable   Using algebraic techniques to isolate the variable on one side of the equation with coefficient 1.

algebraic expression   A mathematical set of instructions (containing constant numbers, variables, and the operations among them) that indicates the sequence in which to perform the computations.

arithmetic sequence   A list of numbers in which consecutive numbers share a common difference.

associative property of addition   For all numbers a, b, and c, $(a + b) + c = a + (b + c)$. For example, $(2 + 3) + 4 = 5 + 4 = 9$ and $2 + (3 + 4) = 2 + 7 = 9$.

associative property of multiplication   For all numbers a, b, and c, $ab(c) = a(bc)$. For example, $(2 \cdot 3) \cdot 4 = 6 \cdot 4 = 24$ and $2 \cdot (3 \cdot 4) = 2 \cdot 12 = 24$.

average   The sum of a collection of numbers, divided by how many numbers are in the collection.

average rate of change   The ratio of the change in the output (dependent) variable to the change in the input (independent) variable.

binomial   A polynomial with exactly two terms.

coefficient   A number (constant) that multiplies a variable.

common factor   A factor contained in each term of an algebraic expression.

commutative property of addition   For all numbers a and b, $a + b = b + a$, which means that changing the order of the addends does not change the result.

commutative property of multiplication   For all numbers a and b, $ab = ba$, which means that changing the order of the factors does not change the result.

completely factored form   An algebraic expression written in factored form where none of its factors can themselves be factored any further.

constant   A number or symbol in an algebraic expression whose value is understood to remain fixed and does not change.

constant of proportionality (or constant of variation)   In a direct variation problem, the following statements are equivalent. The number represented by k is called the constant of proportionality or constant of variation.

a. y varies directly as x

b. y is directly proportional to x

c. $y = kx$ for some constant k.

decay factor   Ratio of new (decreased) value to the original value.

decreasing function   As the x-values increase, the corresponding y-values decrease; the graph falls from left to right; For example, a linear function with a negative slope.

degree of a monomial in a single variable   The exponent on its variable. If the monomial is a constant, the degree is zero.

degree of a polynomial   The highest degree among all its terms.

denominator   The number written below the fraction bar that indicates the number of equal sized parts into which a unit has been divided.

dependent variable   Another name for the output variable of a function.

difference   The result of subtracting one number (or expression) from another.

direct variation   A relationship defined algebraically by $y = k \cdot x$, where k is a constant, in which whenever x *increases* by a multiplicative factor (e.g., doubles), y also *increases* by the same factor.

distributive property   The property of multiplication over addition (or subtraction) that states that $a(b + c) = ab + ac$.

dividend   In a division problem, the number that is divided into parts. For example, in $27 \div 9 = 3$, the number 27 is the dividend, 9 is the divisor, and 3 is the quotient.

divisor   In a division problem, the number that divides the dividend. For example, in $27 \div 9 = 3$, the number 27 is the dividend, 9 is the divisor, and 3 is the quotient.

domain   The set (collection) of all possible input values for a function.

equation   A statement that two algebraic expressions are equal.

equivalent expressions   Two algebraic expressions are equivalent if they always produce identical outputs when given the same input value.

equivalent fractions   Fractions that represent the same ratio. For example, $\frac{4}{6}$ and $\frac{10}{15}$ both represent the ratio $\frac{2}{3}$. Divide both the numerator and denominator of $\frac{4}{6}$ by 2 to obtain $\frac{2}{3}$. Similarly, divide both the numerator and denominator of $\frac{10}{15}$ by 5 to obtain $\frac{2}{3}$.

error   *See* goodness-of-fit measure.

exponent   The superscript n on an expression of the form b^n. When a number or expression is used as a repeated multiplicative factor such as $b \cdot b \cdot b \cdot b$, the product can be expressed in the compact form b^4. In this case, the superscript 4 is called an exponent and indicates the number of repeated factors. If $b \neq 0$, the expression b^0 is defined to have value 1. If n is a positive integer, b^n represents the product $b \cdot b \cdot b \cdots b$ (n factors of b) and b^{-n} represents the reciprocal $\frac{1}{b^n}$.

exponential function   A function defined by an equation of the form $y = a \cdot b^x, b > 0$, in which the input variable becomes the exponent. For example, $y = 4 \cdot 5^x$.

extrapolation   The process of evaluating a function by choosing input values outside a given range of values.

factor, a   When two or more algebraic expressions or numbers are multiplied together to form a product, those individual expressions or numbers are called the factors of that product. For example, the product $5 \cdot a \cdot (b + 2)$ contains the three factors 5, a, and $b + 2$.

Fibonacci sequence   A sequence of numbers in which each number is generated by adding the two numbers that precede it.

formula   An algebraic statement describing the relationship among a group of variables.

function   A rule (given in tabular, graphical, or symbolic form) that relates (assigns) to any permissible input value exactly one output value.

function terminology and notation   The name of a functional relationship between input x and output y. You say that y is a function of x, and you write $y = f(x)$.

geometric sequence   A list of numbers in which consecutive numbers share a common ratio.

goodness-of-fit   A measure of the difference between the actual data values and the values produced by the model.

graph   A collection of ordered pairs that have been plotted as points on a rectangular coordinate system.

greatest common factor   The largest factor that exactly divides each term in an expression.

growth factor   Ratio of new (increased) value to the original value.

horizontal axis   The horizontal number line of a rectangular coordinate system. When graphing a relationship between two variables, the input (independent) variable values are referenced on this axis.

horizontal intercept   A point where a line or curve crosses the horizontal axis.

improper fraction   A fraction in which the absolute value of the numerator is greater than or equal to the absolute value of the denominator.

increasing function   As x increases, the corresponding y-values increase also; the graph rises; For example, a linear function with a positive slope.

independent variable   Another name for the input variable of a function.

inequality   A mathematical statement that one quantity is greater than (or less than) another quantity.

input   Replacement values for a variable in an algebraic expression, table, or function. The value that is listed first in a relationship involving two variables.

integers   The set of positive and negative counting numbers and zero.

interpolation   Th process of evaluating a function by choosing input values within a given range of values.

inverse variation   A relationship defined algebraically by $y = \dfrac{k}{x}$, where k is a constant, in which whenever x *increases* by a multiplicative factor (e.g., doubles), y *decreases* by the same factor (e.g., y is halved).

least common denominator (LCD)   The smallest number that is a multiple of each denominator in two or more fractions.

like terms   Terms that contain identical variable factors (including exponents) and may differ only in their numerical coefficients.

line of best fit   When measured or observed data pairs are plotted, the line whose points lie closest to the plotted points.

linear function   A function defined by an equation of the form $f(x) = mx + b$, where m and b are constants in which m is the slope of the line and b its y-intercept. The graph of a linear function is a straight line.

lowest terms   Phrase describing a fraction whose numerator and denominator have no factors in common other than 1.

mean   The arithmetic average of a set of data.

method of least squares   A statistical procedure for determining a line of best fit from a set of data pairs.

mixed number   The sum of an integer and a fraction, written in the form $a\dfrac{b}{c}$, where a is the integer, and $\dfrac{b}{c}$ is the fraction.

monomial   A single term consisting of either a number (constant), a variable, or a product of variables raised to a positive integer exponent, along with a coefficient.

negative slope   The slope associated with a decreasing line whose y-values decrease as you move to the right along the line.

numerator   The number written above the fraction bar specifying the number of equal parts that are being counted.

numerical method   Guess, check, and repeat approach.

order of operations convention   The universally accepted rules for the order in which to perform each of the arithmetic operations contained in a given expression or computation.

ordered pair   Two numbers or symbols, separated by a comma and enclosed in a set of parentheses. An ordered pair can have several interpretations, depending on its context. Common interpretations are as the coordinates of a point in the plane and as an input/output pair of a function.

origin   The point at which the horizontal and vertical axes intersect.

output   Values produced by evaluating an algebraic expression, table, or function. The value that is listed second in a relationship involving two variables.

parabola   The graph of a quadratic function (second-degree polynomial function). The graph is $\cup$-shaped, opening upward or downward.

point-slope form   The symbolic rule for a line in which the slope, m, and a point (x_1, y_1) are known: $y - y_1 = m(x - x_1)$.

polynomial   An algebraic expression formed by adding and/or subtracting monomials.

positive slope   The slope associated with an increasing line whose y-values increase as you move to the right along the line.

practical domain   The domain determined by the situation being studied.

practical range   The range determined by the situation being studied.

product   The result of performing a multiplication of two or more factors.

proportional reasoning   The thought process by which a known ratio is applied to one piece of information to determine a related, but yet unknown, second piece of information.

proportion   An equation stating that two ratios are equal.

quadrants   The four regions of the plane separated by the vertical and horizontal axes. The quadrants are numbered from I to IV. Quadrant I is located in the upper right with the numbering continuing counterclockwise. Therefore, quadrant IV is located in the lower right.

quadratic equation   Any equation that can be written in the form $ax^2 + bx + c = 0, a \neq 0$.

quadratic function   A function defined by an equation of the form $y = ax^2 + bx + c$, $a \neq 0$.

quotient   The result of performing a division operation.

range   The set (collection) of all output values for a function.

rate   A comparison formed by dividing of two quantities that have different units of measurement.

rational number   A number that can be written in the form $\frac{a}{b}$, where a and b are integers and b is not zero.

ratio   A quotient that compares two similar numerical quantities, such as "part" to "whole."

reciprocal   Two numbers are reciprocals of each other if their product is 1. Obtain the reciprocal of a fraction by switching the numerator and denominator. For example, the reciprocal of $\frac{3}{4}$ is $\frac{4}{3}$.

regression line   A line that best approximates a set of data points as shown on a scatterplot.

relative change   The comparison of the actual change to the original value. Relative change is measured by the ratio

$$\text{relative change} = \frac{\text{actual change}}{\text{original value}}.$$

Since relative change is frequently reported as a percent, it is often called *percent change*.

relative error   The ratio of the measured error to the exact value.

scaling   The process of assigning a fixed distance between adjacent tick marks on a coordinate axis.

scatterplot   A set of points in the plane whose coordinate pairs represent input/output pairs of a data set.

scientific notation   A concise way for expressing very large or very small numbers as a product—(a number between 1 and 10) · (the appropriate power of 10).

slope of a line   A measure of the steepness of a line. It is calculated as the ratio of the rise (vertical change) to the run (horizontal change) between any two points on the line.

slope-intercept form   The form $y = mx + b$ of a linear function in which m denotes the slope and b denotes the y-intercept.

solving an inequality   The process of determining the values of the variable that makes the inequality a true statement.

square root   The square root of a nonnegative number N is a number M whose square is N. The symbol for square root is $\sqrt{}$. For example, $\sqrt{9} = 3$ because $3^2 = 9$.

substitution method   A method used to solve a system of two linear equations. Choose one equation and rewrite it to express one variable in terms of the other. In the other equation, replace the chosen variable with its expression and solve to determine the value of the second variable. Substitute the value in either original equation to determine the value of the first variable.

sum   The result of performing an addition.

symbolic rule   A shorthand code that indicates a sequence of operations to be performed on the input variable, x, to produce the corresponding output variable, y.

system of linear equations   Two linear equations considered together.

terms   Parts of an algebraic expression separated by the addition, $+$, and subtraction, $-$, symbols.

trinomial   A polynomial with exactly three terms.

unit analysis   The process of using measurement units to solve proportion problems by setting up and performing a sequence of multiplications and/or divisions so the appropriate measurement units cancel, leaving the desired measurement unit of the result. Also called dimensional analysis.

variable   A quantity that takes on specific numerical values and will often have a unit of measure (e.g., dollars, years, miles) associated with it.

verbal rule   A statement that describes in words the relationship between input and output variables.

vertical axis   The vertical number line of a rectangular coordinate system. When graphing a relationship between two variables, the output (dependent) variable values are referenced on this axis.

vertical intercept   The point where a line or curve crosses the vertical axis.

weighted average   An average in which some numbers count more heavily than others.

zero-product rule   The algebraic principle that states if a and b are real numbers such that $a \cdot b = 0$, then either a or b, or both, must be equal to zero.

Index

Geometric Formulas

Perimeter and Area of a Triangle, and
Sum of the Measures of the Angles

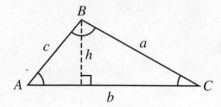

$$P = a + b + c$$
$$A = \tfrac{1}{2}bh$$
$$A + B + C = 180°$$

Pythagorean Theorem

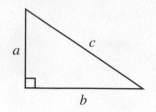

$$a^2 + b^2 = c^2$$

Perimeter and Area of a Rectangle

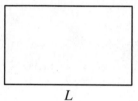

$$P = 2L + 2W$$
$$A = LW$$

Perimeter and Area of a Square

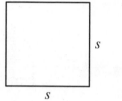

$$P = 4s$$
$$A = s^2$$

Area of a Trapezoid

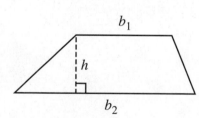

$$A = \tfrac{1}{2}h(b_1 + b_2)$$

Circumference and Area of a Circle

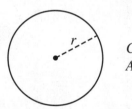

$$C = 2\pi r$$
$$A = \pi r^2$$

Volume and Surface Area of a Rectangular Solid

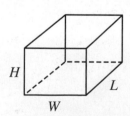

$$V = LWH$$
$$SA = 2LW + 2LH + 2WH$$

Volume and Surface Area of a Sphere

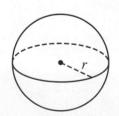

$$V = \tfrac{4}{3}\pi r^3$$
$$SA = 4\pi r^2$$

Volume and Surface Area of a Right
Circular Cylinder

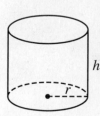

$$V = \pi r^2 h$$
$$SA = 2\pi r^2 + 2\pi rh$$

Volume and Surface Area of a
Right Circular Cone

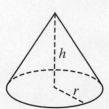

$$V = \tfrac{1}{3}\pi r^2 h$$
$$SA = \pi r^2 + \pi rl$$